VOLUME EIGHTY SIX

ADVANCES IN
PARASITOLOGY
Malaria Control and Elimination Programme in the People's Republic of China

SERIES EDITOR

D. ROLLINSON
Life Sciences Department
The Natural History Museum,
London, UK
d.rollinson@nhm.ac.uk

J. R. STOTHARD
Department of Parasitology
Liverpool School of Tropical
Medicine Liverpool, UK
r.stothard@liv.ac.uk

EDITORIAL BOARD

M. G. BASÁÑEZ
Professor in Parasite Epidemiology,
Department of Infectious Disease
Epidemiology Faculty of Medicine
(St Mary's Campus), Imperial
College, London, London, UK

S. BROOKER
Wellcome Trust Research Fellow
and Professor, London School of
Hygiene and Tropical Medicine,
Faculty of Infectious and Tropical,
Diseases, London, UK

R. B. GASSER
Department of Veterinary Science,
The University of Melbourne,
Parkville, Victoria, Australia

N. HALL
School of Biological Sciences,
Biosciences Building, University of
Liverpool, Liverpool, UK

R. C. OLIVEIRA
Centro de Pesquisas Rene Rachou/
CPqRR - A FIOCRUZ em Minas
Gerais, Rene Rachou Research
Center/CPqRR - The Oswaldo Cruz
Foundation in the State of Minas
Gerais-Brazil, Brazil

R. E. SINDEN
Immunology and Infection
Section, Department of Biological
Sciences, Sir Alexander Fleming
Building, Imperial College of
Science, Technology and
Medicine, London, UK

D. L. SMITH
Johns Hopkins Malaria Research
Institute & Department of
Epidemiology, Johns Hopkins
Bloomberg School of Public Health,
Baltimore, MD, USA

R. C. A. THOMPSON
Head, WHO Collaborating Centre
for the Molecular Epidemiology
of Parasitic Infections, Principal
Investigator, Environmental
Biotechnology CRC (EBCRC), School
of Veterinary and Biomedical
Sciences, Murdoch University,
Murdoch, WA, Australia

X.-N. ZHOU
Professor, Director, National
Institute of Parasitic Diseases,
Chinese Center for Disease Control
and Prevention, Shanghai, People's
Republic of China

VOLUME EIGHTY SIX

ADVANCES IN
PARASITOLOGY

Malaria Control and Elimination Programme in the People's Republic of China

Edited by

X.-N. ZHOU
*National Institute of Parasitic Diseases,
Chinese Center for Disease Control and Prevention,
Shanghai, People's Republic of China*

R. KRAMER
Global Health Institute, Duke University, USA

W.-Z. YANG
*Chinese Center for Disease Control and Prevention,
Beijing, People's Republic of China*

AMSTERDAM • BOSTON • HEIDELBERG • LONDON
NEW YORK • OXFORD • PARIS • SAN DIEGO
SAN FRANCISCO • SINGAPORE • SYDNEY • TOKYO
Academic Press is an imprint of Elsevier

Academic Press is an imprint of Elsevier
32 Jamestown Road, London NW1 7BY, UK
525 B Street, Suite 1800, San Diego, CA 92101-4495, USA
225 Wyman Street, Waltham, MA 02451, USA
The Boulevard, Langford Lane, Kidlington, Oxford OX5 1GB, UK

First edition 2014

Copyright © 2014 Elsevier Ltd. All rights reserved.

No part of this publication may be reproduced or transmitted in any form or by any means, electronic or mechanical, including photocopying, recording, or any information storage and retrieval system, without permission in writing from the publisher. Details on how to seek permission, further information about the Publisher's permissions policies and our arrangements with organizations such as the Copyright Clearance Center and the Copyright Licensing Agency, can be found at our website: www.elsevier.com/permissions.

This book and the individual contributions contained in it are protected under copyright by the Publisher (other than as may be noted herein).

Notices
Knowledge and best practice in this field are constantly changing. As new research and experience broaden our understanding, changes in research methods, professional practices, or medical treatment may become necessary.

Practitioners and researchers must always rely on their own experience and knowledge in evaluating and using any information, methods, compounds, or experiments described herein. In using such information or methods they should be mindful of their own safety and the safety of others, including parties for whom they have a professional responsibility.

To the fullest extent of the law, neither the Publisher nor the authors, contributors, or editors, assume any liability for any injury and/or damage to persons or property as a matter of products liability, negligence or otherwise, or from any use or operation of any methods, products, instructions, or ideas contained in the material herein.

ISBN: 978-0-12-800869-0
ISSN: 0065-308X

For information on all Academic Press publications
visit our website at http://store.elsevier.com

CONTENTS

Contributors — xi
Preface — xvii

1. **Historical Patterns of Malaria Transmission in China** — 1
   Jian-Hai Yin, Shui-Sen Zhou, Zhi-Gui Xia, Ru-Bo Wang, Ying-Jun Qian, Wei-Zhong Yang, Xiao-Nong Zhou

   1. Introduction — 2
   2. Background — 2
   3. Historical Patterns of Malaria in China — 3
   4. Looking Forward — 12
   5. Conclusions — 15
   Acknowledgements — 15
   References — 15

2. **Feasibility and Roadmap Analysis for Malaria Elimination in China** — 21
   Xiao-Nong Zhou, Zhi-Gui Xia, Ru-Bo Wang, Ying-Jun Qian, Shui-Sen Zhou, Jürg Utzinger, Marcel Tanner, Randall Kramer, Wei-Zhong Yang

   1. Introduction — 22
   2. Feasibility Assessment at the National Level — 23
   3. Correlation Between Incidence Patterns and Interventions in Four Target Provinces — 28
   4. Feasibility Analysis of Malaria Elimination in China — 31
   5. Phase-Based Malaria Elimination Strategies — 36
   6. Conclusions and Recommendations — 39
   Acknowledgements — 42
   References — 42

3. **Lessons from Malaria Control to Elimination: Case Study in Hainan and Yunnan Provinces** — 47
   Zhi-Gui Xia, Li Zhang, Jun Feng, Mei Li, Xin-Yu Feng, Lin-Hua Tang, Shan-Qing Wang, Heng-Lin Yang, Qi Gao, Randall Kramer, Tambo Ernest, Peiling Yap, Xiao-Nong Zhou

   1. Introduction — 48
   2. Background of Two Provinces — 51
   3. Malaria Transmission Patterns, Particularly in 2004–2012 — 53

4.	History of Malaria Control	61
5.	Malaria Interventions from Control to Elimination	63
6.	Challenges for Malaria Elimination	68
7.	Conclusions	74
	Acknowledgements	76
	References	76

4. Surveillance and Response to Drive the National Malaria Elimination Programme — 81

Xin-Yu Feng, Zhi-Gui Xia, Sirenda Vong, Wei-Zhong Yang, Shui-Sen Zhou

1.	Introduction	82
2.	Background	83
3.	Surveillance in the National Malaria Elimination Programme	86
4.	Response Strategy in the National Malaria Elimination Programme	94
5.	Challenges	101
6.	Conclusions and Recommendations	103
	Acknowledgements	103
	References	104

5. Operational Research Needs Toward Malaria Elimination in China — 109

Shen-Bo Chen, Chuan Ju, Jun-Hu Chen, Bin Zheng, Fang Huang, Ning Xiao, Xia Zhou, Tambo Ernest, Xiao-Nong Zhou

1.	Introduction	110
2.	Methods	111
3.	Research Challenges in the Stage of Malaria Elimination	112
4.	Research Priority toward Malaria Elimination in China	120
5.	Conclusions and Recommendations	126
	Acknowledgements	127
	References	127

6. Approaches to the Evaluation of Malaria Elimination at County Level: Case Study in the Yangtze River Delta Region — 135

Min Zhu, Wei Ruan, Sheng-Jun Fei, Jian-Qiang Song, Yu Zhang, Xiao-Gang Mou, Qi-Chao Pan, Ling-Ling Zhang, Xiao-Qin Guo, Jun-Hua Xu, Tian-Ming Chen, Bin Zhou, Peiling Yap, Li-Nong Yao, Li Cai

1.	Background	137
2.	Introduction	138

3.	Pilot Project of the National Malaria Elimination Programme	143
4.	Results and Achievements	158
5.	Conclusions	180
	Acknowledgements	180
	References	180

7. Surveillance and Response Strategy in the Malaria Post-elimination Stage: Case Study of Fujian Province — 183

Fa-Zhu Yang, Peiling Yap, Shan-Ying Zhang, Han-Guo Xie, Rong Ouyang, Yao-Ying Lin, Zhu-Yun Chen

1.	Introduction	184
2.	Background	185
3.	Surveillance after Malaria Elimination	195
4.	Response in Post-elimination	197
5.	Experiences and Lessons	201
6.	Conclusions	202
	Acknowledgements	202
	References	202

8. Preparation of Malaria Resurgence in China: Case Study of Vivax Malaria Re-emergence and Outbreak in Huang-Huai Plain in 2006 — 205

Hong-Wei Zhang, Ying Liu, Shao-Sen Zhang, Bian-Li Xu, Wei-Dong Li, Ji-Hai Tang, Shui-Sen Zhou, Fang Huang

1.	Introduction	206
2.	Background	206
3.	Vivax Malaria Re-emergence and Outbreak in Huang-Huai Plain	209
4.	Characteristics of Malaria Re-emergence and Outbreak in Huang-Huai Plain	213
5.	Factors of Malaria Re-emergence and Outbreak in Huang-Huai Plain	215
6.	Response to Re-emergence and Outbreak in Huang-Huai Plain	218
7.	Effects on Response to Re-emergence and Outbreak	223
8.	Challenge of Malaria Elimination in Huang-Huai Plain	224
9.	Conclusions	225
	Acknowledgements	226
	References	226

9. Preparedness for Malaria Resurgence in China: Case Study on Imported Cases in 2000–2012   231

Jun Feng, Zhi-Gui Xia, Sirenda Vong, Wei-Zhong Yang, Shui-Sen Zhou, Ning Xiao

1. Introduction   232
2. Background   232
3. Epidemiological Situation   234
4. High Risk Areas and Risk Factors   241
5. Strategies on Prevention of Malaria Reintroduction   247
6. Case Studies   257
7. Challenges   258
8. The Way Forward   260
Acknowledgements   261
References   261

10. Preparation for Malaria Resurgence in China: Approach in Risk Assessment and Rapid Response   267

Ying-Jun Qian, Li Zhang, Zhi-Gui Xia, Sirenda Vong, Wei-Zhong Yang, Duo-Quan Wang, Ning Xiao

1. Background   268
2. Malaria in China   268
3. Risk Determinants of Secondary Transmission by Imported Malaria   271
4. Response to Reintroduction of Malaria   274
5. Research Needs in the National Malaria Elimination Programme   280
6. Discussions   281
7. Conclusions   282
Acknowledgements   283
References   283

11. Transition from Control to Elimination: Impact of the 10-Year Global Fund Project on Malaria Control and Elimination in China   289

Ru-Bo Wang, Qing-Feng Zhang, Bin Zheng, Zhi-Gui Xia, Shui-Sen Zhou, Lin-Hua Tang, Qi Gao, Li-Ying Wang, Rong-Rong Wang

1. Introduction   290
2. Epidemiological Justification for Rounds of GFATM Malaria Programmes   291
3. General Information about Rounds   292
4. Programme Management   296

	5. Programme Input	300
	6. Main Programme Activities and Output	302
	7. Achievements and Impacts	310
	8. Looking Forward	315
	9. Conclusions	316
	Acknowledgements	317
	References	317
12.	**China–Africa Cooperation Initiatives in Malaria Control and Elimination**	**319**

Zhi-Gui Xia, Ru-Bo Wang, Duo-Quan Wang, Jun Feng, Qi Zheng, Chang-Sheng Deng, Salim Abdulla, Ya-Yi Guan, Wei Ding, Jia-Wen Yao, Ying-Jun Qian, Andrea Bosman, Robert David Newman, Tambo Ernest, Michael O'leary, Ning Xiao

	1. Background	320
	2. Existing China–Africa Collaboration on Malaria Control	322
	3. The Challenges and Needs for Malaria Control in Africa	323
	4. Potential Opportunity and Contribution to Enhance the Partnership	328
	5. Collaborative Research Scopes	332
	6. The Way Forward	333
	Annex 1: Antimalaria Centers in Africa Established by Chinese Government in 2007–2009	334
	Acknowledgements	335
	References	335

Index	*339*
Contents of Volumes in This Series	*347*

CONTRIBUTORS

Salim Abdulla
Ifakara Health Institute, Dar es Salaam, Tanzania

Andrea Bosman
Global Malaria Programme, World Health Organization, Geneva, Switzerland

Li Cai
Shanghai Municipal Center for Disease Control & Prevention, Shanghai, People's Republic of China

Jun-Hu Chen
National Institute of Parasitic Diseases, Chinese Center for Disease Control and Prevention; Key Laboratory of Parasite and Vector Biology of the Chinese Ministry of Health; WHO Collaborating Centre for Malaria, Schistosomiasis and Filariasis, Shanghai, People's Republic of China

Shen-Bo Chen
National Institute of Parasitic Diseases, Chinese Center for Disease Control and Prevention; Key Laboratory of Parasite and Vector Biology of the Chinese Ministry of Health; WHO Collaborating Centre for Malaria, Schistosomiasis and Filariasis, Shanghai, People's Republic of China

Tian-Ming Chen
Haiyan County Center for Disease Control & Prevention, Haiyan, People's Republic of China

Zhu-Yun Chen
Fujian Center for Disease Control and Prevention, Fujian, People's Republic of China

Chang-Sheng Deng
Guangzhou University of Traditional Chinese Medicine, Guangdong, People's Republic of China

Wei Ding
National Institute of Parasitic Diseases, Chinese Center for Disease Control and Prevention; Key Laboratory of Parasite and Vector Biology, MOH; WHO Collaborating Centre for Malaria, Schistosomiasis and Filariasis, Shanghai, People's Republic of China

Tambo Ernest
Centre for Sustainable Malaria Control, Faculty of Natural and Environmental Science; Center for Sustainable Malaria Control, Biochemistry Department, Faculty of Natural and Agricultural Sciences, University of Pretoria, Pretoria, South Africa

Sheng-Jun Fei
Songjiang District Center for Disease Control & Prevention, Shanghai, People's Republic of China

Jun Feng
National Institute of Parasitic Diseases, Chinese Center for Disease Control and Prevention; Key Laboratory of Parasite and Vector Biology, MOH; WHO Collaborating Centre for Malaria, Schistosomiasis and Filariasis, Shanghai, People's Republic of China

Xin-Yu Feng
National Institute of Parasitic Diseases, Chinese Center for Disease Control and Prevention; Key Laboratory of Parasite and Vector Biology, MOH; WHO Collaborating Centre for Malaria, Schistosomiasis and Filariasis, Shanghai, People's Republic of China

Qi Gao
Jiangsu Provincial Institute of Parasitic Diseases, Wuxi, People's Republic of China

Ya-Yi Guan
National Institute of Parasitic Diseases, Chinese Center for Disease Control and Prevention; Key Laboratory of Parasite and Vector Biology, MOH; WHO Collaborating Centre for Malaria, Schistosomiasis and Filariasis, Shanghai, People's Republic of China

Xiao-Qin Guo
Songjiang District Center for Disease Control & Prevention, Shanghai, People's Republic of China

Fang Huang
National Institute of Parasitic Diseases, Chinese Center for Disease Control and Prevention; Key Laboratory of Parasite and Vector Biology of the Chinese Ministry of Health; WHO Collaborating Centre for Malaria, Schistosomiasis and Filariasis, Shanghai, People's Republic of China

Chuan Ju
National Institute of Parasitic Diseases, Chinese Center for Disease Control and Prevention; Key Laboratory of Parasite and Vector Biology of the Chinese Ministry of Health; WHO Collaborating Centre for Malaria, Schistosomiasis and Filariasis, Shanghai, People's Republic of China

Randall Kramer
Duke Global Health Institute, Duke University, Durham, NC, USA

Mei Li
National Institute of Parasitic Diseases, Chinese Center for Disease Control and Prevention; Key Laboratory of Parasite and Vector Biology, MOH; WHO Collaborating Centre for Malaria, Schistosomiasis and Filariasis, Shanghai, People's Republic of China

Wei-Dong Li
Anhui Center for Disease Control and Prevention, Hefei, People's Republic of China

Yao-Ying Lin
Fujian Center for Disease Control and Prevention, Fujian, People's Republic of China

Ying Liu
Henan Center for Disease Control and Prevention, Zhengzhou, People's Republic of China

Xiao-Gang Mou
Anji County Center for Disease Control & Prevention, Anji, People's Republic of China

Robert David Newman
Global Malaria Programme, World Health Organization, Geneva, Switzerland

Michael O'leary
World Health Organization, China Representative Office, Beijing, People's Republic of China

Rong Ouyang
Fujian Center for Disease Control and Prevention, Fujian, People's Republic of China

Qi-Chao Pan
Shanghai Municipal Center for Disease Control & Prevention, Shanghai, People's Republic of China

Ying-Jun Qian
National Institute of Parasitic Diseases, Chinese Center for Disease Control and Prevention; Key Laboratory of Parasite and Vector Biology, MOH; WHO Collaborating Centre for Malaria, Schistosomiasis and Filariasis, Shanghai, People's Republic of China

Wei Ruan
Zhejiang Provincial Center for Disease Control & Prevention, Hangzhou, People's Republic of China

Jian-Qiang Song
Zhabei District Center for Disease Control & Prevention, Shanghai, People's Republic of China

Ji-Hai Tang
National Institute of Parasitic Diseases, Chinese Center for Disease Control and Prevention; Key Laboratory of Parasite and Vector Biology, Ministry of Health; WHO Collaborating Centre for Malaria, Schistosomiasis and Filariasis, Shanghai, People's Republic of China

Lin-Hua Tang
National Institute of Parasitic Diseases, Chinese Center for Disease Control and Prevention; Key Laboratory of Parasite and Vector Biology, Ministry of Health; WHO Collaborating Centre for Malaria, Schistosomiasis and Filariasis, Shanghai, People's Republic of China

Marcel Tanner
Department of Epidemiology and Public Health, Swiss Tropical and Public Health Institute; University of Basel, Basel, Switzerland

Jürg Utzinger
Department of Epidemiology and Public Health, Swiss Tropical and Public Health Institute; University of Basel, Basel, Switzerland

Sirenda Vong
World Health Organization, China Representative Office, Beijing, People's Republic of China

Duo–Quan Wang
National Institute of Parasitic Diseases, Chinese Center for Disease Control and Prevention; Key Laboratory of Parasite and Vector Biology, MOH; WHO Collaborating Centre for Malaria, Schistosomiasis and Filariasis, Shanghai, People's Republic of China

Shan-Qing Wang
Hainan Provincial Center for Disease Control and Prevention, Haikou, People's Republic of China

Li-Ying Wang
National Health and Family Planning Commission of the People's Republic of China, Beijing, People's Republic of China

Ru-Bo Wang
National Institute of Parasitic Diseases, Chinese Center for Disease Control and Prevention; Key Laboratory of Parasite and Vector Biology, Ministry of Health; WHO Collaborating Centre for Malaria, Schistosomiasis and Filariasis, Shanghai, People's Republic of China

Rong-Rong Wang
National Health and Family Planning Commission of the People's Republic of China, Beijing, People's Republic of China

Zhi-Gui Xia
National Institute of Parasitic Diseases, Chinese Center for Disease Control and Prevention; Key Laboratory of Parasite and Vector Biology, MOH; WHO Collaborating Centre for Malaria, Schistosomiasis and Filariasis, Shanghai, People's Republic of China

Ning Xiao
National Institute of Parasitic Diseases, Chinese Center for Disease Control and Prevention; Key Laboratory of Parasite and Vector Biology of the Chinese Ministry of Health; WHO Collaborating Centre for Malaria, Schistosomiasis and Filariasis, Shanghai, People's Republic of China

Han-Guo Xie
Fujian Center for Disease Control and Prevention, Fujian, People's Republic of China

Jun-Hua Xu
Zhabei District Center for Disease Control & Prevention, Shanghai, People's Republic of China

Bian-Li Xu
Henan Center for Disease Control and Prevention, Zhengzhou, People's Republic of China

Wei-Zhong Yang
Chinese Preventive Medicine Association, Beijing, People's Republic of China; Chinese Center for Disease Control and Prevention, Beijing, People's Republic of China

Heng-Lin Yang
Yunnan Provincial Institute of Parasitic Diseases, Pu-er, People's Republic of China

Fa-Zhu Yang
Fujian Center for Disease Control and Prevention, Fujian, People's Republic of China

Li-Nong Yao
Zhejiang Provincial Center for Disease Control & Prevention, Hangzhou, People's Republic of China

Jia-Wen Yao
National Institute of Parasitic Diseases, Chinese Center for Disease Control and Prevention; Key Laboratory of Parasite and Vector Biology, MOH; WHO Collaborating Centre for Malaria, Schistosomiasis and Filariasis, Shanghai, People's Republic of China

Peiling Yap
Department of Epidemiology and Public Health, Swiss Tropical and Public Health Institute, Basel, Switzerland; University of Basel, Basel, Switzerland

Jian-Hai Yin
National Institute of Parasitic Diseases, Chinese Center for Disease Control and Prevention; Key Laboratory of Parasite and Vector Biology, MOH; WHO Collaborating Centre for Malaria, Schistosomiasis and Filariasis, Shanghai, People's Republic of China

Yu Zhang
Haiyan County Center for Disease Control & Prevention, Haiyan, People's Republic of China

Ling-Ling Zhang
Zhejiang Provincial Center for Disease Control & Prevention, Hangzhou, People's Republic of China

Shan-Ying Zhang
Fujian Center for Disease Control and Prevention, Fujian, People's Republic of China

Hong-Wei Zhang
Henan Center for Disease Control and Prevention, Zhengzhou, People's Republic of China

Shao-Sen Zhang
National Institute of Parasitic Diseases, Chinese Center for Disease Control and Prevention; Key Laboratory of Parasite and Vector Biology, Ministry of Health; WHO Collaborating Centre for Malaria, Schistosomiasis and Filariasis, Shanghai, People's Republic of China

Li Zhang
National Institute of Parasitic Diseases, Chinese Center for Disease Control and Prevention; Key Laboratory of Parasite and Vector Biology, MOH; WHO Collaborating Centre for Malaria, Schistosomiasis and Filariasis, Shanghai, People's Republic of China

Qing-Feng Zhang
National Institute of Parasitic Diseases, Chinese Center for Disease Control and Prevention; WHO Collaborating Centre for Malaria, Schistosomiasis and Filariasis; Key Laboratory of Parasite and Vector Biology, Ministry of Health, Shanghai, People's Republic of China

Bin Zheng
National Institute of Parasitic Diseases, Chinese Center for Disease Control and Prevention; Key Laboratory of Parasite and Vector Biology of the Chinese Ministry of Health; WHO Collaborating Centre for Malaria, Schistosomiasis and Filariasis, Shanghai, People's Republic of China

Qi Zheng
National Institute of Parasitic Diseases, Chinese Center for Disease Control and Prevention; Key Laboratory of Parasite and Vector Biology, MOH; WHO Collaborating Centre for Malaria, Schistosomiasis and Filariasis, Shanghai, People's Republic of China

Shui-Sen Zhou
National Institute of Parasitic Diseases, Chinese Center for Disease Control and Prevention; Key Laboratory of Parasite and Vector Biology, MOH; WHO Collaborating Centre for Malaria, Schistosomiasis and Filariasis, Shanghai, People's Republic of China

Xiao-Nong Zhou
National Institute of Parasitic Diseases, Chinese Center for Disease Control and Prevention; Key Laboratory of Parasite and Vector Biology, MOH; WHO Collaborating Centre for Malaria, Schistosomiasis and Filariasis, Shanghai, People's Republic of China

Xia Zhou
National Institute of Parasitic Diseases, Chinese Center for Disease Control and Prevention; Key Laboratory of Parasite and Vector Biology of the Chinese Ministry of Health; WHO Collaborating Centre for Malaria, Schistosomiasis and Filariasis, Shanghai, People's Republic of China

Bin Zhou
Anji County Center for Disease Control & Prevention, Anji, People's Republic of China

Min Zhu
Shanghai Municipal Center for Disease Control & Prevention, Shanghai, People's Republic of China

PREFACE

 MALARIA CONTROL AND ELIMINATION PROGRAMME IN THE PEOPLE'S REPUBLIC OF CHINA

The purpose of this special issue is to tell the story of China's successful control of malaria in recent decades and its careful planning to move the country to an elimination phase (Tambo et al., 2012; Zhou et al., 2014). China has made considerable progress on slowing malaria transmission over the past 60 years, but few health researchers and programme managers outside of China are aware of how this has been achieved.

Historical evidence on the prevalence of malaria goes back 4000 years. Over 30 million annual cases are estimated to have occurred prior to the establishment of the People's Republic of China in 1949 (Tang, 2009; Chen, 2014). After 1949, a number of initiatives were undertaken to control malaria, but incidence remained at more than 24 million through the early 1970s. Through sustained efforts in different control phases, the number of cases dramatically declined to less than 15,000 by 2009. At the same time, the scope of endemic areas was greatly narrowed. Falciparum malaria was eliminated by this time except in Hainan and Yunnan provinces (Yin et al., 2014). In 2010 China took the large step of initiating the National Malaria Elimination Programme (NMEP), described in the government document of 'Action Plan for Malaria Elimination in China' co-issued by 13 ministries in 2010. In the context of achieving the Millennium Development Goals as well as the global goal to eradicate malaria, China committed to eliminating malaria by 2020 (Gao et al., 2011).

Many challenges remain for China to achieve its elimination goal. This collection of papers discusses how China is moving from the control stage to the elimination stage. After reviewing the history of malaria control, papers in this volume discuss how to prepare for resurgence, how to deal with border issues and how to engender greater international cooperation (Figure 1).

First, Chapters 1 and 2 present national historical patterns of transmission of malaria as well as a feasibility analysis and a road map analysis on malaria elimination (Yin et al., 2014; Zhou et al., 2014). The first stage of control from 1950 to 1980 was characterised by high malaria prevalence

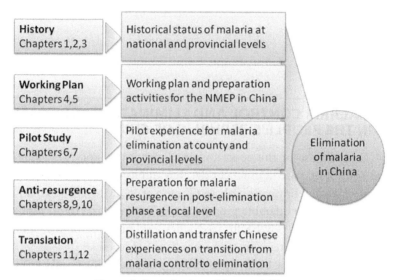

Figure 1 Organisation of the special issue.

with fluctuating peaks roughly every 10 years. During the second stage after 1980, there were slight rebounds but no incidence peaks due to sustained control efforts and improved capacity. This significant reduction of malaria incidence both in the control stage was demonstrated by the experiences at the provincial level in Hainan and Yunnan in Chapter 3 (Xia et al., 2014b). Chapter 3 discusses the situation in Yunnan and Hainan, which have experienced high malaria transmission with falciparum and vivax malaria epidemic year round. This is due to ecological conditions well suited for transmission and very efficient vectors, *Anopheles minimus* and *An. dirus*. Understanding progress and challenges in these provinces is critical not only to further elimination here but also in the entire country.

Second, Chapters 4 and 5 provide a view of preparations to launch the NMEP (Chen et al., 2014; Feng et al., 2014b). Chapter 4 introduces the working plan for surveillance and response systems in the malaria elimination phase, with the key indicators for an elimination plan. Imported cases provide particular challenges in the NMEP. To address this issue, sentinel surveillance was carried out in 2013 in Anhui, Henan, Zhejiang and Shaanxi provinces. At the same time, 13 provinces carried out screening of workers returning from abroad, and found 737 cases among 4358 persons screened. Chapter 5 analyses the challenges and needs of operational research in the NMEP. In particular, gaps were identified, which could hinder the progress of China's NMEP. Therefore, priorities for operational research were recommended, including development of new screening tools, better diagnostic tool, entomological studies linking environmental and climate factors and integrated, multipronged strategies for malaria elimination in the People's Republic of China.

Third, Chapters 6 and 7 introduce an intervention strategy leading to transition from control to elimination phases in the lower epidemic regions, such as Shanghai, Zhejiang and Fujian provinces (Yang et al., 2014; Zhu et al., 2014). Chapter 6 reviews the surveillance strategy carried out in four pilot sites in Shanghai and Zhejiang provinces, to provide the guidelines for malaria elimination assessment at the national level. Chapter 7 presents the experiences with surveillance and response in the malaria postelimination phase in Fujian province, where no local case has been detected since 2002.

Fourth, there were many instances of malaria resurgence in different parts of the world historically following a period of successful suppression of malaria transmission. This is most often the result of a subsequent weakening of control programmes (Cohen et al., 2012). Chapters 8, 9 and 10 examine how programme managers in China can prepare for resurgence in the postelimination phase. Chapter 8 examines what can happen when malaria re-emerges after successful elimination (Zhang et al., 2014). There was a serious outbreak of vivax malaria in 2006 in the Huang-Huai plain in Central China. It was subsequently controlled with mass drug administration and case management. Chapters 9 and 10 examine imported malaria from Southeast Asia and Africa as a potential cause of resurgence (Feng et al., 2014a; Qian et al., 2014). The chapters draw a number of lessons to avoid future outbreaks with a particular emphasis on aggressive screening, ongoing malaria education and maintaining the ability to mount rapid responses.

Finally, as in many countries in the world, successful efforts in malaria control and elimination depend critically on strong intersectoral collaborations, clear intervention strategies supported by efficient health information systems, active involvement of communities and innovative operational research. Success stories described in Chapter 11 were gleaned from five projects supported by the Global Fund. With the efficient use of resources provided by the Global Fund, China has successfully transitioned from malaria control to malaria elimination (Wang et al., 2014). Therefore, China's considerable success in malaria control over the past 60 years presents opportunities to transfer knowledge to the African countries that are still faced with a significant burden of the disease. Chapter 12 examines knowledge transfer opportunities, especially in diagnostic systems, drug delivery systems, improved reporting and related capacity building (Xia et al., 2014a). However, a well-coordinated, culturally sensitive approach is needed. China's experience suggests that pilot projects are the best way to test the transfer of the Chinese experience to the African context.

Randall Kramer
Ning Xiao
Xiao-Nong Zhou

ACKNOWLEDGEMENTS

This special issue was supported by Chinese National Science and Technology Major Project (grant No. 2012ZX10004-220) and China UK Global Health Support Programme (grant no. GHSP-CS-OP1, -OP2, and -OP3).

REFERENCE

Chen, C., 2014. Development of antimalarial drugs and their application in China: a historical review. Infect. Dis. Poverty 3, 9.

Chen, S.B., Ju, C., Chen, J.H., Zheng, B., Huang, F., Xiao, N., Zhou, X., Tambo, E., Zhou, X.N., 2014. Operational research needs toward malaria elimination in China. Adv. Parasitol. 86, 109–127.

Cohen, J.M., Smith, D.L., Cotter, C., Ward, A., Yamey, G., Sabot, O.J., Moonen, B., 2012. Malaria resurgence: a systematic review and assessment of its causes. Malar. J. 11, 122.

Gao, Q., 2011. Opportunities and challenges of malaria elimination in China. Chin. J. Schisto. Control 23, 347–349 (in Chinese).

Feng, J., Xia, Z.G., Vong, S., Yang, W.Z., Zhou, S.S., Xiao, N., 2014a. Preparedness for malaria resurgence in China: case study on imported cases in 2000–2012. Adv. Parasitol 86, 229–64.

Feng, X.Y., Xia, Z.G., Vong, S., Yang, W.Z., Zhou, S.S., 2014b. Surveillance and response to drive the national malaria elimination programme. Adv. Parasitol 86, 81–108.

Qian, Y.J., Zhang, L., Xia, Z.G., Vong, S., Yang, W.Z., Wang, D.Q., Xiao, N., 2014. Preparation for malaria resurgence in China: approach in risk assessment and rapid response. Adv. Parasitol 86, 265–84.

Tang, L.H., 2009. Malaria in China: from control to elimination. Int. Med. Parasitic. Dis. 36, 8 (in Chinese).

Tambo, E., Adedeji, A.A., Huang, F., Chen, J.H., Zhou, S.S., Tang, L.H., 2012. Scaling up impact of malaria control programmes: a tale of events in Sub-Saharan Africa and People's Republic of China. Infect. Dis. Poverty 1, 7.

Wang, R.B., Zhang, Q.F., Zheng, B., Xia, Z.G., Zhou, S.S., Tang, L.H., Gao, Q., Wang, L.Y., Wang, R.R., 2014. Transition from control to elimination: impact of the 10-year global fund project on malaria control and elimination in China. Adv. Parasitol 86, 285–314.

Xia, Z.G., Wang, R.B., Wang, D.Q., Feng, J., Zheng, Q., Deng, C.S., Abdulla, S., Guan, Y.Y., Ding, W., Yao, J.W., Qian, Y.J., Bosman, A., Newman, R.D., Tambo, E., O'leary, M., Xiao, N., 2014a. China–Africa cooperation initiatives in malaria control and elimination. Adv. Parasitol 86, 215–336.

Xia, Z.G., Zhang, L., Feng, J., Li, M., Feng, X.Y., Tang, L.H., Wang, S.Q., Yang, H.L., Gao, Q., Kramer, R., Tambo, E., Yap, P., Zhou, X.N., 2014b. Lessons from malaria control to elimination: case study in Hainan and Yunnan provinces. Adv. Parasitol 86, 47–80.

Yang, F.Z., Yap, P., Zhang, S.Y., Xie, H.G., Ouyang, R., Lin, Y.Y., Chen, Z.Y., 2014. Surveillance and response strategy in the malaria post-elimination stage: case study of Fujian Province. Adv. Parasitol 86, 181–202.

Yin, J.H., Zhou, S.S., Xia, Z.G., Wang, R.B., Qian, Y.J., Yang, W.Z., Zhou, X.N., 2014. Historical patterns of malaria transmission in China. Adv. Parasitol 86, 1–20.

Zhang, H.W., Liu, Y., Zhang, S.S., Xu, B.L., Li, W.D., Tang, J.H., Zhou, S.S., Huang, F., 2014. Preparation of malaria resurgence in China: case study of vivax malaria re-emergence and outbreak in Huang-Huai plain in 2006. Adv. Parasitol 86, 203–28.

Zhou, X.N., Xia, Z.G., Wang, R.B., Qian, Y.J., Zhou, S.S., Utzinger, J., Tanner, T., Kramer, R., Yang, W.Z., 2014. Feasibility and roadmap analysis for malaria elimination in China. Adv. Parasitol 86, 21–46.

Zhu, M., Ruan, W., Fei, S.J., Son, J.Q., Zhang, Y., Mao, X.G., Pan, Q.C., Zhang, L.L., Guo, X.Q., Xu, J.H., Chen, T.M., Zhou, B., Yao, L.N., Cai, L., 2014. Approaches to the evaluation of malaria elimination at county level: case study in the Yangtze River Delta Region. Adv. Parasitol 86, 133–80.

CHAPTER ONE

Historical Patterns of Malaria Transmission in China

Jian-Hai Yin[1], Shui-Sen Zhou[1,*], Zhi-Gui Xia[1], Ru-Bo Wang[1], Ying-Jun Qian[1], Wei-Zhong Yang[2], Xiao-Nong Zhou[1,*]

[1]National Institute of Parasitic Diseases, Chinese Center for Disease Control and Prevention; Key Laboratory of Parasite and Vector Biology, MOH; WHO Collaborating Centre for Malaria, Schistosomiasis and Filariasis, Shanghai, People's Republic of China
[2]Chinese Preventive Medicine Association, Beijing, People's Republic of China; Chinese Center for Disease Control and Prevention, Beijing, People's Republic of China
*Corresponding author: E-mail: zss163@hotmail.com, zhouxn1@chinacdc.cn

Contents

1. Introduction	2
2. Background	2
3. Historical Patterns of Malaria in China	3
3.1 First phase: indeterminacy of malaria transmission (1949–1959)	3
3.2 Second phase: outbreak and pandemic transmission (1960–1979)	5
3.3 Third phase: decline with sporadic case distribution (1980–1999)	8
3.4 Fourth phase: low transmission with a re-emergence in Central China (2000–2009)	10
4. Looking Forward	12
4.1 Challenges	13
4.1.1 Imported malaria	*13*
4.1.2 Asymptomatic and low-parasitemia infections	*14*
4.1.3 Resistance of P. falciparum to artemisinin	*14*
4.2 Recommendation	14
5. Conclusion	15
Acknowledgements	15
References	15

Abstract

The historical patterns of malaria transmission in the People's Republic of China from 1949 to 2010 are presented in this chapter to illustrate the changes in epidemiological features and malaria burden during five decades. A significant reduction of malaria incidence has resulted in initiation of a national malaria elimination programme. However, challenges in malaria elimination have been identified. Foci (or hot spots) have occurred in unstable transmission areas, indicating an urgent need for strengthened surveillance and response in the transition stage from control to elimination.

Advances in Parasitology, Volume 86
ISSN 0065-308X
http://dx.doi.org/10.1016/B978-0-12-800869-0.00001-9

1. INTRODUCTION

Malaria is one of the most important tropical diseases, mostly affecting poor and vulnerable groups in tropical and subtropical areas of the world. The disease has caused calamity throughout the history of mankind, with records dating back more than 4000 years ago, and it has seriously affected the socioeconomic development in the People's Republic of China (P.R. China). Since 1949, the number of malaria cases, mainly caused by infections of *Plasmodium falciparum* and *P. vivax* in P.R. China, have declined dramatically through years of efforts. These achievements are attributed to the many effective interventions at different malaria transmission phases. A national malaria elimination programme (NMEP) was launched in P.R. China in 2010, with a goal of eliminating malaria by 2020. This chapter presents a comprehensive review of the historic patterns of malaria transmission in P.R. China so that the reader can understand the perspectives of NMEP initiation.

2. BACKGROUND

Great achievements in the global battle against malaria have been made in recent decades, with tremendous funding for malaria research and innovative measures, such as rapid diagnostic tests (RDTs), long-lasting insecticidal nets and artemisinin-based combination therapy (ACT). According to the *World Malaria Report*, in 2013 (WHO, 2013), global malaria mortality and incidence rates were reduced by about 45% and 29% compared to the year 2000, respectively. However, malaria epidemiology changed significantly due to different factors, such as vector control measures and increasing population flows, resulting in several new challenges to meet the terminal goal of malaria eradication (Bhatia et al., 2013; Cotter et al., 2013; Kitua et al., 2011). Thus, malaria is still one of the greatest public health problems around the world, and there is still a long way to go to overcome this ancient foe.

Malaria has been recorded in Chinese traditional medicine books, and the prevalence of the disease dates back about 4000 years ago in the history of P.R. China. The disease spreads widely, especially in rural areas, and outbreaks occurred frequently in the past. Before 1949, it was estimated that more than 30 million cases of malaria occurred nationwide each year, and mortality was around 1% (Zhou, 1985). However, through sustained efforts in different control phases, malaria has declined dramatically, to less than 15,000 cases in 2009 (Zhou et al., 2011a), and the endemic areas have shrunk greatly. Moreover, the transmission of malaria species has changed; no autochthonous malariae malaria and ovale malaria were reported in recent years, and falciparum malaria was only reported in Yunnan (Xia et al., 2012; Xia et al., 2013; Zhou et al., 2011b). Vivax malaria is the predominant species in P.R. China, with a significant reduction compared to the historical data (Tang et al., 2012; Zhou, 1991). The main malaria vector mosquitoes are *Anopheles sinensis*, *An. dirus*, *An. anthropophagus* and *An. minimus* (MOH, 2007), while *An. jeyporiensis candidiensis*, *An. pseudowillmori* and *An. sacharovi* also play roles in malaria transmission in some specific localities (Wu et al., 2009; Wu et al., 2012; Wu et al., 2013; Zhou, 1991). To fulfil Millennium Development Goals and achieve global eradication of malaria, the Chinese government initiated the NMEP in 2010, with the goal of eliminating malaria nationwide by 2020. Although tremendous successes in malaria control have been achieved, challenges still exist in the transition stage from control towards elimination.

The main objective of this chapter is to comprehensively review malaria transmission patterns, including malaria cases, vector mosquitoes and specific strategies, in the different malaria transmission phases in P.R. China, to synthesize the experiences and lessons worthy to be learnt by other malaria endemic areas. We also identify the potential challenges that need to be addressed in order to reach the goal of malaria elimination nationally and eradication globally.

3. HISTORICAL PATTERNS OF MALARIA IN CHINA

From the foundation of P.R. China in 1949 to the launch of NMEP in 2010, the malaria transmission in the country can be principally divided into four phases, according to the transmission profiles, namely: (1) indeterminacy of malaria transmission (1949–1959), (2) outbreak and pandemic transmission (1960–1979), (3) decline with sporadic case distribution (1980–1999), and (4) low transmission with a re-emergence in central China (2000–2009) (Figure 1.1).

3.1 First phase: indeterminacy of malaria transmission (1949–1959)

Although malaria was highly prevalent nationwide, the information on parasite species, vectors, epidemiology, and geographic/demographic distributions were not well addressed in this time period because of the country's poor economy and low capacity. However, many professional institutes were established in succession at national and provincial levels, and malaria prevalence was investigated in some endemic areas for professional training and pilot studies. The first five-year National Malaria Control Programme (NMCP) was issued with two epoch-making events: malaria was designated as a notifiable disease and a malaria-reporting mechanism was implemented nationwide (MOH, 1956).

Many investigations and surveys were carried out to understand the malaria profile and to determine epidemiological factors. Baseline data on *Plasmodium* species and vectors were collected from trials (Tang et al., 2012; Zhou, 1981). The country was divided into four different zones on the basis of the extensive epidemiological surveys (Tang et al., 2012; Zhou, 1981; Zhou, 1991). Zone 1 was the tropical and subtropical areas covering the southern parts of Yunnan, most parts of Guangdong (including Hainan Island), Guangxi, and southeastern Fujian. In Zone 1, *P. falciparum* was the predominant species; *P. vivax* was common, while *P. malariae* was scattered and *P. ovale* was also suspected in the southwestern Yunnan. The season of malaria transmission was from March to December. Zone 2 included Guizhou, Hunan, Jiangxi, Hubei, Zhejiang and Shanghai, northern Yunnan, parts of Tibet, Guangxi, Guangdong, Gansu, Shanxi and Henan. *P. vivax* was the main species, and the season of malaria transmission was from May to November, with a peak from August to October. Zone 3 covered Shandong, Liaoning, Jilin, Heilongjiang, Beijing and Tianjin, most parts of Hebei and Shaanxi, parts of Shanxi, Henan, Jiangsu, Anhui and Xinjiang. Only vivax malaria was prevalent in this zone, with the transmission season lasting from June to November (with the peak from August to October). Zone 4 consisted of malaria-free areas including the cold high-altitude areas, dry desert, plateau, etc.

An. sinensis was found to be widely distributed in most parts of the country; it was the only malaria vector or the preponderant one in many provinces, such as Gansu, Liaoning, Shanxi, Shaanxi, Shandong, Henan, Shanghai, Jiangsu and Zhejiang Provinces. It was also the main vector in Anhui, Hubei, Fujian, Hunan and Sichuan provinces, with *An. minimus* also playing an important role in malaria transmission in these areas. *An. minimus* was the main vector in the mountainous areas of Guangxi and the high or hyper-high transmission areas in Guizhou. *A. sinensis* was the main vector in the plains of Guangxi and the northwestern and southern of Guizhou. Also, *An. minimus* was the preponderant vector in Guangdong, Hainan, Yunnan and Jiangxi provinces. Other mosquitoes such as *An. dirus* in Hainan island (a part of Guangdong province then) and *An. jeyporiensis candidiensis* in several southern provinces and *An. sacharovi* in Xinjiang also played a part role in malaria transmission in local areas (Tang et al., 2012; Zhou, 1991).

3.2 Second phase: outbreak and pandemic transmission (1960–1979)

The transmission of malaria in this period was unstable and hyperendemic (Figure 1.1(a)). A pandemic transmission of vivax malaria occurred in the Huang-huai plain (Sang et al., 2011) and central China, including Jiangsu, Shandong, Henan, Anhui and Hubei provinces, because of the interruption of malaria control activities attributed by a natural disaster in the beginning of 1960s and political unrest in 1967. In addition, malarial outbreaks caused by population movements were recorded in Hainan, Zhejiang, Fujian and Yunnan, etc. (Tang et al., 2012; Zhou, 1991). During the 20 years, more than 18,000,000 malaria cases were accumulatively reported from each of five central parts of China, including Henan, Jiangsu, Shandong, Anhui and Hubei provinces. More than 1,000,000 cases were individually reported from the other 10 provinces including Jiangxi, Zhejiang, Sichuan, Hunan, Fujian, Guangdong, Hebei, Guangxi, Shanghai and Yunnan, and approximately 1,000,000 cases were form other endemic areas (Tang et al., 2012). Annual deaths suffered from malaria numbered in the hundreds, with the highest total of approximately 2050 in 1963 (Figure 1.2(a)). The country was stratified into four transmission areas according to malaria prevalence (Figure 1.3(a)).

The pandemic transmission in the central parts was caused by *An. Sinensis*, with its wide distribution, large population, and outdoor-biting behaviour. Nevertheless, malaria prevalence in Hainan and the southeastern coastal areas decreased because of the shrinking distribution of *An. minimus*. During this phase, *An. anthropophagus* was identified as an important malaria vector in the some southern and central areas with a high transmission capacity. Meanwhile, *An. dirus* become the main vector in Hainan with the elimination of *An. minimus* (Tang et al., 2012).

Specific strategies were formulated to control outbreaks and pandemics, mainly based on the species of vector in particular areas. Vector control interventions, such as insecticide treated net (ITN), combined with case management were the primary strategies implemented in the southern areas, where *An. minimus* was the predominant vector (MOH, 1964; Tang et al., 2012). The northern parts of China, where *An. sinensis* was the key vector, adopted integrated measures, including environment improvement, a radical treatment that administration with primaquine plus pyrimethamine/quinine, and prophylactic chemotherapy with pyrimethamine in high-transmission settings.

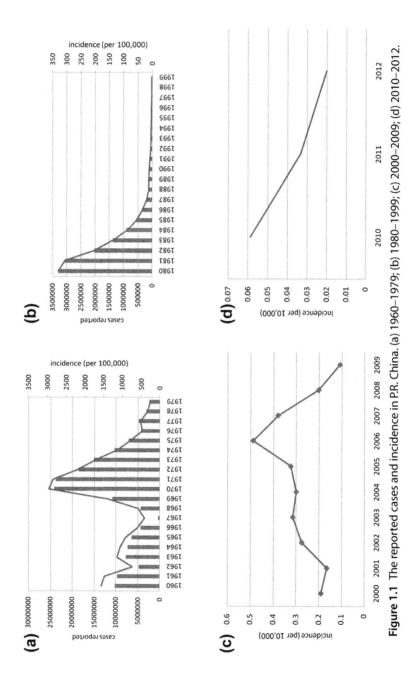

Figure 1.1 The reported cases and incidence in P.R. China. (a) 1960–1979; (b) 1980–1999; (c) 2000–2009; (d) 2010–2012.

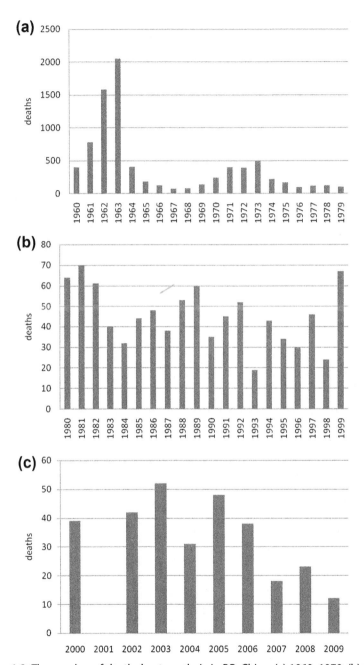

Figure 1.2 The number of death due to malaria in P.R. China. (a) 1960–1979; (b) 1980–1999; (c) 2000–2009.

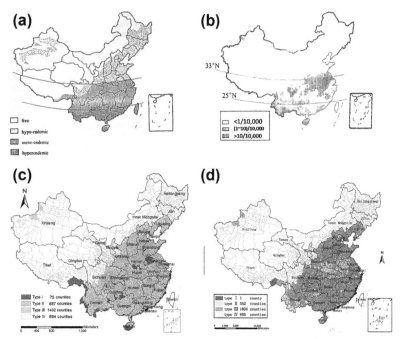

Figure 1.3 The geographical stratification of malaria based on the incidences in P.R. China: (a) 1979 (Tang et al., 1991); (b) 1985 (Tang et al., 2012); (c) 2007–2009 (Yin et al., 2013); (d) 2010–2012 (Yin et al., 2013).

3.3 Third phase: decline with sporadic case distribution (1980–1999)

In this phase, malaria prevalence was declining gradually, with the exception of the year 1989 and 1994 (MoH Expert Advisory Committee on Malaria, 1993; MoH Expert Advisory Committee on Malaria, 1994; MoH Expert Advisory Committee on Malaria, 1995; MoH Expert Advisory Committee on Malaria, 1998; MoH Expert Advisory Committee on Malaria, 1999; MoH Expert Advisory Committee on Malaria, 2000; MoH Expert Advisory Committee on Parasitic Diseases, 1988; MoH Expert Advisory Committee on Parasitic Diseases, 1989; MoH Expert Advisory Committee on Parasitic Diseases, 1990; MoH Expert Advisory Committee on Parasitic Diseases, 1991; MoH Expert Advisory Committee on Parasitic Diseases, 1992; MoH Expert Advisory Committee on Parasitic Diseases, 1996; MoH Expert Advisory Committee on Parasitic Diseases, 1997; CoMS Malaria Commission, 1983; CoMS Malaria Commission, 1984; CoMS Malaria Commission, 1985; CoMS Malaria Commission, 1986; CoMS Malaria

Commission, 1987) (Figure 1.1(b)). The case reduction rates were reported of approximately 30–43% during 1982–1988 and approximately 15–25% during 1990–1996, except for 1994 because of increasing population movement towards southern parts of China with high malaria transmission (MoH Expert Advisory Committee on Malaria, 1995). In the first 10 years of this period, more than 40% of cases were reported from central China including Jiangsu, Shandong, Henan, Anhui and Hubei, while in the latter 10 years more than 40% of cases were from the southern China, including Yunnan, Hainan, Guizhou, Guangxi, Guangdong, Fujian, Sichuan and Chongqing. Vivax malaria was the predominant species and was widely distributed in the endemic areas in the period. Falciparum malaria has been limited to Yunnan and Hainan provinces since 1995, although it also occurred in a few provinces in the past. The number of death due to malaria dramatically decreased, with less than 70 deaths reported annually (Tang et al., 1991) (Figure 1.2(b)). Malaria stratification was also updated along with the changing transmission (Figure 1.3(b)).

No significant variation in the species and distribution of vectors was found, but the malaria situation in the areas with different main vectors underwent a great change. In the areas where *An. sinensis* was the dominant vector, malaria transmission steadily decreased due to the shrinking of breeding grounds. However, the transmission was unstable in the areas where *An. anthropophagus* was the main vector, with a higher vector capacity (Pan et al., 1999; Wang et al., 1986). *An. dirus* was mainly distributed in particular areas in Yunnan and Hainan, which greatly contributed to the high prevalence of falciparum malaria in Hainan.

Control measures were tailored to different malaria profiles and specific vectors. At the beginning of this phase, case management combined with vector control interventions were adopted in the areas where vivax malaria was prevalent by *An. sinensis*. ITNs using DDT or pyrethroid insecticides followed by case management were applied in the areas where *An. anthropophagus* or *An. minimus* was the main vector. Environment modifications for breeding sites reduction were added in the areas with *An. dirus*. Residual foci elimination and case surveillance were carried out in areas with relatively low transmission, where the prevalence was less than 5 per 10,000 population. Especially in this phase, the concept and target of malaria eradication were initially proposed in 1983 (MOH, 1983). Subsequent technical guidelines and protocols related to case management, vector control, surveillance, and training were developed, with a target of basically eradicating malaria in particular localities

(MOH, 1984a,b; MOH, 1986; MOH, 1986; MOH, 1989; MOH, 1990; MOH, 1996; MOH et al., 1985).

3.4 Fourth phase: low transmission with a re-emergence in Central China (2000–2009)

Malaria transmission presented an unstable pattern from 2000 to 2009, with the highest incidence of 0.49 per 10,000 in 2006 and the lowest incidence of 0.11 per 10,000 in 2009 (MoH Expert Advisory Committee on Malaria, 2001; Sheng et al., 2003; Zhou et al., 2005; Zhou et al., 2006a; Zhou et al., 2006b; Zhou et al., 2007; Zhou et al., 2008; Zhou et al., 2009; Zhou et al., 2011) (Figure 1.1(c)). Moreover, there still existed a large number of suspected malaria cases nationwide, with no data available for the suspected cases in 2000 and 2001. Less than 100 deaths due to malaria were reported in each year (Figure 1.2(c)). A new malaria transmission map was drawn for further stratification leading to malaria control and elimination (Figure 1.3(c)).

In particular, since 2001, a re-emergence unpredictably occurred in central parts of China along the Huang-Huai River, including Anhui, Henan, Hubei and Jiangsu provinces. Anhui Province alone accounted for more than 50% of the country's total cases. Even so, Yunnan Province maintained its rank in the top three for numbers and incidence rates (Figure 1.4). Vivax malaria accounted for a majority of cases. Locally-transmitted falciparum malaria declined significantly and was limited to Yunnan and Hainan provinces (Figure 1.5).

An. sinensis, *An. dirus*, *An. anthropophagus,* and *An. minimus* were still the main malaria vectors in this period (MOH, 2007), while *An. pseudowillmori* was identified as the key malaria transmiting vector in Motuo County of Tibet (Wu et al., 2009). The intensity of *An. anthropophagus* and *An. minimus* were much lower, so it was difficult to capture in the most endemic areas (Wang et al., 2013). New geographic distributions of *An. anthropophagus* were identified in three provinces of Henan, Shandong and Liaoning (Gao et al., 2002; Ma et al., 2000; Shang et al., 2007).

Specific strategies were adopted in particular malaria transmission areas and for the control of re-emergence. Comprehensive measures were intensively implemented in the high-transmission areas of borders of Yunnan and the central southern mountainous counties of Hainan to control or eliminate falciparum malaria, supported by the first round of the Global Fund to Fight AIDS, Tuberculosis and Malaria programme (GFATM).

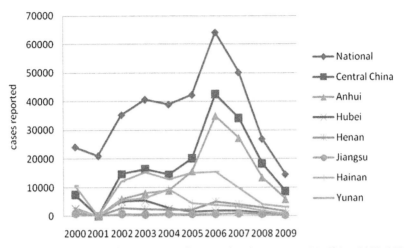

Figure 1.4 The changes of the number of reported malaria cases in P.R. China, 2000–2009.

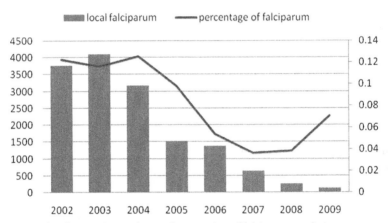

Figure 1.5 The changes in number and proportion of falciparum malaria cases reported in P.R. China, 2002–2009.

The main activities involved in various aspects of malaria control which included early diagnosis and appropriate treatment, focal vector control interventions, malaria management among mobile populations, health education and promotion, and surveillance, monitoring and evaluation. The successful application and implementation of the fifth round of the GFATM greatly contributed to the rollback of re-emergence in central China and made a strong base for the NMEP.

Table 1.1 Epidemiological features of four phases of malaria in P.R. China from 1949 to 2009

Phase (period)	Epidemiological features	Control strategy
Indeterminacy of malaria transmission phase (1949–1959)	Seriously prevalent, information about malaria species and vectors unclear nationwide	Notifiable disease enrolment, malaria-reporting mechanism, malaria control trials, baseline surveys of *Plasmodium* and vectors, etc.
Outbreak and pandemic transmission phase (1960–1979)	Unstable and hyperendemic, vivax malaria pandemic in the Huang-huai plain and central China	ITNs combined with case management; integrated measures composed of environment improvements, radical treatment with primaquine plus pyrimethamine/quinine, prophylactic chemotherapy with pyrimethamine, etc.
Decline phase with sporadic case distribution (1980–1999)	Decreasing, sporadic distribution	Case management, vector control with ITNs, environment modification, case surveillance, personnel training, etc.
Low transmission phase with a re-emergence in the central China (2000–2009)	Unstable but low transmission, unpredictable re-emergence in the central China	Early diagnosis and appropriate treatment, focal vector control interventions, malaria management among mobile populations, health education and promotion, and surveillance, monitoring and evaluation, etc.

4. LOOKING FORWARD

With tremendous investments and intensive interventions, malaria has been effectively controlled in China (Figure 1.1(d), Table 1.1), with autochthonous cases being dramatically reduced to limited localities (Xia et al., 2012; Xia et al., 2013; Zhou et al., 2011b) (Figure 1.6). In recent years, locally transmitted malaria was limited to a few counties, from 303 counties in 18 provinces in 2010, to 160 counties in 12 provinces in 2011, down to 41 counties in 5 provinces in 2012 (Yin et al., 2013) (Figure 1.3(d)). However, almost all of counties in the whole country have reported the importation of malaria mainly due to labours back from other countries, covering 651 counties in 23 provinces in 2010, 760 counties in 26 provinces in 2011, and 598 counties in

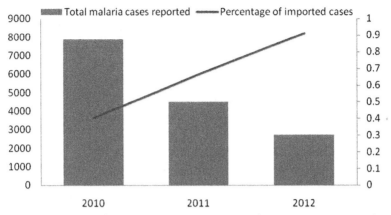

Figure 1.6 The trends of the number and proportion of malaria cases reported in P.R. China, 2010–2012.

29 provinces in 2012. Only 15 counties – 11 counties of Anhui, 3 of Yunnan, and 1 of Tibet – had no imported cases in the years of 2010, 2011 and 2012. Therefore, there still have technical and managerial challenges in the transition from control towards elimination (Zheng et al., 2013; Tambo et al., 2014a).

4.1 Challenges

As efforts toward malaria elimination progress, more and more areas will be investigated by timely experts for surveillance and response after onsets of malaria cases in order to interruption of malaria transmission promptly in the NMEP. This will result in a shortage of human resources and finances for malaria surveillance and response (Yin et al., 2013), including vector surveillance (Tambo et al., 2014b). It is an essential to sustain personnel training and finances, as well as formulate proposals for postelimination malaria surveillance. It must pay much more attention to the following three scenarios, while keeping routine surveillance for malaria cases and mosquito vectors.

4.1.1 Imported malaria

Imported malaria has become an important challenge with an increasing proportion of total cases, accounting for more than 90% in 2012 (Figure 1.6). There has been a remarkable upward trend in recent years, with the increasing movement of labourers returning from Africa and frequent movement cross the borders. Moreover, not only dominant falciparum malaria from African countries and vivax malaria from Asian countries, but also some malariae and ovale malaria, have been reported with an increasing pattern as well (Xia et al., 2012). The transmission settings, especially the distribution of malaria vectors, cannot be changed fundamentally following the reduction of malaria transmission. It is necessary to monitor whether domestic mosquitoes in P.R. China could be infected with the imported *Plasmodium* spp. In addition, current measures on imported cases and foci-related investigation should refer to the diagnosis, treatment, and management of imported malaria cases. In addition, domestic multisectoral cooperation, international multilateral collaboration, and information harmonization should be sustained and strengthened.

4.1.2 Asymptomatic and low-parasitemia infections

Little data about asymptomatic or low-parasitemia individuals in P.R. China can be retrieved. These populations account for most infections from *Anopheles* mosquitoes and are difficult to identify due to low-parasitemia (Ganguly et al., 2013; Harris et al., 2010; Starzengruber et al., 2014; Zoghi et al., 2012). It is also an important challenge related to transmission interruption. China lacked malaria-sensitive diagnostic tools for asymptomatic patients and patients with low levels of parasitemia. Although RDTs, polymerase chain reaction, and other techniques have been used in NMEP, most cases were tested based on the results of microscopy (Zheng et al., 2013). Therefore, development of much more sensitive and cost-effective techniques for malaria diagnosis is essential for the NMEP in P.R. China.

4.1.3 Resistance of P. falciparum to artemisinin

Although *P. falciparum* resistance to artemisinin is restricted to the Greater Mekong Subregion including Cambodia, Laos, Myanmar, Thailand and Vietnam, different day-3 positivity rates among malaria patients treated with artemether-lumefantrine were reported in South America (Ariey et al., 2014; Liu, 2014; Miotto et al., 2013; WHO, 2014). The consequence of widespread resistance to artemisinin will be catastrophic (Bhatia et al., 2013). Routine monitoring of *P. falciparum* resistance to artemisinin is especially necessary to carry out in Yunnan border areas, although no artemisinin-resistant cases have been found so far (Starzengruber et al., 2012).

4.2 Recommendation

To achieve the final goal of eliminating malaria in China, sustained efforts are essential to transfer from the control stage to the elimination stage. Therefore, efforts to accelerate the transition stage from control to elimination are recommended in the following five fields:
- Improving surveillance technology with more sensitivity of malaria detection as well as modeling transmission patterns supported by the

essential database (ERACGoM mal 2011; malEra Consultative Group on Monitoring et al., 2011; Zhou et al., 2013; Liu et al., 2012; Xia et al., 2013).
- Strengthening the health system through multisectorial cooperation, which enforces the response action to malaria foci (Alonso and Tanner, 2013; ERACGoHS mal, 2011).
- Sustaining the NMEP investments to further improve the quality of the programme activities, such as monitoring and evaluation, health education, etc. (Alonso and Tanner, 2013; Brijnath et al., 2014; Koenker et al., 2014; ERACGoIS mal, 2011; malEra Consultative Group on Monitoring, 2011).
- Enforcing capacity building, with both professional institutions and personnel working in NMEP (Alonso et al., 2011).
- Encouraging more activities in research and development in order to cope with emerging issues, such as artemisinin resistance, identification of parasite origins, G6PD deficiency, community-based interventions, etc. (Alonso and Tanner, 2013; Brijnath et al., 2014; Liu, 2014; ERACGoD mal, 2011).

5. CONCLUSION

Malaria incidence in P.R. China has been reduced to the lowest level in the history. However, the NMEP goal of malaria elimination will not be achieved by 2020 if the efforts of the consistent phase-specific interventions are not intensified enough. Major challenges include malaria surveillance and response, which should be tailored to local settings in the transition stage from control to elimination, particularly for the increasing patterns of imported malaria cases, asymptomatic and low-parasitemia infections, and artemisinin resistance. It is suggested that the activities of the NMEP, including multilateral cooperation, investments in NMEP, health systems, and personnel capacity, research and development, have to be strengthened in both the elimination and postelimination stages to finally win the battle against malaria.

ACKNOWLEDGEMENTS
This project was supported by the National S & T Major Programme (grant no. 2012ZX10004220), the National S & T Supporting Project (grant no. 2007BAC03A02) and the China UK Global Health Support Programme (grant no. GHSP-CS-OP1).

REFERENCES
Ariey, F., Witkowski, B., Amaratunga, C., Beghain, J., Langlois, A.C., Khim, N., Kim, S., Duru, V., Bouchier, C., Ma, L., et al., 2014. A molecular marker of artemisinin-resistant *Plasmodium falciparum* malaria. Nature 505, 50–55.
Alonso, P.L., Tanner, M., 2013. Public health challenges and prospects for malaria control and elimination. Nat. Med. 19, 150–155.

Alonso, P.L., Brown, G., Arevalo-Herrera, M., Binka, F., Chitnis, C., Collins, F., Doumbo, O.K., Greenwood, B., Hall, B.F., Levine, M.M., et al., 2011. A research agenda to underpin malaria eradication. PLoS Med. 8, e1000406.

Bhatia, R., Rastogi, R.M., Ortega, L., 2013. Malaria successes and challenges in Asia. J. Vector Borne Dis. 50, 239–247.

Brijnath, B., Butler, C.D., McMichael, A.J., 2014. In an interconnected world: joint research priorities for the environment, agriculture and infectious disease. Infect. Dis. Poverty 3, 2.

Cotter, C., Sturrock, H.J., Hsiang, M.S., Liu, J., Phillips, A.A., Hwang, J., Gueye, C.S., Fullman, N., Gosling, R.D., Feachem, R.G., 2013. The changing epidemiology of malaria elimination: new strategies for new challenges. Lancet 382, 900–911.

Expert Advisory Committee on Malaria MoH, 1993. Malaria situation in the People's Republic of China in 1992. Chin. J. Parasitol. Parasit. Dis. 11, 161–164.

Expert Advisory Committee on Malaria MoH, 1994. Malaria situation in the People's Republic of China in 1993. Chin. J. Parasitol. Parasit. Dis. 12, 161–164.

Expert Advisory Committee on Malaria MoH, 1995. Malaria situation in the People's Republic of China in 1994. Chin. J. Parasitol. Parasit. Dis. 13, 161–164.

Expert Advisory Committee on Malaria MoH, 1998. Malaria situation in the People's Republic of China in 1997. Chin. J. Parasitol. Parasit. Dis. 16, 161–163.

Expert Advisory Committee on Malaria MoH, 1999. Malaria situation in the People's Republic of China in 1998. Chin. J. Parasitol. Parasit. Dis. 17, 193–195.

Expert Advisory Committee on Malaria MoH, 2000. Malaria situation in the People's Republic of China in 1999. Chin. J. Parasitol. Parasit. Dis. 18, 129–131.

Expert Advisory Committee on Parasitic Diseases MoH, 1988. Malaria situation in China, 1987. Chin. J. Parasitol. Parasit. Dis. 6, 241–244.

Expert Advisory Committee on Parasitic Diseases MoH, 1989. Malaria situation in China, 1988. Chin. J. Parasitol. Parasit. Dis. 7, 241–244.

Expert Advisory Committee on Parasitic Diseases MoH, 1990. Malaria situation in China, 1989. Chin. J. Parasitol. Parasit. Dis. 8, 241–244.

Expert Advisory Committee on Parasitic Diseases MoH, 1991. Malaria situation in the People's Republic of China in 1990. Chin. J. Parasitol. Parasit. Dis. 9, 250–253.

Expert Advisory Committee on Parasitic Diseases MoH, 1992. Malaria situation in the People's Republic of China in 1991. Chin. J. Parasitol. Parasit. Dis. 10, 161–165.

Expert Advisory Committee on Parasitic Diseases MoH, 1996. Malaria situation in the People's Republic of China in 1995. Chin. J. Parasitol. Parasit. Dis. 14, 169–172.

Expert Advisory Committee on Parasitic Diseases MoH, 1997. Malaria situation in the People's Republic of China in 1996. Chin. J. Parasitol. Parasit. Dis. 15, 129–132.

Expert Advisory Committee on Malaria MoH, 2001. Malaria situation in the People's Republic of China in 2000. Chin. J. Parasitol. Parasit. Dis. 19, 257–259.

Gao, Q., Cooper, R.D., Zhou, H.Y., Pan, B., Yang, W., Guo, C.K., Huang, G.Q., Li, F.H., Li, J.L., Shen, B.X., et al., 2002. Genetic identification of *Anopheles anthropophagus* and *Anopheles sinesis* by PCR-RFLP. Chin. J. Zoonoses 18, 39–42.

Ganguly, S., Saha, P., Guha, S.K., Biswas, A., Das, S., Kundu, P.K., Maji, A.K., 2013. High prevalence of asymptomatic malaria in a tribal population in eastern India. J. Clin. Microbiol. 51, 1439–1444.

Harris, I., Sharrock, W.W., Bain, L.M., Gray, K.A., Bobogare, A., Boaz, L., Lilley, K., Krause, D., Vallely, A., Johnson, M.L., et al., 2010. A large proportion of asymptomatic *Plasmodium* infections with low and sub-microscopic parasite densities in the low transmission setting of Temotu Province, Solomon Islands: challenges for malaria diagnostics in an elimination setting. Malar. J. 9, 254.

Kitua, A., Ogundahunsi, O., Lines, J., Mgone, C., 2011. Conquering malaria: enhancing the impact of effective interventions towards elimination in the diverse and changing epidemiology. J. Global Infect. Dis. 3, 161–165.

Koenker, H., Keating, J., Alilio, M., Acosta, A., Lynch, M., Nafo-Traore, F., 2014. Strategic roles for behaviour change communication in a changing malaria landscape. Malar. J. 13, 1.

Liu, D.Q., 2014. Surveillance of antimalarial drug resistance in China in the 1980s-1990s. Infect. Dis. Poverty 3, 8.

Liu, J., Yang, B., Cheung, W.K., Yang, G., 2012. Malaria transmission modelling: a network perspective. Infect. Dis. Poverty 1, 11.

MOH, 2007. Malaria Control Manual. People's Medical Publishing House, Beijing.

MOH, 1956. Malaria Control Programme.

MOH, 1964. Technical Proposal for Malaria Control in China.

MOH, 1983. Malaria control programme in China, 1983–1985.

MOH, 1984a. Malaria control management regulation.

MOH, 1984b. The criteria and assessment for malaria control and basic eradication (Trial).

Malaria Commission CoMS, 1983. Epidemiological status of malaria in the People's Republic of China in 1981 and suggestions for control. J. Parasitol. Parasit. Dis. 1, 2–4.

Malaria Commission CoMS, 1984. Malaria control and morbidity in China in 1982. J. Parasitol. Parasit. Dis. 2, 1–2.

Malaria Commission CoMS, 1985. The malaria situation in 1984 in the People's Republic of China. J. Parasitol. Parasit. Dis. 3, 241–243.

Malaria Commission CoMS, 1986. The malaria situation in China, 1985. J. Parasitol. Parasit. Dis. 4, 241–243.

Malaria Commission CoMS, 1987. Malaria situation in China. 1986. Chin. J. Parasitol. Parasit. Dis. 5, 241–243.

MOH, 1986a. Malaria Control Programme in China, 1986–1990.

MOH, 1986b. The Interim Criteria for Malaria Control, Basic Eradication, Eradication and Basic Eradication of Falciparum Malaria. .

MOH, 1989. Technical Proposal for Malaria Control.

MOH, 1990. Malaria Control Programme in China, 1991–1995.

MOH, 1996. Malaria Control Programme in China, 1996–2000.

MOH, MOPS, MOEP, MOWC, MOA, 1985. Interim Measures To Manage Malaria In Floating Populations.

Ma, Y.J., Qu, F.Y., Cao, Y.C., Yang, B.J., 2000. On molecular identification and taxonomic status of *Anopheles lesteri* and *Anopheles anthropophagus* in China (Diptera: Culicidae). Chin. J. Parasitol. Parasit. Dis. 18, 325–328.

Miotto, O., Almagro-Garcia, J., Manske, M., Macinnis, B., Campino, S., Rockett, K.A., Amaratunga, C., Lim, P., Suon, S., Sreng, S., et al., 2013. Multiple populations of artemisinin-resistant *Plasmodium falciparum* in Cambodia. Nat. Genet. 45, 648–655.

mal ERACGoM, 2011. A research agenda for malaria eradication: modeling. PLoS Med. 8, e1000403.

malEra Consultative Group on Monitoring E, Surveillance, 2011. A research agenda for malaria eradication: monitoring, evaluation, and surveillance. PLoS Med. 8, e1000400.

mal ERACGoHS, Operational R, 2011. A research agenda for malaria eradication: health systems and operational research. PLoS Med. 8, e1000397.

mal ERACGoIS, 2011. A research agenda for malaria eradication: cross-cutting issues for eradication. PLoS Med. 8, e1000404.

mal ERACGoD, Diagnostics, 2011. A research agenda for malaria eradication: diagnoses and diagnostics. PLoS Med. 8, e1000396.

Pan, B., Zhu, T.H., Li, Z.Z., Xu, R.H., Xu, Y.Y., Wu, X.G., Lin, R.X., Wu, C.G., Yang, Z.H., Ynag, W.S., 1999. Studies on distribution, ecological feature, malaria transmission effect and control measure of Anopheles anthropophagus in Guangdong province. Chin. J. Vector Biology Control 10, 374–378.

Sang, L.Y., Xu, B.L., Su, Y.P., 2011. Malaria Endemics and Control in Henan Province. Zhongyuan Publishing Media Group, No. 66 Zhengjingwulu, Zhengzhou.

Sheng, H.F., Zhou, S.S., Gu, Z.C., Zheng, X., 2003. Malaria situation in the People's Republic of China in 2002. Chin. J. Parasitol. Parasit. Dis. 21, 193–196.

Shang, L.Y., Chen, J.S., Li, D.F., LI P, Su YP., Liu, H., 2007. Studies on distribution, ecological feature and malaria transmission effect of *Anopheles anthropophagus* in Henan province, China. J. Pathog. Biol. 2, 304–306.

Starzengruber, P., Fuehrer, H.P., Ley, B., Thriemer, K., Swoboda, P., Habler, V.E., Jung, M., Graninger, W., Khan, W.A., Haque, R., et al., 2014. High prevalence of asymptomatic malaria in south-eastern Bangladesh. Malar. J. 13, 16.

Starzengruber, P., Swoboda, P., Fuehrer, H.P., Khan, W.A., Hofecker, V., Siedl, A., Fally, M., Graf, O., Teja-Isavadharm, P., Haque, R., et al., 2012. Current status of artemisinin-resistant falciparum malaria in South Asia: a randomized controlled artesunate monotherapy trial in Bangladesh. PLoS One 7, e52236.

Tambo, E., Ai, L., Zhou, X., Chen, J.H., Hu, W., Bergquist, R., et al., 2014a. Surveillance-response systems: the key to elimination of tropical diseases. Infect. Dis. Poverty 3, 17.

Tambo, E., Ugwu, E.C., Ngogang, J.Y., 2014b. Need of surveillance response systems to combat Ebola outbreaks and other emerging infectious diseases in African countries. Infect. Dis. Poverty 3, 29.

Tang, L.H., Xu, L.Q., Chen, Y.D., 2012. Parasitic Disease Control and Research in China. Beijing Science & Technology Press, Beijing.

Tang, L.H., Qian, H.L., Xu, S.H., 1991. Malaria and its control in the People's Republic of China. Southeast Asian J. Tropical Med. Public Health 22, 467–476.

Wu, S., Pan, J.Y., Wang, X.Z., Zhou, S.S., Zhang, G.Q., Liu, Q., Tang, L.H., 2009. *Anopheles pseudowillmori* is the predominant malaria vector in Motuo County, Tibet Autonomous Region. Malar. J. 8, 46.

Wu, S., Huang, F., Zhou, S.S., Tang, L.H., 2012. Study on malaria vectors in malaria endemic areas of Tibet Autonomous Region. Chin. J. Schistosomiasis Control 24, 711–713.

Wu, S., Huang, F., Wang, D.Q., Xu, G.J., Jiang, W.K., Zhou, S.S., Tang, L.H., Wang, H.J., Zhuoma, Y.J., Yong, J., et al., 2013. Ecological behavior comparison between *Anopheles pseudowillmori* and *A. willmori* in villages with malaria outbreaks in Motuo County, Tibet Autonomous Region. Chin. J. Schistosomiasis Control 25, 362–366.

Wang, H., Gao, C.K., Huang, H.M., Liu, C.F., Qian, H.L., Lin, S.Y., Zhu, G.S., Zhao, S.Q., Zhen, J.J., 1986. Geographic distribution of *Anopheles lesteri anthropophagus* and its role in malaria transmission in Guangxi. Chin. J. Parasitol. Parasit. Dis. 5, 104–106.

Wang, W.M., Cao, J., Zhou, H.Y., Li, J.L., Zhu, G.D., Gu, Y.P., Liu, Y.B., Cao, Y.Y., 2013. Seasonal increase and decrease of malaria vector in monitoring sites of Jiangsu province in 2005-2010. China Trop. Med. 13, 152–155.

WHO, 2013. World Malaria Report 2013. WHO Press, Avenue Appia, 1211 Geneva 27, Switzerland.

WHO, 2014. Update on Artemisinin Resistance.

Xia, S., Allotey, P., Reidpath, D.D., Yang, P., Sheng, H.F., Zhou, X.N., 2013. Combating Infect. Dis. Poverty: a year on. Infect. Dis. Poverty 2, 27.

Xia, Z.G., Yang, M.N., Zhou, S.S., 2012. Malaria situation in the People's Republic of China in 2011. Chin. J. Parasitol. Parasit. Dis. 30, 419–422.

Xia, Z.G., Feng, J., Zhou, S.S., 2013. Malaria situation in the People's Republic of China in 2012. Chin. J. Parasitol. Parasit. Dis. 31, 413–418.

Yin, J.H., Yang, M.N., Zhou, S.S., Wang, Y., Feng, J., Xia, Z.G., 2013. Changing malaria transmission and implications in China towards National Malaria Elimination Programme between 2010 and 2012. PLoS One 8, e74228.

Yin, W., Dong, J., Tu, C., Edwards, J., Guo, F., Zhou, H., et al., 2013. Challenges and needs for China to eliminate rabies. Infect. Dis. Poverty 2, 23.

Zhou, Z.J., 1985. Current status of malaria in China. In: The Asia and Pacific Conference on Malaria: 1985, pp. 31–39.
Zhou, S.S., Wang, Y., Xia, Z.G., 2011a. Malaria situation in the People's Republic of China in 2009. Chin. J. Parasitol. Parasit. Dis. 29, 1–3.
Zhou, S.S., Wang, Y., Li, Y., 2011b. Malaria situation in the People's Republic of China in 2010. Chin. J. Parasitol. Parasit. Dis. 29, 401–403.
Zhou, Z.J., 1991. Malaria Control and Research in China. People's Medical Publishing House, Beijing.
Zhou, Z.J., 1981. The malaria situation in the People's Republic of China. Bull. World Health Organ. 59, 931–936.
Zhou, S.S., Tang, L.H., Sheng, H.F., 2005. Malaria situation in the People's Republic of China in 2003. Chin. J. Parasitol. Parasit. Dis. 23, 385–387.
Zhou, S.S., Tang, L.H., Sheng, H.F., Wang, Y., 2006a. Malaria situation in the People's Republic of China in 2004. Chin. J. Parasitol. Parasit. Dis. 24, 1–3.
Zhou, S.S., Wang, Y., Tang, L.H., 2006b. Malaria situation in the People's Republic of China in 2005. Chin. J. Parasitol. Parasit. Dis. 24, 401–403.
Zhou, S.S., Wang, Y., Tang, L.H., 2007. Malaria situation in the People's Republic of China in 2006. Chin. J. Parasitol. Parasit. Dis. 25, 439–441.
Zhou, S.S., Wang, Y., Fang, W., Tang, L.H., 2008. Malaria situation in the People's Republic of China in 2007. Chin. J. Parasitol. Parasit. Dis. 26, 401–403.
Zhou, S.S., Wang, Y., Fang, W., Tang, L.H., 2009. Malaria situation in the People's Republic of China in 2008. Chin. J. Parasitol. Parasit. Dis. 27 (457), 455–456.
Zoghi, S., Mehrizi, A.A., Raeisi, A., Haghdoost, A.A., Turki, H., Safari, R., Kahanali, A.A., Zakeri, S., 2012. Survey for asymptomatic malaria cases in low transmission settings of Iran under elimination programme. Malar. J. 11, 126.
Zheng, Q., Vanderslott, S., Jiang, B., Xu, L.L., Liu, C.S., Huo, L.L., Duan, L.P., Wu, N.B., Li, S.Z., Xia, Z.G., et al., 2013. Research gaps for three main tropical diseases in the People's Republic of China. Infect. Dis. poverty 2, 15.
Zhou, X.N., Bergquist, R., Tanner, M., 2013. Elimination of tropical disease through surveillance and response. Infect. Dis. poverty 2, 1.

CHAPTER TWO

Feasibility and Roadmap Analysis for Malaria Elimination in China

Xiao-Nong Zhou[1,2,*], Zhi-Gui Xia[1,2], Ru-Bo Wang[1,2], Ying-Jun Qian[1,2], Shui-Sen Zhou[1,2], Jürg Utzinger[3,4], Marcel Tanner[3,4], Randall Kramer[5], Wei-Zhong Yang[6,*]

[1]National Institute of Parasitic Diseases, Chinese Center for Disease Control and Prevention, Shanghai, People's Republic of China
[2]Key Laboratory of Parasite and Vector Biology, MOH; WHO Collaborating Centre for Malaria, Schistosomiasis and Filariasis, Shanghai, People's Republic of China
[3]Department of Epidemiology and Public Health, Swiss Tropical and Public Health Institute, Basel, Switzerland
[4]University of Basel, Basel, Switzerland
[5]Duke Global Health Institute, Duke University, Durham, NC, USA
[6]Chinese Preventive Medicine Association, Beijing, People's Republic of China; Chinese Center for Disease Control and Prevention, Beijing, People's Republic of China
*Corresponding authors: E-mail: zhouxn1@chinacdc.cn; yangwz@chinacdc.cn

Contents

1. Introduction	22
2. Feasibility Assessment at the National Level	23
2.1 Data sources	24
2.2 Assessment indicators	24
2.2.1 Malaria transmission risk index	25
2.2.2 Malaria elimination capacity index	25
2.2.3 Risk of malaria transmission in a population	25
3. Correlation Between Incidence Patterns and Interventions in Four Target Provinces	28
3.1 Historical transmission pattern of four provinces	28
3.2 Correlation between malaria incidence and interventions	29
3.3 Effective intervention during the transition stage from control to elimination	30
4. Feasibility Analysis of Malaria Elimination in China	31
4.1 Transmission risks	31
4.2 Malaria elimination capacities	31
4.2.1 Technical capacity	32
4.2.2 Resources capacity	32
4.3 Feasibility analysis for malaria elimination at the national level	33
4.3.1 Feasibility analysis employed by MTRI and MECI	33
4.3.2 Feasibility analysis employed by malaria incidence, MTRI and MECI	33
4.3.3 Feasibility analysis with geographic variations	33
5. Phase-Based Malaria Elimination Strategies	36
5.1 Classification of elimination phases	36
5.2 Strategy formulations in each stage	37

6. Conclusions and Recommendations 39
   6.1 Conclusions 39
   6.2 Recommendations 40
Acknowledgements 42
References 42

Abstract

To understand the current status of the malaria control programme at the county level in accordance with the criteria of the World Health Organisation, the gaps and feasibility of malaria elimination at the county and national levels were analysed based on three kinds of indicators: transmission capacity, capacity of the professional team, and the intensity of intervention. Finally, a roadmap for national malaria elimination in the People's Republic of China is proposed based on the results of a feasibility assessment at the national level.

1. INTRODUCTION

With the articulation of the United Nations Millennium Development Goals, global malaria control has received growing attention from the international community (Committee and Secretariat, 2012; Dye et al., 2013). Considerable progress has been made in global malaria control, with malaria morbidity and mortality effectively reduced in some countries, and hence, malaria-endemic areas have shrunk (Bhaumik, 2013). With the progressive move from malaria control to elimination, the World Health Organisation (WHO) released guidelines to support countries aiming for malaria elimination (Kelly et al., 2012; Mendis et al., 2009; WHO, 2007). According to WHO guidelines, a blood slide positivity rate (SPR) among febrile patients lower than 5% is an indicator of preelimination, while an annual parasite incidence (API) among the at-risk population for 3 consecutive years without local infection below 1 per 1000 indicates achievement of the elimination stage (Clements et al., 2013; Cotter et al., 2013).

A serious threat to public health, malaria is a major parasitic disease hindering socioeconomic development in the People's Republic of China (P.R.China) (Diouf et al., 2014; Laurentz, 1946; Tang et al., 1991). Since the founding of P.R. China in 1949, governments at all levels attached great importance to prevention and control of malaria and made remarkable achievements (Zhou, 1981). For example, the incidence of malaria was reduced from more than 24 million cases in the early 1970s to tens of thousands in the late 1990s, which greatly narrowed the scope of endemic areas (Tang et al., 1991). Malaria due to *Plasmodium falciparum* infection was eliminated, with the exception of Yunnan and Hainan provinces. After 2000, malaria reemerged in some areas

(Gao et al., 2012; Zhou et al., 2012). However, with the implementation of the 2006–2015 National Malaria Control Programme (NMEP), increased support provided by the central and local governments improved the malaria control status. By 2009, the national incidence of malaria dropped to 14,000 cases, and the incidence rate was lower than 1 per 10,000 in 95% of the counties (cities, districts) in 24 malaria-endemic provinces (autonomous regions and municipalities) (Zhou et al., 2011). The incidence rate was above 1 per 10,000 in only 87 counties, indicating that P.R. China has marched into the preparation stage of elimination (Yin et al., 2013b).

However, malaria elimination still faces many challenges in P.R. China. First, some areas of the country are still seriously endemic, such as border areas of Yunnan province adjacent to Myanmar (Bi et al., 2013). Second, elimination of malaria is a new public health task, so the government has no previous experience and weak capacity in the low- or medium-endemic areas, especially in resource-constrained settings (Xu and Liu, 2012). Third, in a global economy with a mobile population due to trade, tourism and labour migration, there are constant opportunities for imported malaria cases, which are likely to generate local transmission (Pindolia et al., 2012). Fourth, technical bottlenecks are another obstacle on the way to malaria elimination (Yan et al., 2013; Yin et al., 2013a). For example, microscopy could not meet the demands of population surveillance in the malaria elimination stage because it is labour- and time-consuming (Zheng et al., 2013).

It is essential to understand the current status of the malaria control programme in P.R. China, the relevant criteria of WHO, as well as the gaps and feasibility of malaria elimination at county and national levels (Anthony et al., 2012; Maharaj et al., 2012; Moonasar et al., 2013). This will provide more information for determining the roadmap of the national malaria elimination campaign in P.R. China based on feasibility assessments, with an emphasis on potential transmission risks, the capacity of the professional team and the intensity of intervention (Cotter et al., 2013; Kidson and Indaratna, 1998; Moore et al., 2008; Yang et al., 2010).

2. FEASIBILITY ASSESSMENT AT THE NATIONAL LEVEL

A feasibility assessment was an essential step before the initiation of a NMEP to understand the current status and future potential risks, which provide the basis for formulating the goals of the NMEP for a specific time frame (Diouf et al., 2014; El-Moamly, 2013; Zofou et al., 2014). A retrospective survey was conducted using data collected from the NMEP to draw the malaria pattern, potential transmission risks and institutional capacities. In-depth analyses were performed for the relationship between incidence and interventions at the provincial level to demonstrate the feasibility of malaria elimination in different settings and to come forward with a national strategy for P.R. China.

2.1 Data sources

The data were collected from several sources. Information on malaria incidence during 1950–2010 was collected from the national database on infectious diseases. Demographic data were extracted from the *China Statistical Yearbook* (NBSC, 2012). Malaria case information during 2004–2011 (e.g. age, sex and occupation) was collected from the national information and reporting system for infectious diseases (Wang et al., 2008). The *National Annual Report on Schistosomiasis, Malaria and Echinococcosis*, published by the Chinese Center for Disease Control and Prevention (Jin et al., 2006), provided malaria incidence and intervention data collected in each province by year during 2004–2010. Intervention data included information on radical treatment in the pretransmission stage (RTPT), indoor residual spraying (IRS), populations protected by insecticide-treated nets (ITNs), and training of microscopists, vector control staff, epidemiologists and other specialists (Zhao et al., 2013; Xia et al., 2014).

In addition, a database was established using data collected from 24 endemic provinces in 2010 through a questionnaire survey, including transmission risks and elimination capacities at the county level (Zhou et al., 2014). The information on transmission risks included data on malaria transmission risk with morbidity and vector species, while the data on elimination capacities included such information as whether a leading group was established, whether an implementation plan was issued, funding, personnel, supervisions, microscopic examination stations, active case detections, coverage of the reporting system, foci investigations and health education.

2.2 Assessment indicators

All quantitative data indicators were divided into five categories after being weighted by the county population. Those assigned values revealed the different levels of importance (El-Moamly, 2013). Four determinants were used in the assessment, including the incidence of malaria in each county, the value of the transmission capacity of vector species, the malaria transmission risk index for each county and the malaria elimination capability index. The actual value of the last two indices—employing the malaria transmission risk index (MTRI) and malaria elimination capacity index (MECI)—were converted to a uniform dispersion ratio of magnitude through standardised deviation, deviation normalised value = (variable value—min)/(max—min), so that those two indices could be compared among all counties equally (Zhou et al., 2014). For instance, the MECI assigned different weights to each county based on the ability to eliminate malaria weighted by a grade value of each indicator, of which the weighted values are summerised in Table 2.1.

By employing the MTRI, the MECI, and malaria incidence in 2010, a plot with three dimensions was generated to make a quantitative assessment of the potential risks to achieve the goal of the NMEP (Zhou et al., 2014). The different categories of potential risks were mapped at the county level, which will help to formulate the elimination strategy at the national level.

2.2.1 Malaria transmission risk index

The MTRI was defined the potential risk level for malaria transmission in a specific area. In one county, the MTRI = VTI × LPR, where VTI is the vectorial transmission index (defined as the vectorial capacity for transmission of malaria by *Anopheles* mosquito, the main local mosquito species for transmission of malaria) and LPR is the local potential risk for malaria transmission (defined as the potential malaria transmission risk at local level or county level, based on the malaria prevalence in the previous 3 years from 2007 to 2009). The detailed calculations of those indices as well as the weight values are listed in Table 2.1 (Zhou et al., 2014).

2.2.2 Malaria elimination capacity index

The MECI is defined as the local administrative ability devoted to the NMEP, weighted by the ability for data management, intervention work plan in 2010, organisational management, surveillance system, inspection, monitoring and evaluation, mobilisation, prevention for imported cases and financial input (Table 2.1). The calculation formula is as follows:

MECI = data management × 5 + intervention work-plan in 2010 × 6 + organisational management × 4 + surveillance system × 8 + inspection × 6 + mobilisation × 5 + monitoring and evaluation × 8 + prevention for imported cases × 10 + financial input × 9 (Zhou et al., 2014).

2.2.3 Risk of malaria transmission in a population

Malaria case distribution during 2004–2011 is shown in Figure 2.1. During 2004–2011, a total of 237,513 cases of malaria were reported, among which 158,206 were male (66.6%) and 79,307 were female (33.4%). The

Table 2.1 The calculation of indices and their weighted values

Index	Definition	Weight value[a]
Vectorial transmission index (VTI)	VTI is defined as the vectorial capacity for transmission of malaria by the *Anopheles* mosquito, the main local mosquito species for transmission of malaria. These mosquito species in P.R. China include *Anopheles dirus*, *An. jeyporiensis candidiensis*, *An. minimus*, *An. lesteri anthropophagus*, *An. sinensis* and *An. pseudowillmori*	*An. dirus*: 6 *An. jeyporiensis candidiensis*: 5 *An. minimus*: 4 *An. lesteri anthropophagus*: 3 *An. sinensis*: 2 *An. pseudowillmori*: 1
Local potential risk for malaria transmission (LPR)	LPR is defined as the potential malaria transmission risk at the local level (or a county), based on the judgement of malaria prevalence in the previous 3 years (from 2007 to 2009). A total of four strata were classified: stratum 1 is the area where annual malaria prevalence is more than 1 per 100,000 in each of the 3 years; stratum 2 is the area where annual malaria prevalence is more than 1 per 100,000 in at least 1 during 3 years; stratum 3 is the area where malaria prevalence is less than 1 per 100,000 during 3 years; and stratum 4 is the area where no local cases were found during 3 years.	Stratum 1: 10 Stratum 2: 7 Stratum 3: 4 Stratum 4: 1
Malaria elimination capacity index (MECI)	MECI is defined as the local administrative ability devoted to the NMEP, weighted by ability in data management, intervention work plan in 2010, organisational management, surveillance system, inspection, mobilisation, monitoring and evaluation, prevention for imported cases and financial input.	Data management: 5 Intervention work plan in 2010: 6 Organisational management: 4 Surveillance system: 8 Inspection: 6 Mobilisation: 5 Monitoring and evaluation: 8 Prevention for imported cases: 10 Financial input: 9

[a] Note: Weight values were granted based on the importance of achieving the goal of the NMEP, based on Delphi analysis (details of the Delphi analysis are reported elsewhere).

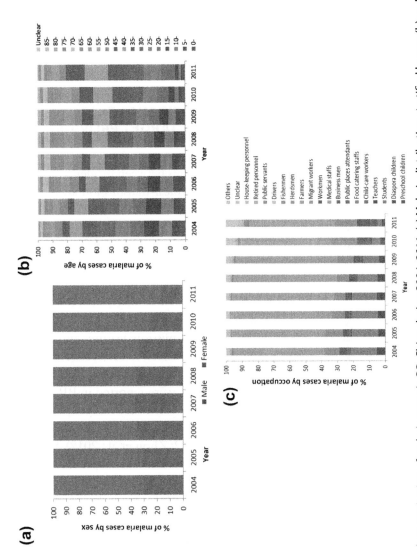

Figure 2.1 The characterisation of malaria cases in P.R. China during 2004–2011. (a) Malaria distribution stratified by sex; (b) malaria distribution stratified by age; (c) malaria distribution stratified by occupation.

ratio of male to female was 1.99:1. Among the reported 237,513 cases, the 10- to 50-year-age group was predominant, accounting for 66.7% of the total number, while the proportion of children under 5 years of age was only 3.3%. The proportion of the 20- to 45-year-old age group tended to increase. The majority of cases were farmers, students and migrant workers, accounting for 61.6%, 14.9% and 9.0% of the total number, respectively, which was 85.5% in total. Among them, the proportion of workers and government staff increased over time from 2004 to 2011.

3. CORRELATION BETWEEN INCIDENCE PATTERNS AND INTERVENTIONS IN FOUR TARGET PROVINCES

To understand the impact of the intervention on the reduction of malaria incidence at the local level, correlations between incidence patterns and interventions were calculated in two typical transmission areas at the provincial level.

3.1 Historical transmission pattern of four provinces

Four typical provinces – Anhui, Henan, Yunnan and Hainan – were selected to represent two typical patterns of malaria transmission in P.R. China: (1) high transmission areas with *P. vivax* singly in central P.R. China; and (2) high transmission areas with mixed *P. vivax* and *P. falciparum* in southern P.R. China. Two transmission features during the preelimination stage were presented with historical data collected from 2004 to 2010 (Zhang et al., 2014). First, the general transmission pattern showed that a total of 244,836 malaria cases were reported in P.R. China from 2004 to 2010, and 88% of all cases were from four target provinces: Anhui (accounting for 44% of the total cases), Henan (8%), Yunnan (26%) and Hainan (10%) provinces (Pan et al., 2012; Zhou et al., 2012). Second, only *P. vivax* malaria was distributed in Anhui and Henan provinces during the last decade, while both *P. vivax* and *P. falciparum* malaria occurred in Hainan and Yunnan provinces (Lin et al., 2009; Xu and Liu, 2012).

P. vivax malaria incidence was highest in Hainan (0.098%) and in Anhui (0.064%) in 2004 and in 2006, and then gradually declined annually to 0.09 and 0.28 per 10,000 in 2010, respectively. In Yunnan and Henan, the peaks of *P. vivax* malaria incidence were in 2005 and 2006, respectively, which were substantially below that in Hainan and Anhui provinces (Xia et al., 2012; Xia et al., 2014). Similarly, incidences in the two provinces decreased to 0.60 and 0.11 per 10,000 in 2010 (Figure 2.2). The incidence of *P. falciparum* malaria in Hainan was relatively higher than that in Yunnan

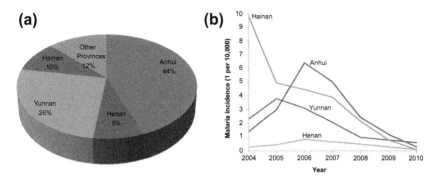

Figure 2.2 Malaria incidence in the four top provinces of Anhui, Henan, Yunnan and Hainan from 2004 to 2010. (a) Proportion of total number of malaria cases by provinces; (b) *P. vivax* malaria incidence for each of four provinces (Xia et al., 2014).

during 2004–2005, but it dropped quickly to zero local cases in 2010. However, the local incidence of *P. falciparum* malaria in Yunnan remained at a certain level with little reduction, to 0.03 per 10,000 (Lin et al., 2009; Wang et al., 2012; Xiao et al., 2012).

3.2 Correlation between malaria incidence and interventions

The correlation between incidence and interventions in four target provinces was calculated based on the time-series cross-sectional (TSCS) data model (Gmel et al., 2001; Reibling, 2013). The detailed process for TSCS data modelling is described in Appendix 1. Results showed that three scenarios could be observed.

First, generally, the TSCS data model showed that provincial differences in the annual incidence of malaria were mainly due to two interventions: RTPT for populations who were either historical cases or at-risk and the application of ITNs. The incidence of malaria differed among the four provinces, mainly because of antimalarial treatment variations: Yunnan was the highest, followed by Hainan, with Henan being the lowest (F = 15.19, $p < 0.001$, $R^2 = 0.724$). As a result of the use of ITNs, Hainan was the highest, followed by Yunnan, whilst Henan showed lowest (F = 16.77, $p < 0.001$, $R^2 = 0.689$) (Xia et al., 2014). These data indicate that both RTPT and ITNs are important interventions explaining the observed reduction of malaria transmission in Yunnan and Hainan provinces (Cao et al., 2013).

Second, in the correlation analysis between *P. vivax* malaria incidence and various interventions, results showed that two interventions – RTPT among historical patients or at-risk populations and microscopy training – influence the variation of *P. vivax* malaria incidence among the four provinces. For *P. vivax* malaria treatment (F = 14.53, $p < 0.001$, $R^2 = 0.721$), Hainan is the highest, followed by Yunnan and then Anhui, with Henan being the lowest. For microscopy training ($F = 11.53, p < 0.001, R^2 = 0.609$), Hainan is the highest and Henan is the lowest (Xia et al., 2014). These data indicate that antimalaria treatment and capacity in diagnosis are the key factors in the control of *P. vivax* malaria (Xu et al., 2002).

Third, in the correlation analysis between *P. falciparum* malaria incidence and intervention measures, two interventions were significant factors: microscopy training ($F = 11.06, p < 0.001, R^2 = 0.870$) and vector control training ($F = 15.28, p < 0.001, R^2 = 0.895$) (Xia et al., 2014). These data indicate that measures integrating treatment with strengthened capacity in diagnosis and vector control are of importance in the control of *P. falciparum* malaria transmission (Tambo et al., 2012).

3.3 Effective intervention during the transition stage from control to elimination

To understand the changes of intervention in the transmission stage from control to elimination, we consider a case study of the two typical patterns of malaria transmission in P.R. China. In the first type of transmission pattern, malaria transmission in Henan province was at the lowest level (0.11 per 10,000) of incidence in history (in the year 2010), reaching the threshold of entering the elimination stage according to WHO criteria. In the second type of transmission pattern in Hainan province, no *P. falciparum* malaria cases were found and *P. vivax* malaria incidence was 0.09 per 10,000 in 2010, which also reached the threshold of entering the elimination stage according to WHO criteria (Committee and Secretariat, 2013a).

In Henan province, where only *P. vivax* malaria is transmitted, the interventions of RTPT for historical patients and at-risk populations and the application of ITNs were effective measures for clearing the infection sources. These interventions had spatial and temporal characteristics based on the results of correlation analysis (Chen et al., 2012; Liu and Xu, 2006; Xu et al., 2006). In addition, to further reduce the incidence, strengthened training to improve the diagnostic capacity of malaria microscopy helps in the timely detection and treatment of malaria so as to control malaria transmission (Bi et al., 2012; Fernando et al., 2013; Huang et al., 1988).

In Hainan province, where both *P. vivax* and *P. falciparum* malaria were transmitted in recent years, an RTPT intervention for malaria patients and the at-risk population, together with strengthened capacity for microscopy diagnosis and vector control, were effective measures (Dapeng et al.,

1996). Specifically for vector control, the application of ITNs was effective, while IRS was ineffective. This is because the major mosquito species for transmission of malaria is *An. dirus*, which is mainly found outdoors in the mountainous area of Hainan province (Wang et al., 2013).

In conclusion, RTPT for historical patients and at-risk populations during the pretransmission season (normally from February to April) and application of ITNs were effective measures in the transition stage from control to elimination of malaria. The cost-effectiveness of the interventions will be improved when the capacity for diagnosis and vector control are strengthened at the same time (Roy et al., 2013; Wang et al., 2012; Zheng et al., 2013).

4. FEASIBILITY ANALYSIS OF MALARIA ELIMINATION IN CHINA

To achieve malaria elimination in a large country like P.R. China, there are many challenges to address. Those challenges must be identified and addressed with the progress of the NMEP (El-Moamly, 2013). Therefore, a feasibility analysis of the potential to achieve the goal of the malaria elimination is essential at the beginning of the NMEP, with a focus on the natural and biological risks in the transmission and intervention capacities at the local level (Clements et al., 2013).

4.1 Transmission risks

The feasibility analysis of transmission risks was conducted in a total of 2147 counties targeting elimination in P.R. China based on the goals of the NMEP (Cao et al., 2013; Yang et al., 2012). The MTRI (see Section 2.2.1) varied from 0 to 60, with an average of 50.8. Geographically, higher MRTIs tended to be distributed in the south, whereas lower MRTIs were located in north P.R. China.

When we analyse the MRTI for four typical transmission provinces, its average value is the highest in Hainan province (23.6), followed by Yunnan (22.9), Anhui (9.0) and Henan (5.6) provinces.

4.2 Malaria elimination capacities

The MECI (see Section 2.2.2) for each of 2147 counties was composed of nine variables. The average MECI value among these 2147 counties was 221.4, varying from 23 to 895. Among the four provinces studied, it was the highest in Yunnan province (915.1), followed by Anhui (363.2), Hainan (349.8) and Henan (313.4) provinces. All of these MECIs exceeded the

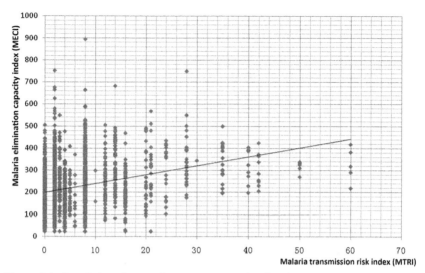

Figure 2.3 Correlation diagram between county-level malaria transmission risk index (MTRI) and malaria elimination capacity index (MECI) for 2147 counties (Zhou et al., 2014).

average value of the whole country. Among the nine variables, two types could be identified: the technical capacity, which can be improved by training, and the resource capacity, which can be increased through the political awareness of policymakers. Figure 2.3 shows that the MECI in 40% of the counties was below average, indicating that the capacities in those counties should be strengthened in order to achieve the goal of the NMEP by 2020.

4.2.1 Technical capacity
The technical capacity included five variables: (i) data management; (ii) intervention work plan in 2010; (iii) inspection; (iv) monitoring and evaluation; and (v) prevention for imported cases. The best way to increase the technical capacity is to strengthen the training activities or technical practice in technical organisations involved in malaria elimination, such as county centres for disease control and prevention, township hospitals and village clinics.

4.2.2 Resources capacity
Resources capacity covered four variables: (i) organisational management; (ii) surveillance systems; (iii) mobilisation; and (iv) financial input. In the areas that previously experienced large epidemics of malaria, county governments can easily maintain the necessary capacity, but it is normally

difficult to increase or maintain the awareness of local governments in the counties where transmission was at low levels for a long time. Therefore, in order to maintain this kind of resource capacity, regulations to mitigate the potential risks of malaria transmission locally in each county are likely to be enforced.

4.3 Feasibility analysis for malaria elimination at the national level

4.3.1 Feasibility analysis employed by MTRI and MECI

The correlation diagram between MTRI and MECI for each county is shown in Figure 2.3, where 85% of counties have MTRI below 20 and MECI below 30. Therefore, counties where MTRI is above 20 need to be monitored rigorously.

4.3.2 Feasibility analysis employed by malaria incidence, MTRI and MECI

A total of 2147 counties were mapped in a three-dimensional figure with four categories (in different colours) of counties based on malaria incence (MI), MTRI and MECI (Figure 2.4).

Figure 2.4 shows that MI rates in the majority of counties are zero, although their MTRI varied from 0% to 20% and MECI varied from 0% to 30%. A few of the counties located in the blue zone are of higher incidence (MI > 1 per 1000 and <1 per 10,000), which is a potential risk area, probably due to the introduction of imported cases from other regions or countries.

4.3.3 Feasibility analysis with geographic variations

When all counties were mapped using these four categories, the spatial distribution of each type of county provided information for decision makers to formulate elimination strategies with certain resources in each region (Figure 2.5). Figure 2.5 indicates that the counties in stratum 1 are distributed along either county borders or provincial borders. Therefore, for malaria elimination in the last stage, close attention must be paid to the border areas, with enhanced capacity building for sustained surveillance and response.

In historical reports on the progress of the national malaria control campaign, malaria stratification has been carried out at the county level. For example, according to the research conducted by Ho Qi and Feng Lanzhou in 1958, malaria endemic areas can be divided into four

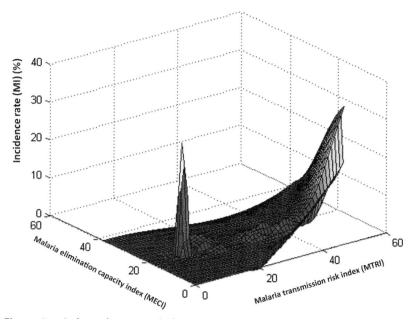

Figure 2.4 A three-dimensional plot employing three variables (MTRI, MECI and MI) showing four strata of counties among all 2147 studied counties (Zhou et al., 2014). Stratum 1 areas are shown in red, located in the top right, with a high local vector capacity (MTRI = 50–60), of which malaria elimination capacity is at the middle level (MECI around 20) and malaria incidence is higher (MI > 1 per 10,000). Stratum 2 areas are shown in green, where high transmission risk (MTRI = 50–60) was present, the local antimalarial capability was at the lower or middle level (MECI = 10–30) and the incidence rate was less than that of stratum 1 (MI > 1 per 1000 and < 1 per 10,000). Stratum 3 areas are shown in yellow, where malaria transmission risk index is low (MTRI around 10), MECI is at the middle level (around 20) and the incidence rate is close to that in stratum 2 (MI > 1 per 1000 and <1 per 10,000). Stratum 4 areas are shown in blue, where both the transmission risk and control capacity index are at low levels (MTRI and MECI = 0–40) and malaria incidence rate decreased with the significant decrease of MTRI and MECI. (For interpretation of the references to color in this figure legend, the reader is referred to the online version of this book).

regions based on splenomegaly rate, protozoa species and media distribution, terrain, climate and other factors combined with the latitude and longitude location of endemic areas (Ho and Feng, 1958; Zhou, 1991). In 1965, unstable and stable areas were proposed by Ho Chi (Ho, 1965). In 1995, Liu Zhaofan divided malaria endemic areas into four clusters according to vector distribution and incidence with reference to natural and geographical profiles (Liu et al., 1995b). In 2007, the NMCP

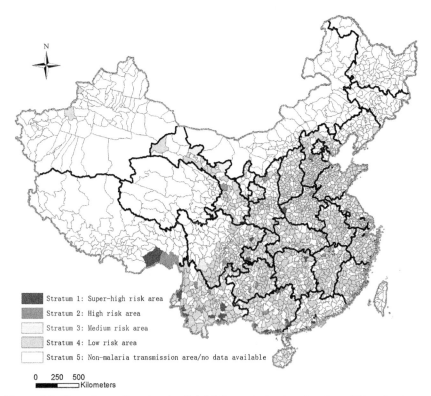

Figure 2.5 Map of counties at potential risk for malaria, showing four different categories of counties with geographic distribution patterns (Zhou et al., 2014).

divided malaria-endemic areas into three areas: (i) high transmission; (ii) unstable transmission; and (iii) under control regions (MOH, 2007).

However, these classification methods are mainly based on vector distribution and the level of malaria incidence. Additionally, malaria stratification was never conducted at the county level in P.R. China. In addition to vector and incidence, malaria transmission risk is also relevant to control capacity (Committee and Secretariat, 2013b; Cui et al., 2012). In the elimination phase, with a dramatic decrease of incidence, it is not adequate to determine malaria risks based on incidence and vector data only.

In this study, considering the transmission risk and control capacities, malaria risk in 2010 was analysed at the county level, based on our proposed new method for malaria risk analysis. This approach could be used in P.R. China, and even other countries, to guide stratification of malaria transmission risk.

5. PHASE-BASED MALARIA ELIMINATION STRATEGIES

The WHO proposed that once malaria incidence is less than 1 case per 1000 population at risk annularly, a malaria elimination programme could be initiated (Bousema et al., 2012; Feachem et al., 2010). Since 1990, P.R. China's average malaria incidence rate at the country level has decreased to 1 per 10,000 annually. However, some locations are still heavily epidemic in specific ecosocial zones, with some outbreaks occurring from time to time. For instance, a malaria outbreak occurred in northern Anhui province in 2006 (Wang et al., 2009; Zhou et al., 2010). The Chinese NMEP was not launched until 2010, with the aim to eliminate malaria nationwide by 2020 (Zheng et al., 2013; Zhou et al., 2011). Nevertheless, taking consideration of the fact that transmission patterns and elimination capacities vary from county to county, malaria elimination strategies need to be identified based on local settings, such as at the county level in different phases of the national programme (Cao et al., 2013; Yang et al., 2012).

5.1 Classification of elimination phases

WHO guidelines divide malaria control into four phases (Moonen et al., 2010; WHO, 2007). Preelimination occurs when the incidence rate is less than 1 per 1000. As shown in this study, from an annual incidence of 1 per 1000 until elimination, malaria control could be further subdivided into the following stages under phases 2 and 3 of WHO classification (or preelimination and elimination phases) stage E1 between SPR<5% in fever patients and 1 per 1000 of incidence rate); stage E2 (incidence rate less than 1 per 1000, and over than 1 per 10,000); stage E3 (incidence rate less than 1 per 10,000); and stage E4 (0 local cases annually).

In stage E1, the annual incidence rate is over 1 per 1000 and transmission capacity is still high; this is equivalent to the preelimination phase in the WHO classification. All stratum 1 areas in the Chinese classification belong to this stage. In this stage E1, transmission control is the main task. In stage E2 the annual incidence rate is between 1 per 1000 and 1 per 10,000 and malaria transmission tends to be stable. This stage E3 is equivalent to the start of elimination phase of the WHO classification and strata 2–3 of the Chinese classification, in which interventions focus on clearing the source of infection by case management and improving control capacities towards elimination. In stage E3, the annual incidence rate is less than 1 per 10,000 and malaria transmission maintains at very low levels, with sporadic distribution of malaria cases. This stage E3 is equivalent to the later elimination phase of the WHO classification and stratum 4 of the Chinese classification, in which the intervention is focused on surveillance and response to preventing imported cases. In stage E4, the annual incidence approaches 0 and almost no local cases occur. This stage E4 is equivalent to the end point of elimination phase of WHO classification and stratum 5 of the Chinese classification, in which the surveillance and response systems is established and aims at prevention of reemergence of malaria due to imported cases (Figures 2.4 and 2.5).

5.2 Strategy formulations in each stage

Elimination strategies differ based on the transmission patterns in different phases of the NMEP. More stages of the elimination programme exist in the classification of the Chinese NMEP compared to the WHO classification. Therefore, more detailed elimination strategies are identified for each stage in the Chinese NMEP. For instance, from preelimination to posttransmission, there are only three phases in the WHO classification—the preelimination phase, the elimination phase and the posttransmission phase—until certification of malaria elimination in 3 years. The elimination phase is normally much longer, usually more than 5–10 years. Therefore, Chinese classification uses five stages, which are based on the indices of malaria transmission and control capability. Stage E1 is equivalent to the preelimination of WHO classification; stage E2 is equivalent to the WHO elimination phase in the early period; stage E3 is equivalent to the WHO elimination phase in the middle period; stage E4 is equivalent to the period of zero local cases in the later WHO elimination phase; and stage E5 is equivalent to the WHO posttransmission phase (Table 2.2).

In accordance with the five strata of malaria endemic areas in P.R. China, the elimination strategy varies from stratum to stratum. For instance, stratum 1 areas (shown in red in Figure 2.6) are mainly located in Yunnan, Hainan and Guizhou. In this important area to eliminate malaria, elimination interventions are comprised of (1) strengthening infection control integrated with vector control measures; and (2) improving control capabilities to reduce the risk of transmission. In stratum 2 (shown in green), these areas are mainly distributed in Yunnan, Hainan, Guizhou, Shaanxi, Hubei, Anhui and other provinces. Due to weak capacities, the elimination interventions are to (1) strengthen training to improve local abilities; and (2) appropriately control infection sources to consolidate malaria control efforts. In stratum 3 (shown in yellow), these areas are mainly scattered in Jiangsu, Zhejiang,

Table 2.2 Comparison of various strategies in the WHO and Chinese classifications

WHO classification	Preelimination	Elimination	Postelimination
Annual incidence	Slide or rapid diagnostic test positivity rate <5% in fever cases	<1 case per 1000 population at risk/year	Zero locally acquired cases for 3 years
WHO strategy	Reinforcing the coverage of good-quality laboratory and clinical services, reporting and surveillance aimed at halting transmission nationwide Perfecting the quality and targeting of case management and vector control operations, and introducing/maintaining activities aimed at consistently reducing the onward transmission from existing cases in residual and new active foci Establishment of a strong surveillance system, with the cooperation of all healthcare providers	Identification and treatment of all malaria reservoir and reduced transmission by vectors with full surveillance for clearing up malaria foci and reducing the number of locally acquired cases to zero Identifying and treating all malaria cases with efficacious antimalarial medicines against liver stage and blood stage parasites, including gametocytes Reducing human–vector contact and the vectorial capacity of the local *Anopheles* mosquito populations in transmission foci by efficacious vector control, personal protection and environmental management methods	Maintain an effective surveillance and response system and strengthen prevention and management of imported malaria to prevent introduced cases and indigenous cases secondary to introduced cases Reduction of vulnerability population Screening of immigrants for malaria and the use of radical treatment in places where importation of malaria is intensive

Chinese classification	Stage E1	Stage E2	Stage E3	Stage E4	Stage E5
Annual incidence	>1/1000	1/1000–1/10,000	<1/10,000	0	0 locally acquired cases for 3 years
Chinese strategy	Strengthening infection control integrated with vector control measures Improving control capabilities to reduce the risk of transmission	Strengthen training to improve local abilities Appropriately control infection source to consolidate malaria control efforts	Strengthen the surveillance-response system Find and treat imported cases earlier	Strengthen surveillance, both active and passive, for early detection of infection sources	Strengthen the surveillance response system to prevent the reintroduction of malaria cases

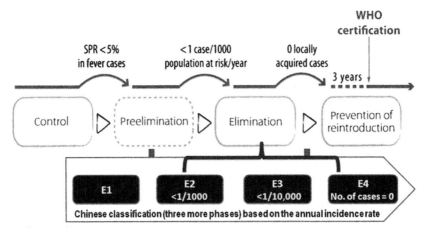

Figure 2.6 Integrated diagram of the malaria elimination stages and phases, indicating the four WHO phases of malaria elimination. Three more phases or stages were classified in our study based on the annual incidence rate: stage E1 is equivalent to the preelimination phase in the WHO classification; stage E2 is equivalent to the WHO elimination phase in the early period; stage E3 is equivalent to the WHO elimination phase in the middle period; stage E4 is equivalent to the WHO phase of zero local cases in the later elimination stage; and stage E5 is equivalent to the WHO posttransmission.

Sichuan, Hunan, Hubei, Henan and other provinces. Few of high incidences occur in these areas, mainly due to the higher numbers of imported malaria cases. Here, the elimination interventions aim to (1) strengthen the surveillance-response system; and (2) locate and treat imported cases earlier (Cao et al., 2013; Yin et al., 2013b). In stratum 4 (in blue), these areas are widely dispersed with sporadic distribution of malaria cases. The main intervention is to strengthen surveillance, both active and passive surveillance, to find the infectious sources earlier.

6. CONCLUSIONS AND RECOMMENDATIONS
6.1 Conclusions

The history of the NMCP in P.R. China can be classified into two stages (MOPH, 1990). The first stage is from 1950 to 1980. Malaria was highly prevalent in this stage, fluctuating every 10 years with three peaks each time (Kung and Huan 1976; Zhou, 1981). The second stage started after 1980. Despite slight rebounds, no incidence peaks were found in this stage due to stable malaria control agencies at all levels, sustained control efforts, increased capacities and sustained interventions that reduced transmission risks significantly (Tang et al., 1991; Zhou, 1991). The setup of the national disease surveillance and reporting system allowed for individual cases to be reported nationwide since 2004 (Liu et al., 1995a). The population distribution of malaria cases nationwide from 2004 to 2011 showed an increasing proportion in the 20- to 45-year-age group, which might be attributed to the occupations of this population, including farming, business and production, which are more susceptible to infections (Tang, 2000; Zhou et al., 2010). Regarding occupational distribution, the proportion of farmers is the greatest but with a declining trend, while the proportions of mobile workers, businessmen and government staff have risen significantly. Particular increases have been noted in mobile workers who migrated back from African countries after working there for years (Jelinek and Muhlberger, 2005; Ming, 2008; Xia et al., 2012; Yin et al., 2013b; Zhang et al., 2010).

The classification approach used in P.R. China's Action Plan for Malaria Elimination (2010–2020) relies on the incidence at the county level during the period of 2006–2008, without considering the capacity of local institutions or vector capacities (Yang et al., 2012). This approach has provided useful information for designing the strategy used in the initiation of the NMEP, but it does not fully reflect malaria transmission risks in the big picture (Qi, 2011). To overcome this gap, our research took comprehensive considerations of transmission risks and control capability (Maude et al., 2011). The results of this classification are helpful for decision makers when assessing risks annually and for professionals involved in the NMEP to find out the key factors that may reduce or increase malaria transmission. In this way, appropriate responses can be easily tailored for local settings and may take place in a shorter time (Liu et al., 2012; Zhou et al., 2013).

6.2 Recommendations

Recommendation 1: Based on the marginal cost-benefit principle, more resources should be allocated and invested in the malaria elimination efforts towards the progress of NMEP (Sabot et al., 2010; Tanner and Hommel, 2010). The best strategy for a country is to maintain elimination efforts pertaining to a certain level of human capacity and to consolidate the achievements of malaria elimination while maintaining a certain level of investment (Alonso and Tanner, 2013).

Recommendation 2: Under the aforementioned circumstances, the following five research priorities are recommended in order to fulfil the optimal goal of eliminating malaria nationwide by 2020 (Greenwood, 2008; Hall and Fauci, 2009; Marsh, 2010; Zheng et al., 2013).

1. Improvement of technology to provide more precise predictions with modelling and geographical information systems, in order to set up or promote active surveillance and response systems to prevent the reestablishment of malaria transmission (Bridges et al., 2012; Zhou et al., 2013).
2. Establishment of a resource bank to use as a repository for the *Plasmodium* parasites and their vectors from the whole country, in order to discover the more specific biomarkers to be used in tracing the different species or strain of parasites, and development of more user-friendly, sensitive and rapid diagnostics for malaria case detection (McMorrow et al., 2011).
3. Development of a surveillance approach to monitor artemisinin drug resistance in the migrant population and insecticide resistance, particularly in the southern border areas, in order to contain the spread of artemisinin resistance worldwide (Huang et al., 2012; Li et al., 2000; Liu, 2014).
4. Acceleration of screening and validating of alternative antimalarial drugs, such as the new formulation of artemisinin-based combination therapies, and screening for more candidates of active compounds or molecules to develop new antimalarial drugs (Anthony et al., 2012; Chen, 2014).
5. Development of a G6PD deficiency screening test for a point-of-care diagnostic for primaquine therapy screening in the NMEP (Domingo et al., 2013; Nie and Zhao, 1999).

Recommendation 3: The classification map of malaria transmission in P.R. China at the county level used in our study needs to be updated every 3 years. The gap between achievement of the NMEP and the true trajectory of malaria transmission needs to be investigated every year. The updated information will provide a clear and accurate picture for the decision makers who provide the resources and formulate the intervention strategies for the NMEP, in order to consolidate the achievements and finally achieve the goal of eliminating malaria in P.R. China by 2020 (Bridges et al., 2012; Smith et al., 2013; White et al., 2009).

Recommendation 4: The quality of NMEP activities must be monitored and evaluated frequently in the periods from 2015 to 2020. The indices of monitoring and evaluation are essential for properly maintaining the quality of the NMEP. It is important to formulate standard indices for monitoring before the evaluation. The standard protocol for surveillance and response, either for malaria elimination or for preventing the reintroduction of malaria, are essential at the county level, both in the elimination stage and in the postelimination stage (Cao et al., 2013; Shah, 2010).

ACKNOWLEDGEMENTS

This work was supported by National Natural Science Foundation of P.R. China (grant no. 81273192), UNICEF/UNDP/World Bank/WHO Special Programme on Research and Training in Tropical Diseases (grant no. 70350) through P.R. China NDI Initiative (Chinese Network on Drug and Diagnostic Innovation), by Chinese National Science and Technology Major Project (grant no. 2012ZX10004-220), the Special Foundation for Technology Research of Science and Technology Research Institute from the Ministry of Science and Technology of P.R. China (grant no. 2011EG150312) and by P.R. China UK Global Health Support Programme (grant no. GHSP-CS-OP1).

REFERENCES

Alonso, P.L., Tanner, M., 2013. Public health challenges and prospects for malaria control and elimination. Nat. Med. 19, 150–155.
Anthony, M.P., Burrows, J.N., Duparc, S., Moehrle, J.J., Wells, T.N., 2012. The global pipeline of new medicines for the control and elimination of malaria. Malar. J. 11, 316.
Bhaumik, S., 2013. Malaria funds drying up: world malaria report 2012. Natl. Med. J. India 26, 62.
Bi, Y., Hu, W., Liu, H., Xiao, Y., Guo, Y., Chen, S., Zhao, L., Tong, S., 2012. Can slide positivity rates predict malaria transmission? Malar. J. 11, 117.
Bi, Y., Hu, W., Yang, H., Zhou, X.N., Yu, W., Guo, Y., Tong, S., 2013. Spatial patterns of malaria reported deaths in Yunnan province, China. Am. J. Trop. Med. Hyg. 88, 526–535.
Bousema, T., Griffin, J.T., Sauerwein, R.W., Smith, D.L., Churcher, T.S., Takken, W., Ghani, A., Drakeley, C., Gosling, R., 2012. Hitting hotspots: spatial targeting of malaria for control and elimination. PLoS Med. 9, e1001165.
Bridges, D.J., Winters, A.M., Hamer, D.H., 2012. Malaria elimination: surveillance and response. Pathog. Global Health 106, 224–231.
Cao, J., Zhou, S.S., Zhou, H.Y., Yu, Y.B., Tang, L.H., Gao, Q., 2013. Malaria from control to elimination in China: transition of goal, strategy and interventions. Chin. J. Schisto. Control 25, 439–443 (in Chinese).
Chen, C., 2014. Development of antimalarial drugs and their application in China: a historical review. Infect. Dis. Poverty 3, 9.
Chen, W.Q., Su, Y.P., Deng, Y., Zhang, H.W., 2012. Epidemiological analysis of imported malaria in Henan province in 2011. Chin. J. Parasitol. Parasit. Dis. 30, 387–390 (in Chinese).
Kung, C.-C., Hung, S.-C., 1976. Malaria control in China, with special reference ot bioenvironmental methods of control. Chin. Med. J. (Engl). 2, 195–202.
Clements, A.C.A., Reid, H.L., Kelly, G.C., Hay, S.I., 2013. Further shrinking the malaria map: how can geospatial science help to achieve malaria elimination? Lancet Infect. Dis. 13, 709–718.
Cotter, C., Sturrock, H.J., Hsiang, M.S., Liu, J., Phillips, A.A., Hwang, J., Gueye, C.S., Fullman, N., Gosling, R.D., Feachem, R.G., 2013. The changing epidemiology of malaria elimination: new strategies for new challenges. Lancet 382, 900–911.
Cui, L., Yan, G., Sattabongkot, J., Chen, B., Cao, Y., Fan, Q., Parker, D., Sirichaisinthop, J., Su, X.Z., Yang, H., Yang, Z., Wang, B., Zhou, G., 2012. Challenges and prospects for malaria elimination in the Greater Mekong subregion. Acta Trop. 121, 240–245.
Dapeng, L., Leyuan, S., Xili, L., Xiance, Y., 1996. A successful control programme for falciparum malaria in Xinyang, China. Trans. R. Soc. Trop. Med. Hyg. 90, 100–102.

Diouf, G., Kpanyen, P.N., Tokpa, A.F., Nie, S., 2014. Changing landscape of malaria in China: progress and feasibility of malaria elimination. Asia-Pacific J. Public Health 26, 93–100.

Domingo, G.J., Satyagraha, A.W., Anvikar, A., Baird, K., Bancone, G., Bansil, P., Carter, N., Cheng, Q., Culpepper, J., Eziefula, C., Fukuda, M., Green, J., Hwang, J., Lacerda, M., McGray, S., Menard, D., Nosten, F., Nuchprayoon, I., Oo, N.N., Bualombai, P., Pumpradit, W., Qian, K., Recht, J., Roca, A., Satimai, W., Sovannaroth, S., Vestergaard, L.S., Von Seidlein, L., 2013. G6PD testing in support of treatment and elimination of malaria: recommendations for evaluation of G6PD tests. Malar. J. 12, 391.

Dye, C., Mertens, T., Hirnschall, G., Mpanju-Shumbusho, W., Newman, R.D., Raviglione, M.C., Savioli, L., Nakatani, H., 2013. WHO and the future of disease control programmes. Lancet 381, 413–418.

El-Moamly, A., 2013. Malaria elimination: needs assessment and priorities for the future. J. Infect. Dev. Ctries. 7, 769–780.

Feachem, R.G., Phillips, A.A., Targett, G.A., Snow, R.W., 2010. Call to action: priorities for malaria elimination. Lancet 376, 1517–1521.

Fernando, S.D., Navaratne, C.J., Galappaththy, G.N., Abeyasinghe, R.R., Silva, N., Wickermasinghe, R., 2013. The importance of accuracy in diagnosis of positive malaria cases in a country progressing towards malaria elimination. J. Global Infect. Dis. 5, 127–130.

Gao, H.W., Wang, L.P., Liang, S., Liu, Y.X., Tong, S.L., Wang, J.J., Li, Y.P., Wang, X.F., Yang, H., Ma, J.Q., Fang, L.Q., Cao, W.C., 2012. Change in rainfall drives malaria re-emergence in Anhui province, China. PLoS One 7, e43686.

Gmel, G., Rehm, J., Frick, U., 2001. Methodological approaches to conducting pooled cross-sectional time series analysis: the example of the association between all-cause mortality and per capita alcohol consumption for men in 15 European states. Eur. Addict. Res. 7, 128–137.

Greenwood, B.M., 2008. Control to elimination: implications for malaria research. Trends Parasitol. 24, 449–454.

Hall, B.F., Fauci, A.S., 2009. Malaria control, elimination, and eradication: the role of the evolving biomedical research agenda. J. Infect. Dis. 200, 1639–1643.

Ho, C., 1965. Studies on malaria in new China. Chin. Med. J. 84, 491–497.

Ho, C., Feng, L.C., 1958. Studies on malaria in new China. Chin. Med. J. 77, 533–551.

Huang, B.P., Ding, H.H., Zhang, J.Y., Zhao, Y.J., Jin, G., Qian, M.Y., Yang, Z.Y., Zhou, H.X., Wang, J.P., Qian, W.Z., Yang, C.X., 1988. Role of microscopy stations for malaria in malaria control in Jiangsu province. Chin. J. Parasitol. Parasit. Dis. 6, 178–181 (in Chinese).

Huang, F., Tang, L., Yang, H., Zhou, S., Liu, H., Li, J., Guo, S., 2012. Molecular epidemiology of drug resistance markers of *Plasmodium falciparum* in Yunnan province, China. Malar. J. 11, 243.

Jelinek, T., Muhlberger, N., 2005. Surveillance of imported diseases as a window to travel health risks. Infect. Dis. Clin. North Am. 19, 1–13.

Jin, S.G., Jiang, T., Ma, J.Q., 2006. Brief introduction of Chinese infectious disease detection report information system. China Digit. Med. 1, 20–22 (in Chinese).

Kelly, G.C., Tanner, M., Vallely, A., Clements, A., 2012. Malaria elimination: moving forward with spatial decision support systems. Trends Parasitol. 28, 297–304.

Kidson, C., Indaratna, K., 1998. Ecology, economics and political will: the vicissitudes of malaria strategies in Asia. Parassitologia 40, 39–46.

Laurentz, F.K., 1946. Malaria control in China. Tex. State J. Med. 42, 386.

Li, Y., Zhu, Y.M., Jiang, H.J., Pan, J.P., Wu, G.S., Wu, J.M., Shi, Y.L., Yang, J.D., Wu, B.A., 2000. Synthesis and antimalarial activity of artemisinin derivatives containing an amino group. J. Med. Chem. 43, 1635–1640.

Lin, H., Lu, L., Tian, L., Zhou, S., Wu, H., Bi, Y., Ho, S.C., Liu, Q., 2009. Spatial and temporal distribution of falciparum malaria in China. Malar. J. 8, 130.

Liu, C., Qian, H., Tang, L., Zheng, X., Gu, Z., Zhu, W., 1995a. Current malaria stratification in China. Chin. J. Parasitol. Parasit. Dis. 13, 8–12 (in Chinese).

Liu, C.F., Qian, H.L., Tang., L.H., Zheng, X., Gu, Z.C., Zhu, W.D., 1995b. Current malaria stratification in China. Chin. J. Parasitol. Parasit. Dis. 1, 8–12 (in Chinese).

Liu, D.Q., 2014. Surveillance of antimalarial drug resistance in China in the 1980s-1990s. Infect. Dis. Poverty 3, 8.

Liu, J., Yang, B., Cheung, W.K., Yang, G., 2012. Malaria transmission modelling: a network perspective. Infect. Dis. Poverty 1, 11.

Liu, X.Z., Xu, B.L., 2006. Malaria situation and evaluation on the control effect in Henan province during 1990-2005. Chin. J. Parasitol. Parasit. Dis. 24, 226–229 (in Chinese).

Maharaj, R., Morris, N., Seocharan, I., Kruger, P., Moonasar, D., Mabuza, A., Raswiswi, E., Raman, J., 2012. The feasibility of malaria elimination in South Africa. Malar. J. 11, 423.

Marsh, K., 2010. Research priorities for malaria elimination. Lancet 376, 1626–1627.

Maude, R.J., White, N.J., White, L.J., 2011. Feasibility of malaria elimination. Lancet 377, 638.

McMorrow, M.L., Aidoo, M., Kachur, S.P., 2011. Malaria rapid diagnostic tests in elimination settings–can they find the last parasite? Clin. Microbiol. Infect. Dis. 17, 1624–1631.

Mendis, K., Rietveld, A., Warsame, M., Bosman, A., Greenwood, B., Wernsdorfer, W.H., 2009. From malaria control to eradication: the WHO perspective. Trop. Med. Int. Health 14, 802–809.

Ming, G., 2008. Malaria cases among those worked in and returned from Uganda. Chin. J. Parasitol. Parasit. Dis. 26, 240 (in Chinese).

MOH, 2007. Malaria Control Manual. People's Medical Publishing House, Beijing (in Chinese).

Moonasar, D., Morris, N., Kleinschmidt, I., Maharaj, R., Raman, J., Mayet, N.T., Benson, F.G., Durrheim, D.N., Blumberg, L., 2013. What will move malaria control to elimination in South Africa? South Afr. Med. J. 103, 801–806.

Moonen, B., Cohen, J.M., Snow, R.W., Slutsker, L., Drakeley, C., Smith, D.L., Abeyasinghe, R.R., Rodriguez, M.H., Maharaj, R., Tanner, M., Targett, G., 2010. Operational strategies to achieve and maintain malaria elimination. Lancet 376, 1592–1603.

Moore, S.J., Min, X., Hill, N., Jones, C., Zaixing, Z., Cameron, M.M., 2008. Border malaria in China: knowledge and use of personal protection by minority populations and implications for malaria control: a questionnaire-based survey. BMC Public. Health 8, 344.

MOPH, 1990. Advisory committee on parasitic diseases. Malaria situation in China, 1989. Chin. J. Parasitol. Parasit. Dis. 8, 241–244 (in Chinese).

NBSC, 2012. National Bureau of Statistics of China. http://www.stats.gov.cn/english/Statisticaldata/AnnualData/ accessed 13 April 2014.

Nie, C., Zhao, S., 1999. A diagnostic kit to screen individuals with glucose-6-phosphate dehydrogenase defect and its application on anti-malaria spot in the countryside. Chin. Med. J. (Engl.) 112, 349–351.

Pan, J.Y., Zhou, S.S., Zheng, X., Huang, F., Wang, D.Q., Shen, Y.Z., Su, Y.P., Zhou, G.C., Liu, F., Jiang, J.J., 2012. Vector capacity of *Anopheles sinensis* in malaria outbreak areas of central China. Parasit. Vectors 5, 136.

Pindolia, D.K., Garcia, A.J., Wesolowski, A., Smith, D.L., Buckee, C.O., Noor, A.M., Snow, R.W., Tatem, A.J., 2012. Human movement data for malaria control and elimination strategic planning. Malar. J. 11, 205.

Qi, G., 2011. Opportunities and challenges of malaria elimination in China. Chin. J. Schisto. Control 23, 347–349 (in Chinese).

Reibling, N., 2013. The international performance of healthcare systems in population health: capabilities of pooled cross-sectional time series methods. Health Policy 112, 122–132.

Roy, M., Bouma, M.J., Ionides, E.L., Dhiman, R.C., Pascual, M., 2013. The potential elimination of *Plasmodium vivax* malaria by relapse treatment: insights from a transmission model and surveillance data from NW India. PLoS Negl. Trop. Dis. 7, e1979.

Sabot, O., Cohen, J.M., Hsiang, M.S., Kahn, J.G., Basu, S., Tang, L., Zheng, B., Gao, Q., Zou, L., Tatarsky, A., Aboobakar, S., Usas, J., Barrett, S., Cohen, J.L., Jamison, D.T., Feachem, R.G., 2010. Costs and financial feasibility of malaria elimination. Lancet 376, 1604–1615.

Shah, N.K., 2010. Assessing strategy and equity in the elimination of malaria. PLoS Med. 7, e1000312.

Smith, D.L., Cohen, J.M., Chiyaka, C., Johnston, G., Gething, P.W., Gosling, R., Buckee, C.O., Laxminarayan, R., Hay, S.I., Tatem, A.J., 2013. A sticky situation: the unexpected stability of malaria elimination. Philosoph. Trans. R. Soc. Lond. Ser. B, Biol. Sci. 368, 20120145.

Tambo, E., Adedeji, A.A., Huang, F., Chen, J.H., Zhou, S.S., Tang, L.H., 2012. Scaling up impact of malaria control programmes: a tale of events in sub-Saharan Africa and People's Republic of China. Infect. Dis. Poverty 1, 7.

Tang, L., 2000. Progress in malaria control in China. Chin. Med. J. (Engl.) 113, 89–92.

Tang, L.H., Qian, H.L., Xu, S.H., 1991. Malaria and its control in the People's Republic of China. Southeast Asian J. Trop. Med. Public Health 22, 467–476.

Tanner, M., Hommel, M., 2010. Towards malaria elimination–a new thematic series. Malar. J. 9, 24.

Wang, D.Q., Xia, Z.G., Zhou, S.S., Zhou, X.N., Wang, R.B., Zhang, Q.F., 2013. A potential threat to malaria elimination: extensive deltamethrin and DDT resistance to *Anopheles sinensis* from the malaria-endemic areas in China. Malar. J. 12, 164.

Wang, G.Z., Wang, S.Q., Hu, X.M., Meng, F., Li, Y.C., Zeng, W., Cai, H.L., 2012. Analysis of malaria epidemic situation in Hainan province, 2010. Chin. J. Schisto. Control 24, 369–370 (in Chinese).

Wang, L., Wang, Y., Jin, S., Wu, Z., Chin, D.P., Koplan, J.P., Wilson, M.E., 2008. Emergence and control of infectious diseases in China. Lancet 372, 1598–1605.

Wang, L.P., Fang, L.Q., Xu, X., Wang, J.J., Ma, J.Q., Cao, W.C., Jin, S.G., 2009. Study on the determinants regarding malaria epidemics in Anhui province during 2004-26. Chin. J. Epidemiol. 30, 38–41 (in Chinese).

White, L.J., Maude, R.J., Pongtavornpinyo, W., Saralamba, S., Aguas, R., Van Effelterre, T., Day, N.P., White, N.J., 2009. The role of simple mathematical models in malaria elimination strategy design. Malar. J. 8, 212.

W.H.O., Secretariat, 2012. Malaria Policy Advisory Committee to the WHO: conclusions and recommendations of September 2012 meeting. Malar. J. 11, 424.

W.H.O., Secretariat, 2013a. Malaria Policy Advisory Committee to the WHO: conclusions and recommendations of March 2013 meeting. Malar. J. 12, 213.

W.H.O., Secretariat, 2013b. Malaria Policy Advisory Committee to the WHO: conclusions and recommendations of September 2013 meeting. Malar. J. 12, 456.

WHO, 2007. Malaria Elimintation: A Field Manual for Low and Moderate Endemic Countries. WHO Press, Geneva.

Xia, Z.G., Xu, J.F., Zhang, S.S., Wang, R.B., Qian, Y.J., Zhou, S.S., Yang, W.Z., Zhou, X.N., 2014. Determination of key interventions for the transition from control to elimination of malaria in China. Chin. J. Schisto. Control 26, 597–600.

Xia, Z.G., Yang, M.N., Zhou, S.S., 2012. Malaria situation in the People's Republic of China in 2011. Chin. J. Parasitol. Parasit. Dis. 30, 419–422 (in Chinese).

Xiao, D., Long, Y., Wang, S., Wu, K., Xu, D., Li, H., Wang, G., Yan, Y., 2012. Epidemic distribution and variation of *Plasmodium falciparum* and *Plasmodium vivax* malaria in Hainan, China during 1995-2008. Am. J. Trop. Med. Hyg. 87, 646–654.

Xu, B.L., Su, Y.P., Shang, L.Y., Zhang, H.W., 2006. Malaria control in henan province, People's Republic of China. Am. J. Trop. Med. Hyg. 74, 564–567.

Xu, J., Liu, H., 2012. The challenges of malaria elimination in Yunnan Province, People's Republic of China. Southeast Asian J. Trop. Med. Public Health 43, 819–824.

Xu, J.W., Yang, H., Yang, Z.Q., Yang, G.C., Ma, X.W., Wang, W.R., Gu, Y.A., Wang, L.B., Yang, X.W., Ma, J., 2002. Cost-effectiveness analysis of the current measures for malaria prevention in Yuanjiang valley, Yunnan province. Chin. J. Parasitol. Parasit. Dis. 20, 238–241 (in Chinese).

Yan, J., Li, N., Wei, X., Li, P., Zhao, Z., Wang, L., Li, S., Li, X., Wang, Y., Li, S., Yang, Z., Zheng, B., Zhou, G., Yan, G., Cui, L., Cao, Y., Fan, Q., 2013. Performance of two rapid diagnostic tests for malaria diagnosis at the China-Myanmar border area. Malar. J. 12, 73.

Yang, G.J., Gao, Q., Zhou, S.S., Malone, J.B., McCarroll, J.C., Tanner, M., Vounatsou, P., Bergquist, R., Utzinger, J., Zhou, X.N., 2010. Mapping and predicting malaria transmission in the People's Republic of China, using integrated biology-driven and statistical models. Geospat. Health 5, 11–22.

Yang, G.J., Tanner, M., Utzinger, J., Malone, J.B., Bergquist, R., Chan, E.Y., Gao, Q., Zhou, X.N., 2012. Malaria surveillance-response strategies in different transmission zones of the People's Republic of China: preparing for climate change. Malar. J. 11, 426.

Yin, J., Xia, Z., Yan, H., Huang, Y., Lu, L., Geng, Y., Xiao, N., Xu, J., He, P., Zhou, S., 2013a. Verification of clinically diagnosed cases during malaria elimination programme in Guizhou Province of China. Malar. J. 12, 130.

Yin, J.H., Yang, M.N., Zhou, S.S., Wang, Y., Feng, J., Xia, Z.G., 2013b. Changing malaria transmission and implications in China towards National Malaria Elimination Programme between 2010 and 2012. PloS One 8, e74228.

Zhang, H.W., Su, Y.P., Zhao, X.D., Yan, Q.Y., Liu, Y., Chen, J.S., 2010. Imported falciparum malaria situation in Henan province during 2005-2009. Chin. J. Parasitol. Parasit. Dis. 28, 476–477 (in Chinese).

Zhao, X.F., Zhang, J.N., Dong, H.J., Zhang, T., Bian, G.L., Sun, Y.W., Yao, M.H., Chen, K.J., Xu, G.Z., 2013. Epidemiological characteristics of malaria in Ningbo City, China 2000-2011. Trop. Biomed. 30, 267–276.

Zheng, Q., Vanderslott, S., Jiang, B., Xu, L.L., Liu, C.S., Huo, L.L., Duan, L.P., Wu, N.B., Li, S.Z., Xia, Z.G., Wu, W.P., Hu, W., Zhang, H.B., 2013. Research gaps for three main tropical diseases in the People's Republic of China. Infect. Dis. Poverty 2, 15.

Zhou, S.S., Huang, F., Wang, J.J., Zhang, S.S., Su, Y.P., Tang, L.H., 2010. Geographical, meteorological and vectorial factors related to malaria re-emergence in Huang-Huai River of central China. Malar. J. 9, 337.

Zhou, S.S., Wang, Y., Xia, Z.G., 2011. Malaria situation in the People's Republic of China in 2009. Chin. J. Parasitol. Parasit. Dis. 29, 1–3 (in Chinese).

Zhou, S.S., Zhang, S.S., Wang, J.J., Zheng, X., Huang, F., Li, W.D., Xu, X., Zhang, H.W., 2012. Spatial correlation between malaria cases and water-bodies in *Anopheles sinensis* dominated areas of Huang-Huai plain, China. Parasit. Vectors 5, 106.

Zhou, X.N., Bergquist, R., Tanner, M., 2013. Elimination of tropical disease through surveillance and response. Infect. Dis. Poverty 2, 1.

Zhou, X.N., Zhang, S.S., Xu, J.F., Xia, Z.G., Wang, R.B., Qian, Y.J., Zhou, S.S., Yang, W.Z., 2014. Risk assessment of malaria transmission for the national malaria elimination in P.R. China. Chin. J. Parasitol. Parasit. Dis. 32, 414–417 (in Chinese).

Zhou, Z.J., 1981. The malaria situation in the People's Republic of China. Bull. World Health Organ. 59, 931–936.

Zhou, Z.J., 1991. The Studies and Control on Malaria in China. People's Medical Publishing House, Beijing (in Chinese).

Zofou, D., Nyasa, R.B., Nsagha, D.S., Ntie-Kang, F., Meriki, H.D., Assob, J.C., Kuete, V., 2014. Control of malaria and other vector-borne protozoan diseases in the tropics: enduring challenges despite considerable progress and achievements. Infect. Dis. Poverty 3, 1.

CHAPTER THREE

Lessons from Malaria Control to Elimination: Case Study in Hainan and Yunnan Provinces

Zhi-Gui Xia[1,#], Li Zhang[1,#], Jun Feng[1], Mei Li[1], Xin-Yu Feng[1], Lin-Hua Tang[1], Shan-Qing Wang[2], Heng-Lin Yang[3], Qi Gao[4], Randall Kramer[5], Tambo Ernest[6], Peiling Yap[7], Xiao-Nong Zhou[1,*]

[1]National Institute of Parasitic Diseases, Chinese Center for Disease Control and Prevention; Key Laboratory of Parasite and Vector Biology, MOH; WHO Collaborating Centre for Malaria, Schistosomiasis and Filariasis, Shanghai, People's Republic of China
[2]Hainan Provincial Center for Disease Control and Prevention, Haikou, People's Republic of China
[3]Yunnan Provincial Institute of Parasitic Diseases, Pu-er, People's Republic of China
[4]Jiangsu Provincial Institute of Parasitic Diseases, Wuxi, People's Republic of China
[5]Duke Global Health Institute, Duke University, Durham, NC, USA
[6]Centre for Sustainable Malaria Control, Faculty of Natural and Environmental Science; Center for Sustainable Malaria Control, Biochemistry Department, Faculty of Natural and Agricultural Sciences, University of Pretoria, Pretoria, South Africa
[7]Department of Epidemiology and Public Health, Swiss Tropical and Public Health Institute, Basel, Switzerland; University of Basel, Basel, Switzerland
*Corresponding author: E-mail: zhouxn1@chinacdc.cn

Contents

1. Introduction	48
2. Background of Two Provinces	51
2.1 Geography, population and health system	51
2.2 Natural factors for malaria transmission	52
2.3 Social-economic factors for malaria transmission	52
2.4 Behavioral factors for malaria transmission	53
3. Malaria Transmission Patterns, Particularly in 2004–2012	53
3.1 Overview of malaria prevalence	53
3.2 Cases by *Plasmodium* species	54
3.3 Spatiotemporal distribution	55
3.4 Demographical characteristics	57
3.5 Cases by acquisition	59
4. History of Malaria Control	61
5. Malaria Interventions from Control to Elimination	63
5.1 Case detection	63
5.2 Antimalarial drug administration	63
5.3 Vector control	64
5.4 Pilot studies	66

[#] Contributed equally to this paper.

5.5 Partnership	67
5.6 Surveillance	68
6. Challenges for Malaria Elimination	68
6.1 Challenges in Hainan province	68
6.1.1 Management of the mobile population	*68*
6.1.2 Prevention of malaria reintroduction	*69*
6.2 Challenges in Yunnan province	70
6.2.1 High malaria transmission on the China–Myanmar border	*70*
6.2.2 Heterogeneity and complexity of malaria epidemiology and vectors	*70*
6.2.3 Management of the mobile population	*71*
6.2.4 Capability building at the ground level	*72*
6.2.5 Prevention of malaria reintroduction	*72*
6.2.6 Monitoring of antimalarial drug resistance	*73*
7. Conclusions	74
Acknowledgements	76
References	76

Abstract

Reduction patterns of *Plasmodium falciparum* and *P. vivax* malaria transmission and the role of an integrated strategy of case management and vector control are compared between different ecological zones. The epidemiology of malaria in Hainan and Yunnan provinces was disparate, even though distinct malaria control strategies have been adapted to different situations based on risk group, vector behaviours, local health infrastructure, and environmental conditions. The island Hainan appears to be victorious in eliminating malaria. However, there is still a long way to go to prevent the reintroduction of malaria in Hainan province and eliminating malaria in the border areas of Yunnan province. This review of the experiences and challenges from malaria control to elimination in Hainan and Yunnan provinces of southern China will provide a basis for the future elimination of malaria in the whole country.

1. INTRODUCTION

Malaria is an important parasitic disease that has severely damaged the Chinese people's life safety and health and slowed the country's economic and social development in the early years of last century. (Chen, 2014). Since the establishment of the People's Republic of China (P.R. China) in 1949, malaria prevention and control in the country has made remarkable achievements with governmental support at all levels. The epidemic area has been greatly reduced, with the number of malaria cases decreasing from more than 24 million in the early 1970s to tens of thousands by the end of the 1990s (Tang, 1999a).

P. falciparum malaria has been eliminated from all areas by 2000, except for the provinces of Yunnan and Hainan (Bi et al., 2013; Tang, 1999a; Tang et al., 2012a).

Due to the complexity of risk factors, the spread of malaria is fast and capricious. Since 2000, the epidemic situation of malaria in P.R. China has become more severe, with outbreaks occurring regularly in some regions (Zhang et al., 2008b; Zhou et al., 2008). This can be mainly explained by weakened prevention and control systems, inadequate fund investments in some regions, and the influence of the mobile population and high endemic situations in some of the neighbouring countries on the border areas of southern China (Zheng et al., 2013; Zofou et al., 2014). Malaria prevalence in the country reached its peak in 2006, when the number of malaria cases was more than 60,000, and the incidence was up to around 5 per 100,000. Due to the presence of geographical conditions naturally suited for transmission, multi-drug resistant *P. falciparum* and highly efficient vectors of *Anopheles minimus* and *An. dirus*, both Hainan and Yunnan are provinces with high malaria transmission, where *P. falciparum and P. vivax* malaria are epidemic throughout the year (Liu, 2014; Zhou et al., 2007).

To accelerate the process for efficient malaria prevention and control in light of the changing malaria situation in China, to ensure people's health and promote social-economic development in epidemic areas, and to position China as the leading developing country in achieving malaria elimination, the *National Malaria Control programme (NMCP) in 2006–2015* was formulated and issued by the Ministry of Health with the following goals (China, 2006):

1. By 2010, malaria will be under control in all counties, except for the border counties of Yunnan and the mountainous counties of central and southern part of Hainan, and 70% of the counties will have achieved basic elimination of malaria.
2. By 2015, malaria will be under control in the border counties of Yunnan and in mountainous counties of central and southern part of Hainan, while the other previously epidemic counties will have achieved basic elimination of malaria, and Hainan province will have eliminated falciparum malaria.

Through implementing the NMCP from 2006 to 2015, the central and local governments have increased support of and investments in malaria control, causing the rising in the number endemic counties to be effectively controlled. By 2009, the number of malaria patients dropped to about 14,000; the incidence in more than 95% of the counties dropped to below 1 per 10,000. To further protect the health of China's population, promote coordinated development of the economy and society, and respond to the Millennium Development Goals on the eradication of malaria globally, the Chinese Ministry of Health, together with the other 12 central governmental sectors, jointly issued the *Chinese Malaria Elimination Action Plan (2010–2020)* that was a sign of launching the National Malaria Elimination Programme (NMEP) in P.R. China (China, 2010). Based on the malaria situation reported in 2006–2008, all the counties of P.R. China were divided into the following four types:

Type 1: There were local malaria infections during those 3 years, and the incidence rate of each year was greater than or equal to 1 per 10,000.

Type 2: There were local infections during those 3 years, and the incidence rate was less than 1 per 10,000 for at least a year.

Type 3: No local infection was reported in three consecutive years.

Type 4: Nonendemic area of malaria.

The following goals were set for the different types of malaria transmission:

1. All Type 3 counties are to achieve malaria elimination by 2015.
2. All Type 2 counties and Yunnan Type 1 counties located outside the border area are to achieve no local malaria infection by 2015 and attain elimination by 2018.
3. Type 1 counties in the border region of Yunnan should have their malaria incidence drop to below 1 per 10,000 by 2015, achieve no local malaria infection by 2017, and attain elimination by 2020.

According to the *Chinese Malaria Elimination Action Plan (2010–2020)*, there were 10 Type 1 counties and 8 Type 2 counties in Hainan province and 19 Type 1 counties, 55 Type 2 counties and 55 Type 3 counties in Yunnan province. In 2012, after nearly 3 years of implementation, there was no local malaria infection for 1 year, no local *P. falciparum* malaria infection for 3 consecutive years, and only 13 imported malaria cases were reported that year in Hainan province. On the other hand, Yunnan province reported 853 malaria cases, but with only 133 indigenous cases that were distributed in 20 counties, and the county incidence rates were lower than 1 per 10,000 in 2012 (Xia et al., 2013).

Because both Hainan and Yunnan are the most serious provinces for malaria transmission in P.R. China before launching the NMEP, their progress in elimination has a significant impact on the overall malaria situation of the country. Therefore, this paper aims to analyse the changing characteristics of *P. falciparum* and *P. vivax* malaria transmission in both provinces, review the successful experiences of malaria control, and identify the risks and challenges for the future. Distillation of such knowledge will not only benefit further malaria elimination in these two provinces but also the whole country.

2. BACKGROUND OF TWO PROVINCES
2.1 Geography, population and health system

Hainan Island, the southernmost province of China, is located at north latitude 18°10′–20°10′ and east longitude 108°37′–111°03′. The region has an area of 33,900 square kilometres with 20 counties and a population of 8.87 million in 2012. The province is characterized by mountains, hills, plateaus and plains. The tropical monsoon and marine climates jointly produce a generally warm temperature, with an annual temperature change of less than 15 °C (59 °F) and rich rainfall of average annual precipitation between 1500 mm and 2000 mm (Wen et al., 2011; Xiao et al., 2010, 2012). Owing to its tropical climate, Hainan's economy is predominantly agricultural, and the island is cultivated extensively with paddy rice, coconuts, palm oil, sisal, tropical fruits, black pepper, coffee, tea and rubber trees. Hainan's health system is composed of hospitals, centres for disease control and prevention, and blood collection centres at different levels. These health and medical institutions play a huge part in malaria control, both historically as well as currently. In Hainan province, the counties are directly under the jurisdiction of provincial authority without a prefecture level. Unlike Yunnan province, almost all the counties are in close proximity with each other and can be reached by car within 3 h from the provincial capital of Hainan.

Yunnan province is located in southwestern China, it includes 129 counties and borders Myanmar, Lao People's Democratic Republic (Lao PDR) and Vietnam, spanning approximately 394,000 square kilometres, with a population of 46.59 million in 2012. It has similar monsoon climates as Hainan province and lowland tropical rain forest vegetation. Temperatures regularly exceed 30 °C (86 °F) in the warmer half of the year in most areas and average annual rainfall ranges from 600 mm to 2300 mm. Yunnan is one of China's relatively underdeveloped provinces with more poverty-stricken counties than the other provinces (Bi & Tong, 2014). However, due to its special geographical location, the province has comparative advantages in exporting agricultural and mineral resources (Feng et al., 2012; Hao et al., 2013; Meng et al., 2010). Health systems, including hospitals, centres for disease control and prevention as well as blood collection centres in Yunnan province at all levels play a key role in dealing with all types of public health threats, including malaria. At the provincial level, the Yunnan Provincial Institute of Parasitic Diseases is responsible for malaria control in the whole province.

2.2 Natural factors for malaria transmission

Hainan province, located in the tropics, has been one of the main malaria transmission areas in P.R. China. Its climate and environment are within the suitable range for the breeding of *An. dirus* and *An. minimus*. Geographically, malaria cases in Hainan province were mainly distributed in patches in the southwest region of the island, while fewer cases were sparsely distributed in the northeast area. Such distributions are largely related to the natural environment of the different counties. The temperature is hotter and the humidity of the mountainous areas is higher in the southwest than in the northeast, making the southwest region more suitable for the growth of anopheline mosquitoes and the transmission of malaria (Wang et al., 2012a; Wang and Cai, 1996).

Yunnan borders Myanmar, Lao PDR and Vietnam, which are highly endemic for malaria. Influenced by the Indian Ocean, the Pacific monsoon climate and its special landforms, Yunnan is well suited for malaria transmission under various climates and special human landscapes (Yang, 2001; Yang et al., 1988). Due to a complex interplay of all these factors, Yunnan has become the most endemic province for malaria in P.R. China, rendering it the most difficult for malaria control in the country. Geographically, complicated malaria distribution and numerous *Anopheles* species are the characteristics of malaria epidemiology in Yunnan province. In the high-risk border area, where the number of malaria cases is largely dependent on climatic conditions, including temperature, relative humidity and rainfall (Bi et al., 2013), the main vector is *An. minimus*. While in the rice-growing area, *An. sinensis* is the main vector; in the high-altitude mountains of the northwest, it is *An. kunmingensis*; and in the valley areas of the northeast, it is *An. anthropophagus* (Dong, 2000).

2.3 Social-economic factors for malaria transmission

In Hainan province, some scholars have found that the culture, malaria knowledge and health organisation conditions at the grassroots level of Li and Miao minorities in high malaria-endemic mountainous areas had a positive effect on local malaria control. Health care service was the main social factor, and carrying out health education and improving the capacities of grassroots health organization in malaria diagnosis and treatment were the critical elements for effective malaria control (Cai et al., 1995; Chen et al., 1995, 1996).

On the other hand, studies in Yunnan province have demonstrated that the main factors for local malaria transmission were mountainous lodging, usage of mosquito nets, house structure, and per capita income of local people (Chen et al., 1998).

2.4 Behavioral factors for malaria transmission

Rubber planting is the main source of family income for Hainan middle-west residents; by having to go to the mountains for such work, they stand a greater chance of contracting malaria. Other studies in Nan-qiao Township of Wan-ning County showed that mountain lodging was the main source of malaria infections, especially *P. falciparum* malaria, and infections from the mountains were highly common in Hainan province (Cai and Weng, 2013). Workers in the mountains had to engage in labour-intensive jobs, lived in dilapidated conditions near anopheline mosquito breeding sites, and lacked access to mosquito nets and health care, making them a high-risk population to be targeted in the NMCP in Hainan (Wu et al., 1995).

In border and minority areas of Yunnan province, residents have relatively low educational levels, special living habits, and frequent movement across the border. The border minority populations lag behind in awareness for malaria, with a lower proportion of people seeking diagnosis within 48 h of fever or using of insecticide-treated mosquito nets, making them the main targeted group for local malaria control (Chen et al., 1998; Cheng et al., 1997).

3. MALARIA TRANSMISSION PATTERNS, PARTICULARLY IN 2004–2012

The epidemiological data of malaria reported through the web-based case reporting system and annual malaria statistics reporting system from 2004 to 2012 were collected and analysed with the following main results.

3.1 Overview of malaria prevalence

An average of 26,674 malaria cases per year were reported nationwide in 2004–2012 (excluding Hong Kong, Macau and Taiwan). From 2004 to 2006, malaria cases increased, with about 38,000 cases each in 2004 and 2005 and around 60,000 cases in 2006. From 2007 to 2012, malaria was effectively controlled in the country, as the reported number of cases and incidence rate declined at an average annual rate of 40.7% and 40.0%, respectively.

Hainan province reported an annual average of 2650 malaria cases in 2004–2012, and its yearly contribution to the total number of cases nationally has declined to 0.4% in 2012, when only 13 cases were reported.

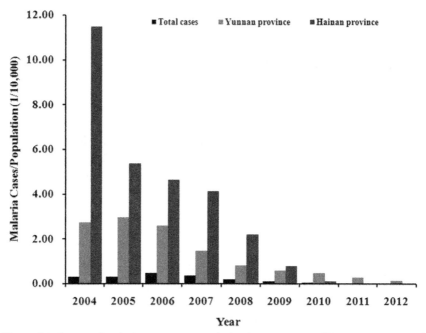

Figure 3.1 Reported malaria cases per population in 2004–2012 of Yunnan, Hainan and nationwide.

Yunnan province reported an annual average of 5984 malaria cases in 2004–2012. The yearly number of cases has been on the decline since 2005. The lowest proportion of malaria infections, accounting for 13.9% of the total number of cases nationwide in 2008, and has since risen to 25.5% in 2012, when a total of 635 cases were reported (Figure 3.1).

3.2 Cases by *Plasmodium* species

In Hainan province, a total of 12,225 (51.3%) *P. vivax* malaria cases, 2061 (8.6%) *P. falciparum* malaria cases and 9567 (40.1%) unclassified malaria cases were reported from 2004 to 2012. The *P. vivax*, *P. falciparum* and unclassified malaria cases declined each year, with average annual rates of 53.3%, 14.3% and 24.9%, respectively. Unfortunately, the proportion of *P. falciparum* malaria saw an increase since 2009. In 2012, the proportion of *P. vivax* and *P. falciparum* malaria stood at 30.0% and 70.0%, respectively.

In the same period, a total of 39,471 (73.3%) *P. vivax* malaria cases, 10,987 (20.4%) *P. falciparum* malaria cases and 3402 (6.3%) unclassified malaria cases were reported, including 108 malaria deaths, in Yunnan province. The *P. vivax*, *P. falciparum* and unclassified malaria cases declined, with average annual rates of 28.6%, 25.7% and 34.3%, respectively. In 2012, the *P. vivax*,

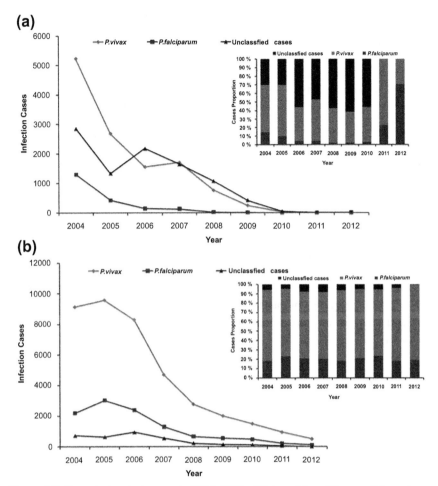

Figure 3.2 Reported malaria cases in 2004–2012 of Hainan (a) and Yunnan (b) provinces by *plasmodium* species.

P. falciparum and unclassified malaria cases accounted for 80.6%, 19.1% and 0.3%, respectively. The highest numbers of total cases, *P. vivax* cases and *P. falciparum* malaria cases from this province were all reported in 2005, accounting for 33.1%, 31.3% and 80.7% of the national figures, respectively (Figure 3.2).

3.3 Spatiotemporal distribution

Geographically, the reported *P. vivax* malaria cases in Hainan province were mainly distributed in the counties of Qiong-zhong (14.9%), Dong-fang (13.8%), Bai-sha (12.4%), Le-dong (11.9%) and Wan-ning (10.2%). The *P. falciparum* malaria cases were mainly found in counties of Dong-fang (29.1%), San-ya (21.6%), Le-dong (15.1%), Ling-shui (8.2%) and Bai-sha (6.3%).

In Yunnan province, the reported *P. vivax* malaria cases were mainly distributed in the prefectures of Bao-shan (29.3%), De-hong (26.2%), Lin-cang (8.2%), Hong-he (7.6%) and Pu-er (7.0%). The *P. falciparum* malaria cases were mainly reported from the prefectures of De-hong (55.7%), Bao-shan (41.2%), Lin-cang (6.6%), Pu-er (5.5%) and Hong-he (4.5%).

Both the geographical distribution and numbers of *P. vivax* and *P. falciparum* malaria cases reported in Hainan and Yunnan provinces have shrunk, but this phenomenon was more significant in Hainan province. There are still many reported malaria cases in areas of Yunnan province, particularly those areas of bordering other malaria endemic countries, such as Myanmar (Figure 3.3 and Figure 3.4).

With regard to seasonal distribution, the reported *P. vivax* and *P. falciparum* malaria cases in Hainan province occurred mainly from May to September, accounting for 48.2% and 61.9%, respectively, of all cases in a year. The cases peaked in August, when the reported numbers of the two species were 13.9% and 14.7%, respectively. These seasonal characteristics of malaria distribution were more obvious from 2004 to 2007 and less evident recently as malaria transmissions were reduced to a very low level and most cases were imported.

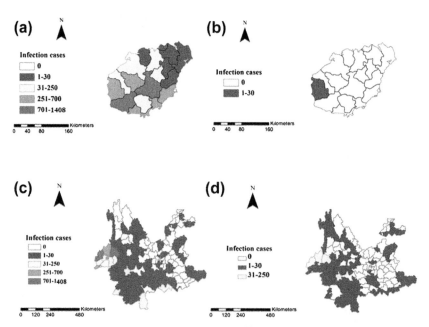

Figure 3.3 Geographical distribution of reported *P. vivax* malaria cases in Hainan province in 2004 (a) and 2012 (b) and Yunnan province in 2004 (c) and 2012 (d), classified by the number of malaria cases in each county.

In Yunnan province, the reported *P. vivax* and *P. falciparum* malaria cases occurred mainly in April to August. The top-five months for *P. vivax* malaria occurrence were June (13.0%), July (12.5%), May (12.3%), August (11.1%) and September (9.1%). For *P. falciparum* malaria cases, these months were May (16.2%), June (14.7%), April (11.2%), July (10.2%) and November (8.5%). Likewise, the seasonal characteristics were also less evident recently in this province (Figure 3.5).

3.4 Demographical characteristics

Among *P. vivax* malaria cases reported from Hainan province between 2004 and 2012, a total of 761 (6.2%) cases were in the age group of <10 years, 7985 (65.3%) cases were in the 10- to 34-year group, 3207 (26.2%) cases were reported for 35–60 years, and 272 (2.2%) cases were for >60 years. In particular, the number of young and middle-aged patients was 3.6 times higher than older patients. Similarly, the number of male patients was 3.4 times higher than that of the females. For *P. falciparum* infections, 178 (8.6%) cases were in the age group of <10 years, 1361 (66.0%) cases were for 10–34 years, 484 (23.5%) cases for 35–60 years, and 38 (1.8%) cases for >60 years. In particular, the number of young and middle-aged

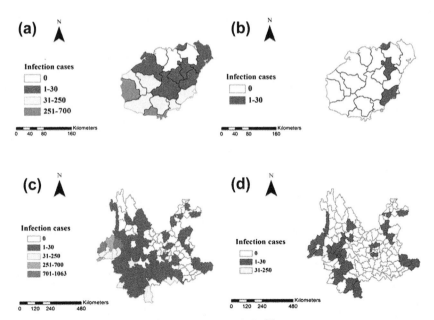

Figure 3.4 Geographical distribution of reported *P. falciparum* malaria cases in Hainan province in 2004 (a) and 2012 (b) and Yunnan province in 2004 (c) and 2012 (d), classified by the number of malaria cases in each county.

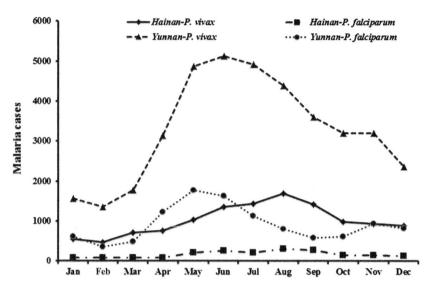

Figure 3.5 The patterns of the number of the reported *P. vivax* and *P. falciparum* malaria cases in 2004–2012 of Hainan and Yunnan provinces by month.

patients was 3.7 times higher than that of older patients, and the number of male patients was 3.3 times higher than that of the females.

For *P. vivax* malaria cases in Yunnan province, a total of 3080 (7.8%) cases were reported in the age group of <10 years, 26,968 (68.3%) cases in 10–34 years, 8786 (22.3%) cases in 35–60 years, and 637 (1.6%) cases in >60 years. The number of young and middle-aged patients was 3.7 times higher than that of older patients. The number of male patients was also 3.4 times higher than that of female patients. With regards to *P. falciparum* infections, 662 (6.0%) cases were in the age group of <10 years, 7595 (69.1%) cases in 10–34 years, 2594 (23.6%) cases in 35–60 years, and 136 (1.2%) cases in >60 years. The number of young and middle-aged patients was 5.0 times higher than that of older patients. The number of male patients with *P. falciparum* malaria was 4.5 times higher than that in the females (Figure 3.6).

Stratification of *P. vivax* malaria cases in Hainan province from 2004 to 2012 according to occupation revealed that infections occurred mainly in farmers (53.1%), followed by migrant peasant labourers (14.4%), students (12.0%) and workers (11.5%). *P. falciparum* malaria cases were also mainly distributed in farmers (54.5%), followed by students (13.8%), migrant peasant labourers (12.1%) and workers (8.6%).

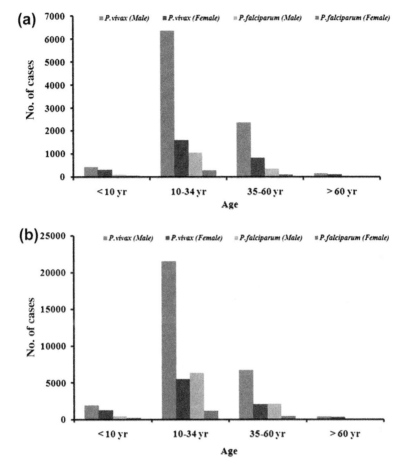

Figure 3.6 The number of the reported *P. vivax* and *P. falciparum* malaria cases in 2004–2012 of Hainan (a) and Yunnan (b) provinces by age group and gender.

In Yunnan province, *P. vivax* malaria cases were highly prevalent in farmers (53.6%), followed by migrant peasant labourers (23.8%), students (8.2%) and workers (3.3%). A similar distribution was observed for *P. falciparum* malaria cases in farmers (54.7%), migrant peasant labourers (26.1%), students (5.6%), and workers (3.6%).

3.5 Cases by acquisition

From 2004 to 2012, a total of 12,210 (99.9%) indigenous and 15 (0.1%) imported *P. vivax* malaria cases were reported in Hainan province. The imported *P. vivax* cases were from Cambodia (5 cases), Myanmar (3 cases),

Pakistan (4 cases) and Indonesia (3 cases). For *P. falciparum* malaria, there were 2046 (99.3%) indigenous cases and 15 (0.7%) imported cases, including 10 cases from Africa, 3 cases from southeast Asia and 2 cases with unclear sources information. There has been no local transmission of *P. falciparum* malaria since 2010 and no local malaria transmission since 2012. Most current cases were imported and only sporadic cases occurred recently in this province (Figure 3.7).

In Yunnan province, a total of 27,411 (69.5%) indigenous and 12,060 (30.6%) imported *P. vivax* malaria cases were reported; the imported cases were mainly from Myanmar (6832 cases, 56.7%). For *P. falciparum* malaria,

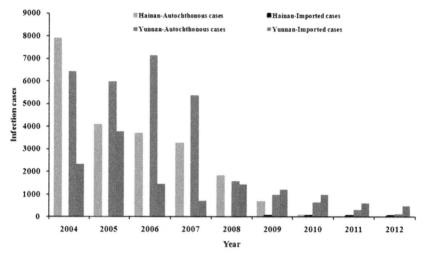

Figure 3.7 The number of the reported *P. vivax* malaria cases in 2004–2012 of Hainan and Yunnan provinces by acquisition.

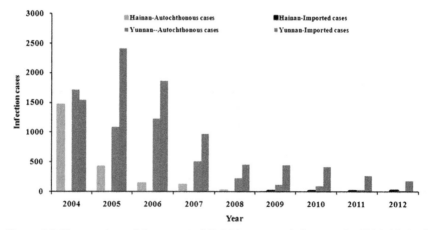

Figure 3.8 The number of the reported *P. falciparum* malaria cases in 2004–2012 of Hainan and Yunnan provinces by acquisition.

there were 5018 (45.7%) indigenous cases and 5969 (54.3%) imported cases; 79.2% of the imported cases were from Africa, including Nigeria (12.3%) and the Republic of Congo (12.3%). The highest proportion (77.7%) of indigenous malaria cases occurred in 2007, and then an increasing proportion of the imported malaria was observed (Figure 3.8). In contrast, a total of 679 imported malaria cases were reported in 2012, accounting for 79.6% of the total number of cases.

4. HISTORY OF MALARIA CONTROL

The history of malaria control in Hainan province can be divided into five stages in the past 60 years (Tang et al., 2012a; Wang, 2000). The first was the stage of investigation and pilot control in 1951–1958, when epidemiological investigations and key control trials were carried out by the National Malaria Research Station and Hainan Malaria Control Dispensary. The second stage in 1959–1966 consisted of comprehensive control, exterminating *An. minimus* and reducing the prevalence of malaria cases. As large-scale malaria campaigns were performed in this stage, using mass chemoprophylaxis and comprehensive insecticide spraying integrated with environmental improvement to control the vectors of *Anopheles* spp., *An. minimus* was basically wiped out by 1964 and a dramatic decline of malaria incidence rate was observed in Hainan Island. Although some success was achieved, this strategy was still slow to proceed as high-quality performance was needed, and the sustainability of such a strategy proved to be difficult.

The third stage, conducted in 1967–1983, relied mainly on vector control and highlighted the containment of chloroquine-resistant *P. falciparum*. The main countermeasures were to strengthen control and research of chloroquine-resistant *P. falciparum*, control infectious sources with mass or targeted drug administration, control vectors with insecticidal nets, and carry out blood tests in fever patients to detect and treat malaria cases. The fourth stage was a point-to-area, joint and comprehensive control stage in 1984–2000. The strategies were to intensify the regular malaria management, keep the former effective interventions, consolidate and enlarge the joint efforts in control of malaria, adopt mass or targeted drug administration to control infectious sources in the high endemic areas, control vectors with insecticidal nets, identify malaria with microscopy and treat febrile patients, carry out malaria surveillance and control among uphill and external mobile populations, establish integrated malaria control trials, and remove the focal zones of malaria.

The fifth stage was carried out in 2001–2012, with further reduced malaria incidence rates; infectious source control was the dominant intervention. The main strategies were to carry out comprehensive and regular malaria management, fulfil blood tests among febrile patients, discover and treat malaria infections and foci in a timely manner, strengthen mobile population management, and implement surveillance on parasitaemia rate in population, antimalarial drug resistance and mosquito vectors. After strong efforts for more than half a century, remarkable achievements have been obtained in Hainan. Annual morbidity of malaria decreased continuously and endemic areas shrunk substantially, such that no high- or super-risk areas exist. Currently, malaria infections are only confined to remote forest zones in some counties of south-central Hainan. The annual incidence of malaria has fallen from 9890.21 per 100,000 in 1995 to 0.11 per 100,000 in 2012, and no malaria deaths were found in eight consecutive years since 2005 (Du, 2001; Huang et al., 2010; Liu et al., 2012; Zeng et al., 2013).

The history of malaria control in Yunnan province can be illustrated in four stages (Li and Wu, 2009; Tang et al., 2012b; Yang, 1995). The first stage in the 1950s was to establish a control system. Prevention teams were orgnized to mobilize the communities to participate in control of malaria and vectors, the malaria control institutions was established to carried out malaria control trials and to explore experiences and methods for control in large areas. The second stage in 1960–1979 was performed to control malaria incidence and decrease malaria mortality with mass preventions and treatments. This included comprehensive and equally important interventions of infectious source control and vector extermination, delivering treatment and antirelapse treatment of malaria patients, and conducting mass chemoprophylaxis in the transmission seasons of endemic areas.

The third stage in 1980–1999 saw the implementation of integrated measurements. Emphasis was placed on prevention of key areas and protection of key populations, highlighting the surveillance of infectious sources and vectors, carrying out blood tests in febrile patients with a testing rate of no less than 5% at the county level and a coverage rate of more than 80% at the village level, identifying and treating malaria patients, eliminating the focal zone, intensifying the management of mobile populations to prevent the import and spread of infectious sources, and establishing vector surveillance sites. The fourth stage in 2000–2012 was conducted to consolidate the achievements of malaria control and improve the surveillance in border areas. It focused on mobile populations and mountainous residents, who were targeted according to the malaria situation and vector distribution in different periods and areas. Integrated strategies of infectious source managements and vector control were applied to further decrease malaria cases and deaths and shrink malaria endemic areas, especially *P. falciparum* malaria, and malaria surveillance and the treatment disposal of imported case were strengthened.

5. MALARIA INTERVENTIONS FROM CONTROL TO ELIMINATION

5.1 Case detection

Blood testing in febrile patients as part of malaria surveillance is the most effective method for the detection of malaria cases. There are some differences in the blood testing procedures from the two provinces. In Hainan province, the malaria microscopy centre in each county's Center for Disease Control and Prevention (CDC) and the malaria microscopy station in each township hospital provided blood test for *Plasmodium* spp. for suspected outpatients with fever. Health workers in villages and farms in malaria endemic areas sampled blood from febrile patients and sent them to health facilities at a higher level at regular intervals for *Plasmodium* detection (Huang et al., 2010; Wang et al., 2003). From 2002 to 2012, a total of 17,32,823 blood tests were done in the province, resulting in a testing rate of 10.2%. A total of 17,522 persons were confirmed with malaria infections, with a positive rate of 0.8%.

In Yunnan province, blood testing was carried out in all medical organizations capable of conducting malaria microscopic examinations, and rapid diagnostic tests (RDTs) for *Plasmodium* antigen detection were used in some qualified medical organizations or village clinics. Active case investigations were done by county CDCs in the transmission seasons in areas including administrative villages with a malaria morbidity of more than 20 per 10,000, and natural villages reporting indigenous cases where no local infections appeared in the past 3 years. Household investigations were used to understand the local malaria situation, and patients with fever or with a history of fever in the past 2 weeks were registered and blood tests were performed on them. From 2002 to 2012, a total of 59,45,383 blood tests were conducted, leading to a testing rate of 1.7%. A total of 74,763 malaria patients were subsequently confirmed, with a positive rate of 1.3% (Figure 3.9).

5.2 Antimalarial drug administration

The first choice of treatment for *P. vivax* malaria infection is chloroquine phosphate (0.6 g single dose for the first day, 0.3 g daily for the next 2 days, total 1.2 g/3 days) combined with primaquine phosphate (22.5 mg single dose/day for 8 days, total 180 mg/8 days) tablets, and antirelapse treatment using primaquine (180 mg, divided into eight doses, one dose daily) given during the nontransmission season for reducing malaria reservoirs entering into the

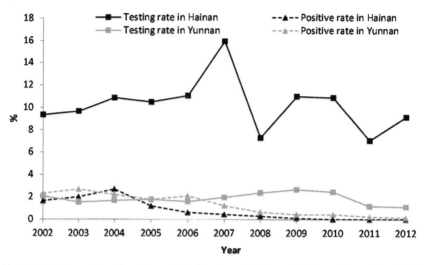

Figure 3.9 Annual blood testing rate in febrile patients and their positive rate in 2002–2012 of Hainan and Yunnan provinces.

transmission season (Chen, 2014). For *P. falciparum* malaria, artemisinin-based combination therapy (ACT) was applied for treatment of uncomplicated cases, while injectable artesunate or artemether is the first choice for severe malaria.

In addition, mass drug administration is one of the interventions in the two provinces for reducing malaria transmission. From 2002 to 2012, a total of 77,206 people in Hainan province received a dose of chloroquine plus primaquine during malaria transmission seasons; in Yunnan province, 14,50,175 doses were administered. For administration of antimalarials in nontransmission seasons, a total of 6,55,287 persons received treatment in Hainan province and 6,62,498 persons in Yunnan province were treated (Figure 3.10 and figure 3.11).

5.3 Vector control

Since the mid-1950s, indoor residual spraying (IRS) had been performed comprehensively 1–2 times per year, mainly using pesticides (dichlorodiphenyl-tricgloroethane or hexachlorocyclohexane) in the province of Hainan. After the 1970s, IRS and distribution of insecticide treated bed-nets(ITN) were focused on key areas or villages using deltamethrin. ITN was implemented once a year before the peak of malaria transmission season, where the bed nets of each natural village or residential group were collected and treated by the professional teams (Cui, 2003). After the 1990s, in order to achieve further reduction of human exposure and vectorial

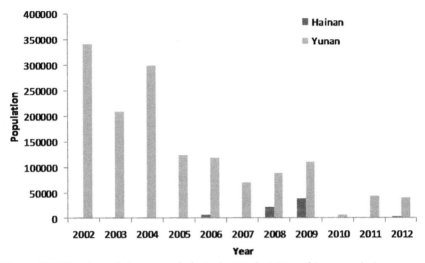

Figure 3.10 Number of chemoprophylaxis doses administered in transmission season in 2002–2012 of Hainan and Yunnan provinces.

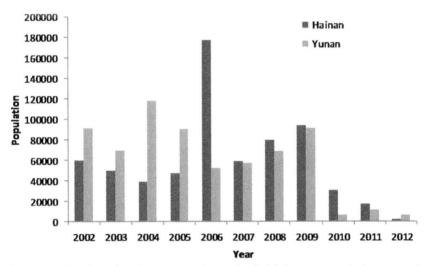

Figure 3.11 Number of patients consuming antimalarials in nontransmission season in 2002–2012 of Hainan and Yunnan provinces.

capacity, the use of bed nets was strengthened through free distribution integrated with health education for residents. From 2002 to 2008, Hainan province distributed 5,68,936 INTs, protecting a population of 9,96,246. In addition, IRS on a total area of 36,64,083 m² was performed, protecting a population of 1,72,828 (Cai and Weng, 2013).

In Yunnan province, IRS was performed 1–2 times at three-monthly intervals in the transmission season using pyrethroid and organophosphate insecticides; their continuous use was limited to no more than 3 years. The first round of IRS was 1 month ahead of the local peak of the transmission season, and the second round was decided based on the locality during the malaria-transmission season. In 2002–2010, a total of 1,68,03,115 and 33,89,998 people were protected by IRS and ITNs, respectively, in the province of Yunnan.

In both Yunnan and Hainan provinces, the Patriotic Health Campaign was combined with the New Rural Communities Construction, where targeted environmental improvements were conducted. Action items included the removal of stagnant water, clearance of ditches, intermittent irrigation, and other measures. Campaigns were used to promote improved indoor ventilation conditions and to change outdoor sleeping habits. Where possible, fish and other biological agents were also used in rice fields, ditches, ponds, and other major water bodies to control the breeding of mosquito larvae (Yang, 2001).

5.4 Pilot studies

In 1991, Hainan translated the successful experience of malaria control in the pilot Mao-yang Township of Wu-zhi-shan County to another 10 townships, each selected from 10 high malaria endemic counties. Besides the increase in equipment and investments such as manpower, materials and financial resources, Hainan formulated unified plans for malaria control and programme management and adopted integrated interventions, such as improving grassroots health organizations and strengthening control efforts. Through 20 consecutive years of operation, promising results were obtained, providing experience for malaria control in high transmission areas, especially against *P. falciparum* malaria in tropical forest areas (Wang et al., 2012b). In the years of 1995–1997, Hainan started another pilot programme in Nan-qiao Township of Wan-ning County. A strategy of controlling the source of infection combined with health education was adopted. The villages and populations at high risk for malaria were targeted; interventions on health education and behavioural changes were performed in those high-risk villages as well as middle and primary schools to improve local knowledge on malaria transmission and self-protection methods. Mass chemoprophylaxis was also administered to residents living in high-risk villages and mountainous lodging, and the treatment of each malaria patient was monitored. Finally, this pilot study achieved the desired effect and provided new experiences for further malaria control in the Hainan mountains (Chen et al., 1996; Chen et al., 1999b). All of these positive pilot experiences solidified the evidence base for effective design and implementation of malaria control strategies and interventions (Hu et al., 2013).

Yunnan also actively explored alternative strategies in malaria control through pilot studies. Malaria prevalence and economic development in Cang-yuan county, which borders Myanmar, were directly affected by malaria epidemics occurring in the border regions of Myanmar. To reduce malaria transmission from the original source, a protection zone on the border was established. Under the support of the Chinese Ministry of Health and the UK Health Unlimited, the pilot project of *Joint Malaria Control on the China–Myanmar Border* was officially launched in 2007 in Cang-yuan county of Yunnan province and six townships of Meng-mao county, as well as Bang-kang and Nan-deng under Wa State in Myanmar, to explore a novel mode of joint malaria control in border areas. Gradually, this mode was expanded to 25 border counties of Yunnan province, and the mechanism was eventually established for joint malaria control on the China–Myanmar border with an aim to reduce malaria morbidity and mortality, which included technical cooperation, information exchange, bilateral training, health education, sharing epidemic information, and providing nets for people in the border areas (Chen et al., 2011; Tang, 1999b).

5.5 Partnership

Hainan has cooperated with relevant research institutes and universities nationwide to carry out a number of malaria studies, due to access to ministerial and provincial awards. In 1990s, Hainan also received funds from the Special Programme for Research & Training in Tropical Diseases (TDR), World Health Organisation (WHO) for health system strengthening and Japan's National Institute of Environmental Sciences. Dozens of domestic and foreign experts and scholars visited Hainan for collaborative research and local malaria control, thus accelerating the process of malaria control in this province. From 2003 to 2012, Hainan conducted three rounds of malaria programmes supported by the Global Fund to Fight AIDS, Tuberculosis and Malaria (GFATM), effectively complementing the local malaria control efforts (Yang et al., 2012).

From 2003 to 2013, Yunnan conducted five rounds of malaria programmes supported by the GFATM. Yunnan province strengthened malaria control in the border region in 2004 and 2005 with support from WHO in six counties, namely Yan-jin, Xin-ping, Yuan-jiang, Hong-he, Yuan-yang and Mo-Jiang, where malaria incidence was high, through provision of standardized malaria management, equipment, antimalarial drugs, health education, technical training and the establishment of a malaria information platform. These activities improved local capacity in malaria control, health education and information delivery. In 2010, Yunnan carried out a WHO-supported malaria control programme to roll back malaria in four counties of Xi-meng, Rui-li, Teng-chong and Cang-yuan. This generated a great breakthrough in malaria control programme management and facilitated the exchange of epidemic information in cross-border regions (Tang et al., 2012b).

5.6 Surveillance

Malaria surveillance is an important component of malaria control and eventual elimination. According to the *National Malaria Surveillance Scheme in 2005–2010* and the *National Surveillance Scheme for Malaria Elimination* in P.R. China, both Hainan and Yunnan provinces set up sentinel sites based on the geographical distribution of malaria incidence and vectors. In 2005–2010, malaria surveillance for the control phase was implemented. Subsequently, malaria surveillance for the elimination phase was initiated. In the control phase, the main purposes of surveillance were to understand the epidemic situation and its influencing factors, master the epidemic patterns and trends, and provide a scientific basis for effective development and evaluation of strategies for malaria control. In the elimination stage, the main purposes of surveillance are to detect malaria cases as well as possible transmission and its influencing factors in a timely manner, assess the control effects and potential risks of malaria transmission, and provide reference and basis for the malaria elimination programme.

Results from the Malaria Information System in 2013 showed that the key indicators for the implementation of the NMEP all reached 100% in Hainan province, including the case reporting rate within 1 day after diagnosis, the case laboratory testing rate, the case laboratory-confirmed rate, the epidemiological case investigation rate within 3 days after diagnosis, and malaria foci clear-up rate within 7 days after case diagnosis. In Yunnan province, only the epidemiological case investigation rate within 3 days after diagnosis was lower than 88%, while all other indicators were above 90%.

6. CHALLENGES FOR MALARIA ELIMINATION

6.1 Challenges in Hainan province

6.1.1 Management of the mobile population

Since reforms on the national economic systems in 1988, malaria control among the mobile population became the focal efforts in Hainan province.

Years of comprehensive malaria prevention and control measures for the Hainan mobile population have achieved good results, as seen through the yearly decline of malaria incidence. Studies have shown that the positive rate of blood tests among the migrant population presenting with fever was higher than not only that of nonmigrant febrile patients, but also the overall positive rate of the province. This suggests that malaria control in Hainan should still have a strong emphasis on the mobile population and efforts should not be reduced in intesity. Malaria control of the mobile population is crucial to prevent the spread of malaria epidemics, develop effective strategies and consolidate antimalarial achievements (Cai, 2002; Lin et al., 2009; Wang et al., 1996). Most mobile populations are from nonimmune communities in nonmalaria endemic areas, have big movement radiuses and relatively weak awareness of self-protection against infectious diseases, and are engaged in agricultural activity in remote and originally high malaria endemic areas, where it is difficult to obtain an early diagnosis and appropriate treatment. All of these factors have brought about a series of difficulties and challenges for successful malaria control and elimination. Therefore, the strengthening of malaria management of the migrant population, especially in the originally high malaria endemic areas of central and southern Hainan, remains the focus for malaria elimination at present and in the future. Consolidation of achievements relating to decades of malaria control in the province might provide new insights on how to solve current challenges (Xu et al., 2007).

6.1.2 Prevention of malaria reintroduction

Hainan has made remarkable achievements in malaria control and elimination. To consolidate the existing achievements and prevent malaria reintroduction and resurgence, the province will adhere to comprehensive strategies of infection source management in combination with vector control and vulnerable population protection. To prevent malaria transmission and outbreaks, the strengthening of monitoring and protection of high-risk groups, including the mobile population, should be continued, and any possible malaria foci should be discovered and eliminated in a timely manner. In the spring and autumn every year, chemoprophylaxis should be provided in a targeted way to reach groups in malaria endemic areas. Routine malaria management should be enhanced through standardized, scientific and institutionalized ways. Finally, it is imperative for all levels of governments to maintain their investments in the above-mentioned prevention strategies so as to sustain the malaria-free status of Hainan province (Wang et al., 2012a).

6.2 Challenges in Yunnan province

6.2.1 High malaria transmission on the China–Myanmar border

Yunnan malaria cases occur mainly in the border region, which is the highest area for the malaria transmission intensively, and the major transmission area of *P. falciparum* in P.R. China. Yunnan has 4,061 km of borderline, and malaria is heavily transmitted in the border areas close to Myanmar. High malaria prevalence in the neighbour countries, frequent movement of the border population and suitable natural settings for malaria transmission are major causes for more severe malaria in the Yunnan border areas. These areas have proved to be more difficult for control than other areas of P.R. China.

There are several specific challenges of malaria control on the China–Myanmar border (Chen et al., 1999a; Li et al., 2011; Yang, 1998; Zhou and Pongsa, 2002). First, it is very difficult to monitor and manage the frequent movement of the local people. According to records in some studies, more than 6,00,000 people cross the Yunnan border annually. With the development of a subeconomic region in Lan-cang and the Mekong River area, the number of mobile individuals and their movement radius will continue to expand. Because the parasitaemia positive rate among border migrants is significantly higher than that of local residents, such movement could result in the constant importation of malaria reservoir into P.R. China. However, there is currently no effective way to manage this population flow. Second, the existence of multiresistant *P. falciparum*, complex malaria vectors and insecticide resistance have made cross-border malaria control more difficult. Third, there is a serious shortage of funding, rendering it difficult to implement normal antimalarial activities. Fourth, there is inadequate awareness of the importance of having and maintaining teams proficient in malaria control at base level. Because malaria control and elimination involves long-term and hard work, many professionals have become weary of the difficulties and challenges ahead. This fatigue, alongside a general trend of decreased malaria burden, has caused unconscious on malaria control among professionals. Facing similar issues, other countries of the Greater Mekong subregion should definitely seek domestic support and international cooperation.

6.2.2 Heterogeneity and complexity of malaria epidemiology and vectors

Malaria control in Yunnan province is more difficult because the vectors are numerous and widely distributed, making the patterns of malaria transmission more complex. Owning to the special terrain and complicated habitats, Yunnan is known for having a three-dimensional climate, housing more *Anopheles* species than the rest of P.R. China. In Yunnan province, the species of *An. minimus*, *An. sinesis*, *An. lesteri* (or *An. anthropophagus*), *An. kunmingensis*, *An. dirus*, *An.jeyporiensis* and *An. nigerrimus* are recognized as being naturally infected by *Plasmodium parasites*. *An. minimus* is the main malaria vector in the valley, reservoir and semi-mountainous areas at altitudes less than 1400 m, south of 27° north latitude. *An. sinesis* and *An. lesteri* are important malaria vectors in the low thermal valley around Jin-sha River, north of 27° north latitude. *An. kunmingensis* is the main malaria vector north of 24° north latitude at altitudes more than 1560 m (Gong et al., 2012). A study showed that the anopheline community on the China–Myanmar border comprises 13 species; *An. minimus* is the dominant species with a high human blood index (Yu et al., 2013). However, the distributions and malaria transmission roles of all the above-mentioned vectors across the border areas are still not clear and require further investigation.

6.2.3 Management of the mobile population

In Yunnan province, there are 12 national ports of first class, eight of second class, 90 border passageways and 103 free points for trade. None of these entry points include a natural barrier or difficult access. Every year, tens of millions of people from Yunnan or other provinces go to the border area for business, construction, mining and other activities. The large size of the mobile population brings great difficulties to malaria control and management (Li, 2006).

At present, the problems in malaria management of mobile populations are as follows. First, there is no specific organization responsible for the task on management of cross-border mobile population, resulting in unstandardized management. Basic demographic and border crossing data are not of high quality. At the same time, local populations focus mainly on economic benefits of mobility, lacking awareness of disease prevention and antimalarial knowledge. Second, rural doctors are not well remunerated, causing them to be less active in the management of mobile populations. Third, the low-quality housing common to the area and poor equipment of the medical teams have proved to be an obstacle to the management of mobile populations. In most rural hospitals, blood tests cannot be carried out. In some hospitals, even though microscopy can be carried out, professionals have limited skills with microscopic examination, so the confirmation rate of malaria is not high. All of these factors could delay the treatment of exogenous malaria, which usually results in the accumulation and spread of infectious source among the mobile populations. The lack of awareness of imported malaria in medical staffs in township hospitals and village clinics, and the insufficient or delayed treatment provided, can result in second or even third generations of malaria cases.

Fourth, unsatisfactory effect and passive position of the overall management of mobile populations were due to insufficient campaigns as well as lack of attention paid by the government, especially to the supervision and urgency for malaria management of mobile populations, which could lead to reduced performance of health professionals. Fifth, networks of joint malaria control are not in place, causing diagnosis and treatment of some mobile patients to be delayed. In short, malaria management in migrant populations, as one of main measures for malaria control, involves complex social system engineering. It is very hard to rely solely on malaria professionals and other health personnel. Strong support from the government and full cooperation of different sectors are definitely required (Tong et al., 2007; Xiong et al., 2009).

6.2.4 Capability building at the ground level

In the stage of malaria elimination, the level of core knowledge on malaria among medical staff at the ground level will directly affect their abilities in discovering, diagnosing and treating the imported malaria cases timely and correctly. In Yunnan province, a relatively sound structure for malaria control system across different levels above the township was established, and in quite a few counties health centres responsible for medical and health care at the village level have also been established through integration of townships and villages. However, the reduced number and aging of technical personnel, increased mobility of staff to other disease areas and the lack of investment from government have led to weakness in malaria control, particularly at levels below provincial levels (Su et al., 2005; Yang, 2001). Therefore, more attention and input should be paid to capacity building at the grassroots level, not only to increase health infrastructure but also improve professional knowledge. Professional and technical trainings should be strengthened and can be performed in batches. The training content should be enriched, and the effectiveness of training must be ensured to improve the capacity of local staffs.

6.2.5 Prevention of malaria reintroduction

Remarkable achievements in malaria prevention and control have been obtained in Yunnan province with the support of domestic and international funds. The incidence of malaria decreased from 249.38 per 10,000 in 1953 to 0.138 per 10,000 in 2012. To achieve the goal of eliminating malaria in Yunnan province by 2020 according to the Chinese NMEP's goal (2010–2020), the following strategies should be further employed on the basis of consolidating previous control results. First, it is necessary to strengthen malaria control and stabilize the malaria epidemic situation in areas having local transmission or high transmission potential. Second, it is to keep malaria rates low in border areas, prevent malaria outbreaks in natural villages, and reduce malaria death cases. Third, it is to actively execute malaria surveillance and response during the postelimination stage, including strengthening of malaria management among mobile populations and treating endogenous and exogenous malaria cases in time, to prevent cases of secondary generation and the re-establishment of malaria transmission (Gao, 2011).

6.2.6 Monitoring of antimalarial drug resistance

In Yunnan province, the scope and degree of *P. falciparum* resistance to antimalarial drugs increased since the 1970s, when a chloroquine-resistant *P. falciparum* strain was found in the Meng-ding township of Geng-ma county. According to some surveys since the 1990s, *P. falciparum* in the border region has produced resistance to chloroquine, amodiaquine, piperaquine, and the sensitivity to artesunate and pyronaridine is less than that observed in the 1980s (Yang, 2001; Zhang et al., 2008a). Due to resistance or reduced sensitivity of *P. falciparum* to a variety of antimalarial drugs, patients infected with *P. falciparum* may not be well treated. Recrudescence and retransmission is an urgent issue for significant control and elimination of *P. falciparum* transmission in the border areas of Yunnan province. The resistance of *P. falciparum* to antimalarial drugs has limited the application of mass drug administration. To control and eliminate malaria effectively, medication should be targeted and combined to delay the emergence of drug resistance.

Artemisinin-based drugs, including ACT, are currently the most effective drugs against *P. falciparum* malaria and have been recommended for worldwide use by WHO. However, because of the irregular application and the cross-resistance of other drugs, the sensitivity of *P. falciparum* to artemisinin has decreased in the border areas of Thailand and Cambodia, and has been spreading to the surrounding countries, such as Vietnam and Myanmar. This has caused great concerns among international academics and disease control agencies (WHO, 2012). Although no ACT-tolerant parasite of *P. falciparum* have been found on the China–Myanmar border (Huang et al., 2012; Wang et al., 2012c), China is facing the risks of introduction and spread of ACT-resistant strains, which will have negative influence on malaria elimination in P.R. China. Therefore, monitoring and surveillance of possible artemisinin resistance through therapeutic efficacy studies under network cooperation of the Great Mekong subregion need to be strengthened.

7. CONCLUSIONS

Owning to their unique natural geographical features, social-economic environments and culture, Hainan and Yunnan have been the provinces most suffered from malaria in P.R. China. The epidemiology of the disease was extremely variable, and different situations required distinct malaria control strategies, which have been adapted to risk groups, vector behaviours, local health infrastructure, and environmental conditions. The building of a preventive and antiepidemic infrastructure and health care system, combined with the training of personnel, have allowed the government to use techniques of mass mobilization to launch programmes of vector control and mass therapy. Provinces and counties were also organized into the regional alliances to facilitate malaria control and surveillance. There was evidence for the efficacy and cost-effectiveness of malaria interventions and a significant increase in funding from all levels of government and the GFATM (2003–2012).

Specifically, the following strategies through the national programmes have played key roles in malaria control:

1. To ensure access to early and accurate diagnosis, followed by prompt, effective and safe treatment through public and private sectors, such as providing qualified microscopy, RDTs and ACTs
2. To ensure full coverage of the population at risk with appropriate vector control measures, such as Long-lasting insecticide-treated nets and IRS
3. To strengthen malaria health education, promotion, and community mobilization efforts and change behaviour, particularly in maximizing utilization of malaria control and elimination services
4. To ensure comprehensive coverage of vulnerable, poor and marginalized populations at high risk for malaria with appropriate malaria interventions
5. To strengthen the malaria surveillance system by improving case reporting, passive and active case detection, entomological and antimalarial resistance monitoring, and ensuring adequate outbreak response capability

6. To provide effective programme management, based on firm leadership commitment, so as to allow high-quality implementation of strategies from malaria control to elimination

With decades of efforts, great progress has been made in these two provinces. In 2012, there was no local infectious malaria for 1 year and no local *P. falciparum* infections for 3 consecutive years in Hainan province. Only 13 malaria cases were reported in that year, all of which were imported. Yunnan province had reported 853 malaria cases, with only 133 local infections in 20 counties, and most of the county incidence rates were lower than 1 per 10,000. However, there is still a long way to go for preventing reintroduction in Hainan province and eliminating malaria in the border areas of Yunnan province.

In recent years, with a growing economy and increasing international exchanges, the number of the imported cases have increased in P.R. China. For example, Hainan province, which is now being developed as an international tourism island, is especially important to solidify the existing malaria elimination efforts on the basis of strengthening the capacity of surveillance and response among mobile population.

The area on the China-Myanmar border is one of the least developed areas, and the population is comprised of ethnic minority groups who are at the highest risk for *P. falciparum* infections because they are often mobile due to their occupations. Control work in the border areas is often hindered by inadequate infrastructure for service delivery, shortages of trained health workers, interruptions in the procurement and delivery of health products, and limited investments in health. It is important to improve living conditions and work opportunities for poor people in their places of residence and to address population movements across borders. Furthermore, people need to be educated about preventive measures and the availability of diagnostic and treatment services, and policies should be reinforced in both public and private health care facilities. There is also a need for adequate coverage with good quality laboratory and clinical services for malaria detection, reporting and surveillance. Also, there should be collaboration between neighbouring countries and organizations to decrease malaria transmission and monitor the therapeutic efficacy of antimalarial drugs in endemic areas of the neighbouring countries (Xu and Liu, 2012). Civil society organizations or non-government organizations (NGOs) present a potentially valuable resource with which to help tackle the growing problem in Great Mekong subregions. These organizations are often able to exercise greater flexibility in accessing hard-to-reach areas, tracking migrants, and implementing cross-border approaches. This enables them to play important roles in assisting the extension, strengthening and community uptake of government health systems and reaching populations that would otherwise be missed. Through these cooperative efforts, malaria can be eliminated in Yunnan province. However, even after local transmission is interrupted, efforts should be maintained to prevent reintroduction and reestablishment of local malaria transmission via the imported cases.

ACKNOWLEDGEMENTS

The project supported by the National S & T Major programme (grant no. 2012ZX10004220), by the National S & T Supporting Project (grant no. 2007BAC03A02) and by China UK Global Health Support Programme (grant no. GHSP-CS-OP1).

REFERENCES

Bi, Y., Tong, S., 2014. Poverty and malaria in the Yunnan province, China. Infect. Dis. Poverty 3, 32.

Bi, Y., Yu, W., Hu, W., Lin, H., Guo, Y., Zhou, X.N., Tong, S., 2013. Impact of climate variability on *Plasmodium vivax* and *Plasmodium falciparum* malaria in Yunnan Province, China. Parasit. Vectors 6, 357.

Cai, H.L., 2002. Analysis of malaria infectious status in the mobile population in hainan province(1996–2000). Chin. Trop. Med. 2, 197–199. (in Chinese)

Cai, M.H., Weng, S.W., 2013. The process of anti-malaria for 60 years and progress of antimalaria after implement global fund malaria project in nanqiao pilot station. Chin. Trop. Med. 13, 245–248.

Cai, X.Z., Deng, D., Wu, K.S., Tang, L.H., Lan, C.X., Gu, Z.C., He, Y.J., Wang, K.A., Wu, D.L., Du, J.W., 1995. A study of human behavior and scioeconomic factors affecting malaria transmission and control in qiongzhong, hainan. Chin. J. Parasitol. Parasit. Dis. 13, 89–93. (in Chinese)

Chen, C., 2014. Development of antimalarial drugs and their application in China: a historical review. Infect. Dis. Poverty 3, 9.

Chen, G.W., LI, H.X., He, Y.Q., Zhou, S., Hou, Z.X., Chen, W.C., Huang, G.Z., Xu, S.Y., 1998. Investigation of behavior, socioeconomic factors affection malaria control in yuanyan county. Chin. J. Parasit. Dis. Control 11, 88–91. (in Chinese)

Chen, G.W., LI, H.X., Xu, S.Y., Wang, L.B., 1999a. The presnt situation and tendency of malaria prevalence in border area of yunnan province. Chin. J. Parasit. Dis. Control 12, 170–172. (in Chinese)

Chen, G.W., Wei, C., Li, H.X., Yang, L.X., Huang, Q., Yan, H.Z., Tian, G.Q., Bai, Z.R., 2011. Malaria epidemic trend and characteristics at monitoring sites in yunnan province in 2008. Chin. J. Parasitol. Parasit. Dis. 29, 55–57. (in Chinese)

Chen, W.J., Shi, P.J., Wu, K.C., Deng, D., Tang, L.H., Cai, X.Z., Gu, Z.C., Si, Y.Z., Li, M.H., Yan, W.X., 1995. An unconditional logistic multi-regression analysis of scioeconomic factors affecting malaria prevalence in mountainous areas of hainan province. Chin. J. Parasit. Dis. Control 8, 161–164. (in Chinese)

Chen, W.J., Wu, K.S., Lan, C.X., Cai, X.Z., Gu, Z.C., Tang, L.H., Tang, X., Pang, X.J., Si, Y.Z., Lin, S.G., He, Y.J., Sheng, H.F., Du, J.W., Wang, G.Z., Zeng, L.H., Cai, Z.F., Deng, D., 1996. An investigation of the relationship between socioeconomic factors and malaria transimission in 10 minority townships in mountainous areas of hainan province. Hainan Med. J. 213–215.

Chen, W.J., Wu, K.S., Lin, M.H., Tang, L.H., Gu, Z.C., wang, S.Q., Lan, C.X., lan, X.H., Li, H.P., Huang, M.S., Chen, X., Sheng, H.F., 1999b. A polit study on malaria control by using new strategy of combing strengthening infection source treatment and health education in mountainous areas of hainan province. Chin. J. Parasitol. Parasit. Dis. 17, 1–4. (in Chinese)

Cheng, F., Lu, Z.X., Zhang, S.Q., 1997. A review of malaria-related social and behavioral risk factors. Med. Soc. 10, 32–34.
Chinese MOH., 2006. National Malaria Control programme during 2006–2015. (in Chinese)
Chinese MOH, 2010. Action Plan of China Malaria Elimination (2010–2020). (in Chinese)
Cui, Y.Q., 2003. The analysis on house spraying and insecticide- treated nets for malaria control. Chin. J. Zoonoses 19, 116–118. (in Chinese)
Dong, X.S., 2000. The malaria vectors and their ecology in yunnan province. Chin. J. Parasit. Dis. Control 13, 144–147. (in Chinese)
Du, J.W., 2001. The trend of malaria in Hainan Province in the past 10 years. Chin. J. Parasit. Dis. Control 14, 309–310. (in Chinese)
Feng, Y., Fu, S., Zhang, H., Li, M., Zhou, T., Wang, J., Zhang, Y., Wang, H., Tang, Q., Liang, G., 2012. Distribution of mosquitoes and mosquito-borne viruses along the China-Myanmar border in Yunnan Province. Jpn. J. Infect. Dis. 65, 215–221.
Gao, Q., 2011. Opportunities and challenges of malaria elimination in China. Chin. J. Schisto. Control 23, 347–349. (in Chinese)
Gong, Z.D., Fu, X.F., Guo, Y.H., 2012. Advances in research on the fauna and diversity of mosquitoes (Culicidae) in Yunnan province. Chin. J. Vector Biol. Control 23, 77–81. (in Chinese)
Hao, M., Jia, D., Li, Q., He, Y., Yuan, L., Xu, S., Chen, K., Wu, J., Shen, L., Sun, L., Zhao, H., Yang, Z., Cui, L., 2013. In vitro sensitivities of *Plasmodium falciparum* isolates from the China–Myanmar border to piperaquine and association with polymorphisms in candidate genes. Antimicrob. Agents Chemother. 57, 1723–1729.
Hu, X.M., Zeng, W., Wang, S.Q., Wang, G.Z., Meng, F., Li, Y.C., 2013. Analysis of malaria surveillance in monitoring sites of Hainan Province from 2006 to 2010. Chin. Trop. Med. 13, 46–49. (in Chinese)
Huang, F., Tang, L., Yang, H., Zhou, S., Liu, H., Li, J., Guo, S., 2012. Molecular epidemiology of drug resistance markers of *Plasmodium falciparum* in Yunnan Province, China. Malar. J. 11, 243.
Huang, J.M., Wang, S.Q., Lin, S.G., 2010. Analysis of results of malarial control, monitoring and infections. Chin. Trop. Med 2, 146–148.
Li, B.F., Shen, J.Y., Chen, Z.W., Zhou, Y.B., Li, Y.X., Yang, F.X., Zhang, Z.F., Zhou, D.L., Guo, X.X., Fang, X.L., Yang, P.M., Yang, Z.G., 2011. Study of the prevalence of malaria infection in individuals returning from abroad at the border areas of Yunnan in 2009. J. Pathog. Biol. 6, 450–452.
Li, S.T., Wu, A.H., 2009. Analysis of results of malaria control measures and effectiveness in Shidian County, Yunnan in 1951~2005. Chin. Trop. Med. 9, 1117. (in Chinese)
Li, Z.H., 2006. Analysis of main factors associated with the prevalence of malaria in Yunnan Province at present. Chin. Trop. Med. 6, 1383–1384. (in Chinese)
Lin, C.F., Zhu, Q.X., Cai, H.L., Wang, S.Q., 2009. Malaria infections in mobile population in Hainan Province in 2004~2008. Chin. Trop. Med. 9, 1666–1667. (in Chinese)
Liu, D.Q., 2014. Surveillance of antimalarial drug resistance in China in the 1980s-1990s. Infect. Dis. Poverty 3, 8.
Liu, Y., Wang, G.Z., Hu, X.M., Cai, H.L., Wang, S.Q., 2012. Status of malaria prevalence in Hainan province from 2006 to 2010. Chin. Trop. Med. 12, 144–145. (in Chinese)
Meng, H., Zhang, R., Yang, H., Fan, Q., Su, X., Miao, J., Cui, L., Yang, Z., 2010. In vitro sensitivity of *Plasmodium falciparum* clinical isolates from the China–Myanmar border area to quinine and association with polymorphism in the Na^+/H^+ exchanger. Antimicrob. Agents Chemother. 54, 4306–4313. (in Chinese)
Su, Y.P., Xu, D.F., Deng, X.L., Liu, D.Q., Zhang, L.X., Zuo, S.L., Chen, G.W., Li, H.X., 2005. Report of malaria epidemic situation and control in yunnan province. Chin. J. Parasit. Dis. Control 18, 161–163. (in Chinese)
Tang, L.H., 1999a. Chinese achievements in malaria control and research. Chin. J. Parasitol. Parasit. Dis. 17, 257–259. (in Chinese)

Tang, L.H., 1999b. Research achievements for malaria control in China. Chin. J. Parasitol. Parasit. Dis. 17, 257–259. (in Chinese)

Tang, L.H., Xu, L.Q., Chen, Y.D., 2012a. Parasitic Disease Control and Research in Hainan Province, Parasitic Disease Control and Research in China. Beijing Science and Technology Press, Beijing. pp. 1296–1300. (in Chinese)

Tang, L.H., Xu, L.Q., Chen, Y.D., 2012b. Parasitic Disease Control and Research in Yunnan Province, Parasitic Disease Control and Research in China. Beijing Science and Technology Press, Beijing. pp. 1370–1373. (in Chinese)

Tong, Y., Wang, X.R., Zhou, Y.B., Li, C.H., Wu, X.H., Wang, Z.Y., Li, C.F., Guo, X.F., Liu, H., Li, L., Lv, S.S., Fu, K.Y., Li, K.H., Tang, Y.Q., 2007. Study on malaria knowledge in border population, Yunnan province. Chin. J. Schisto. Control 19, 59–63. (in Chinese)

Wang, G.Z., Si, Y.Z., Pang, X.J., Cai, H.L., 1996. Analysis on blood examination of febrile cases for detecting malaria among moving population in Hainan province during 1993–1995. J. Prac. Parasit. Dis. 4, 111–113. (in Chinese)

Wang, G.Z., Wang, S.Q., Chen, W.J., Li, C.X., Cai, H.L., 2003. Analysis of results of diagnosis and treatment of malaria patients and blood examination in Hainan Province. Chin. Trop. Med. 3, 584–585. (in Chinese)

Wang, G.Z., Wang, S.Q., Hu, X.M., Meng, F., Li, Y.C., Zeng, W., Cai, H.L., 2012a. Analysis of malaria epidemic situation in Hainan Province, 2010. Chin. J. Schisto. Control 24, 369–370. (in Chinese)

Wang, G.Z., Wang, S.Q., Hu, X.M., Zeng, W., Meng, F., Li, Y.C., Cai, H.L., Lin, S.G., Zhu, Q.X., 2012b. Evaluation of 20- year malaria control effect in pilot areas of Hainan Province. Chin. Trop. Med. 12, 133–136.

Wang, S.Q., 2000. A study on malaria in hainan province. J. Prac. Parasit. Dis. 8, 140. (in Chinese)

Wang, S.Q., Cai, X.Z., 1996. Epidemic situation and control strategies for malaria in Hainan Province. J. Med. Pest Control 12, 24–27. (in Chinese)

Wang, Z., Parker, D., Meng, H., Wu, L., Li, J., Zhao, Z., Zhang, R., Fan, Q., Wang, H., Cui, L., Yang, Z., 2012c. In vitro sensitivity of *Plasmodium falciparum* from China–Myanmar border area to major ACT drugs and polymorphisms in potential target genes. PloS One 7, e30927.

Wen, L., Li, C., Lin, M., Yuan, Z., Huo, D., Li, S., Wang, Y., Chu, C., Jia, R., Song, H., 2011. Spatio-temporal analysis of malaria incidence at the village level in a malaria-endemic area in Hainan, China. Malar. J. 10, 88.

WHO, 2012. Update on Artemisinin Resistance—April 2012. World Health Organization, Geneva.

Wu, K.S., Chen, W.J., Tang, L.H., Deng, D., Lin, M.H., Cai, X.Z., Pan, Y.R., Gu, Z.C., Yan, W.X., Huang, M.S., Zhu, W.D., Sheng, H.F., Chen, X., 1995. A study on behavioural characteristics of staying on the mountain and it relationship with malaria infection in li and miao minorities in Hainan Province. Chin. J. Parasit. Dis. Con. 13, 255–259. (in Chinese)

Xia, Z.G., Feng, J., Zhou, S.S., 2013. Malaria situation in the People's Republic of China in 2012. Chin. J. Parasitol. Parasit. Dis. 31, 413–418.

Xiao, D., Long, Y., Wang, S., Fang, L., Xu, D., Wang, G., Li, L., Cao, W., Yan, Y., 2010. Spatio-temporal distribution of malaria and the association between its epidemic and climate factors in Hainan, China. Malar. J. 9, 185.

Xiao, D., Long, Y., Wang, S., Wu, K., Xu, D., Li, H., Wang, G., Yan, Y., 2012. Epidemic distribution and variation of *Plasmodium falciparum* and *Plasmodium vivax* malaria in Hainan, China during 1995–2008. Am. J. Trop. Med. Hyg. 87, 646–654.

Xiong, L., Yang, H.L., Hu, S.J., Xia, G.H., 2009. Current status and strategy of malaria control in Yunnan Province. Chin. J. Schisto. Control. 21, 147–149. (in Chinese)

Xu, D.Z., Zhang, Y.B., Liu, J.J., Chen, S.Y., Zhang, S.P., 2007. Analysis of measures in surveilance of malaira and control results in Nongken system of Hainan Province. Chin. Trop. Med. 7, 220–222. (in Chinese)

Xu, J., Liu, H., 2012. The challenges of malaria elimination in Yunnan Province, People's Republic of China. Southeast Asian J. Trop. Med. Public Health 43, 819–824.

Yang, C.B., 1995. The improvement on malaria control in Yunnan province. Chin. J. Parasitol. Parasit. Dis. 13, 32–34. (in Chinese)

Yang, H., 1998. Malaria epidemic in border areas of Yunnan Province. J. Dis. Contol Prev. 100–102. (in Chinese)

Yang, H.L., 2001. Malaria control and prevetion in Yunnan province. Chin. J. Parasit. Dis. Con. 14, 54–56. (in Chinese)

Yang, H.L., Du, J.W., Hu, X.M., 2012. Cooperation programme on Malaria in Hainan Province from 2000 to 2011, Malaria Control and Prevention in Hainan Province. Hainan Publishing Press, Hainan. pp. 220–221. (in Chinese)

Yang, Q., Zheng, Z.Y., Zhang, Z.Q., 1988. Malaria epidemic in Yunnan province. Med. Pharm. Yunnan 5, 266–271.

Yu, G., Yan, G., Zhang, N., Zhong, D., Wang, Y., He, Z., Yan, Z., Fu, W., Yang, F., Chen, B., 2013. The Anopheles community and the role of *Anopheles minimus* on malaria transmission on the China–Myanmar border. Parasit. Vectors 6, 264.

Zeng, W., Wang, S.Q., Hu, X.M., Wang, G.Z., Lin, C.F., Meng, F., Li, Y.C., Cai, H.L., 2013. Baseline survey of elimination of malaria in Hainan province. Chin. Trop. Med. 13, 56–58. (in Chinese)

Zhang, J.M., Wu, L.O., Yang, H.L., Duan, Q.X., Shi, M., Yang, Z.Q., 2008a. Malaria prevalence and control in Yunnan province. J. Pathog. Biol. 3, 950–953.

Zhang, W., Wang, L., Fang, L., Ma, J., Xu, Y., Jiang, J., Hui, F., Wang, J., Liang, S., Yang, H., Cao, W., 2008b. Spatial analysis of malaria in Anhui province, China. Malar. J. 7, 206.

Zheng, Q., Vanderslott, S., Jiang, B., Xu, L.L., Liu, C.S., Huo, L.L., Duan, L.P., Wu, N.B., Li, S.Z., Xia, Z.G., Wu, W.P., Hu, W., Zhang, H.B., 2013. Research gaps for three main tropical diseases in the People's Republic of China. Infect. Dis. Poverty 2, 15.

Zhou, S., Pongsa, P., 2002. Analysis of the allocation of malaria control resources in the border areas of Yunnan Province. Chin. Trop. Med. 2, 160–163.

Zhou, S.S., Wang, Y., Fang, W., Tang, L.H., 2008. Malaria situation in the People's Republic of China in 2007. Chin. J. Parasitol. Parasit. Dis. 26, 401–403.

Zhou, S.S., Wang, Y., Tang, L.H., 2007. Malaria situation in the People's Republic of China in 2006. Chin. J. Parasitol. Parasit. Dis. 25, 439–441.

Zofou, D., Nyasa, R.B., Nsagha, D.S., Ntie-Kang, F., Meriki, H.D., Assob, J.C., Kuete, V., 2014. Control of malaria and other vector-borne protozoan diseases in the tropics: enduring challenges despite considerable progress and achievements. Infect. Dis. Poverty 3, 1.

CHAPTER FOUR

Surveillance and Response to Drive the National Malaria Elimination Program

Xin-Yu Feng[1], Zhi-Gui Xia[1], Sirenda Vong[2], Wei-Zhong Yang[3], Shui-Sen Zhou[1,*]

[1]National Institute of Parasitic Diseases, Chinese Center for Disease Control and Prevention; Key Laboratory of Parasite and Vector Biology, MOH; WHO Collaborating Centre for Malaria, Schistosomiasis and Filariasis, Shanghai, People's Republic of China
[2]World Health Organization, China Representative Office, Beijing, People's Republic of China
[3]Chinese Preventive Medicine Association, Beijing, People's Republic of China; Chinese Center for Disease Control and Prevention, Beijing, People's Republic of China
*Corresponding author: E-mail: zss163@hotmail.com

Contents

1. Introduction	82
2. Background	83
2.1 WHO requirements for surveillance in the malaria elimination stage	84
2.2 Malaria surveillance and response in other countries	84
2.3 Malaria surveillance and response in the control and elimination stages in P.R. China	85
3. Surveillance in the National Malaria Elimination Programme	86
3.1 Malaria surveillance system	86
3.2 Malaria surveillance activities	88
3.2.1 Routine surveillance	88
3.2.2 Sentinel surveillance	90
4. Response Strategy in the National Malaria Elimination Programme	94
4.1 1-3-7 strategy	94
4.2 Risk assessment	97
4.3 Case study: imported malaria cases in Shanglin	99
5. Challenges	101
5.1 The sensitivity of surveillance systems needs to be improved	101
5.2 More techniques for surveillance on the imported cases need to be developed	101
5.3 Response mechanism needs to be improved	102
6. Conclusions and Recommendations	103
Acknowledgements	103
References	104

Abstract

The national action plan for malaria elimination in China (2010–2020) was issued by the Chinese Ministry of Health along with other 13 ministries and commissions in 2010. The ultimate goal of the national action plan was to eliminate local transmission of malaria by the end of 2020. Surveillance and response are the most important components driving the whole process of the national malaria elimination programme (NMEP), under the technical guidance used in NMEP. This chapter introduces the evolution of the surveillance from the control to the elimination stages and the current structure of national surveillance system in China. When the NMEP launched, both routine surveillance and sentinel surveillance played critical role in monitoring the process of NMEP. In addition, the current response strategy of NMEP was also reviewed, including the generally developed "1-3-7 Strategy". More effective and sensitive risk assessment tools were introduced, which cannot only predict the trends of malaria, but also are important for the design and adjustment of the surveillance and response systems in the malaria elimination stage. Therefore, this review presents the landscape of malaria surveillance and response in China as well as their contribution to the NMEP, with a focus on activities for early detection of malaria cases, timely control of malaria foci and epidemics, and risk prediction. Furthermore, challenges and recommendations for accelerating NMEP through surveillance are put forward.

1. INTRODUCTION

Currently, malaria is prevalent in 97 countries or regions, with about 3.4 billion people at risk and 470,000–800,000 deaths due to malaria (World Health Organisation (WHO, 2013)). Malaria was also one of major tropical diseases that exerted a great burden on people's health, as well as the social and economic development in the People's Republic of China (P.R. China), particularly in the period from 1950 to 1970s, when annual malaria cases numbered around several millions. Therefore, the government of P.R. China has paid great attention to malaria control during the last six decades. Since the 1980s, the annual incidence rate of malaria has been reduced significantly, with the social economic development in P.R. China. Through efforts at all levels in the country, active prevention and treatment were successfully implemented. The number of counties with annual incidence rate over than 1 per 10,000 was only 75 (3.4%) out of 2189 malaria endemic counties in 2009. Indigenous cases were few and limited to focal localities, which was an initiative basis of the national malaria elimination programme (NMEP) in 2010. The goal of the NMEP is to eliminate malaria in the whole country by 2020 (Ministry of Health, 2010).

Annual malaria incidence further declined with the implementation of the NMEP. Although malaria importation from other countries has increased, with a proportion of 93.3%, only 182 indigenous cases were found among the total 2718 cases in all 31 provinces of P.R. China in 2012 (Xia et al., 2013). Moreover, indigenous cases were focally distributed in limited countries and provinces in the southern and central parts of China, which implies that China has actually stepped into the pre-elimination phase of World Health Organization (WHO) classification, if only malaria incidence is considered (Diouf et al., 2014; Yin et al., 2013). Different from the malaria control stage, with an objective of morbidity and mortality reduction, the elimination programme pursues surveillance and response to all malaria infections and ultimately aims to stop malaria transmission (WHO, 2013). Thus, the main strategies in the pre-elimination require to sensitively identify and clear malaria foci, including residual foci, and to prevent reintroduction of malaria (Guo et al., 2013; Li et al., 2013; Smith et al., 2013). As for the prevention of reintroduction, more attention needs to be paid to particular areas where malaria transmission has been interrupted due to the following reasons: (1) the transmission factors in different types of endemic areas have not fundamentally changed; (2) a rapid increase of the mobile population has resulted in the spread and accumulation of sources of infection (Zhou et al., 2007); and (3) an eco-environmental change has caused climate warming, which can change the composition and density of the vector *Anopheles* spp. population and affect drug resistance of *Plasmodium falciparum* (Guan et al., 2005; Huang et al., 2012a,b; Yang et al., 2007).

2. BACKGROUND

The surveillance and response system is well integrated within the public health system in P.R. China, and all effective and prompt malaria response measures depend on timely and accurate information provided by the surveillance system (Duan et al., 2012; Kelly et al., 2013; Zhou et al., 2013). The relevant information is filtered, verified, stored on dedicated web-based platforms, and then disseminated and analysed by end-users (Barboza et al., 2014; Corberan-Vallet and Lawson, 2014). Depending on epidemic data, vector data and other relevant data acquired through surveillance system, the malaria response system is capable of completing the investigation and verification of individual cases (Chokejindachai and Conway, 2009), screening surrounding populations (Sanders et al., 2014) and implementing relevant vector control strategies (Abeyasinghe et al., 2012; van Bortel et al., 2010).

2.1 WHO requirements for surveillance in the malaria elimination stage

At present, the number of malaria cases detected and reported through the global malaria surveillance system only accounts for a small proportion of the total malaria cases Therefore, WHO urged malaria-endemic countries to strengthen disease surveillance (WHO, 2013), medical and health information and vital registration systems, since the data generated through such systems are critical for assessing and improving the effectiveness of health interventions. In April 2012, to help endemic countries strengthen their malaria surveillance systems, WHO released two operational manuals: *Disease surveillance for malaria control: an operational manual* (WHO, 2012a) and *Disease surveillance for malaria elimination: an operational manual* (WHO, 2012b). These manuals respectively introduced general principles for malaria surveillance in different periods, and gave recommended case definition, key indicators, data recording procedures and guidelines for the establishment of surveillance systems. In addition, these manuals also included templates for recording, reporting and investigating malaria cases.

According to WHO's *Disease surveillance for malaria elimination: an operational manual*, in the malaria elimination stage, the main objectives of surveillance are to find malaria infections (either symptomatic or asymptomatic infections), and to ensure the radical cure of *P. vivax* cases without relapses. In this regard, the implementation of a higher-standard surveillance during the malaria elimination stage is proposed, including examination of malaria parasites for all suspected cases, quality control of the examination, timely and complete case reports, and a comprehensive survey of cases and foci. All the data from examinations and surveys are saved, so that the data can be used for the final assessment and certification of malaria elimination.

2.2 Malaria surveillance and response in other countries

Since the United Nations Millennium Declaration was signed by 189 countries in 2000, many countries have continuously put forward plans to eliminate malaria, including Sri Lanka (Dias et al., 2013; Gunawardena et al., 2014), Turkey (Aydin and Sahin, 2013), Solomon, Vanuatu (Kelly et al., 2013), Botswana (Simon et al., 2013), and South Africa (Maharaj et al., 2012). There are 39 countries that are either planning to eliminate malaria or have entered the elimination or pre-elimination stage (Richard et al., 2009).

Worldwide, disease surveillance mainly relies on the passive surveillance system based on case reports. The surveillance systems cannot monitor all cases and various countries have different surveillance ranges, surveillance indicators, constitutions and sensitivity of surveillance system. In the United States, there are two systems to collect and report malaria cases: the National Malaria Surveillance System and the National Notifiable Diseases Surveillance System. In addition, the Armed Forces Health Surveillance Center also provides relevant information of malaria cases in the military, and usually this information would not be reported to the national health department (CDC, 2013). Such a multichannel reporting system can increase the coverage of malaria cases. In Canada, the surveillance system of infectious diseases was established in 1929; however, the current federal surveillance system only reports information including age and gender of the cases, *Plasmodium* species, mortality rate and possible infections. Thus, it has been suggested to further strengthen the malaria surveillance system in Canada to deal with information of the imported malaria cases (MacLean et al., 2004). In Turkey, Sri Lanka and other countries that are moving towards malaria elimination, case-oriented malaria surveillance systems have been established. Cases detection through doctors or laboratory examinations and vector surveillance were conducted at the same time. Even in some countries where malaria has been eliminated or will be eliminated, the recurrence or outbreak of malaria is possible to roll back after elimination (Galappaththy et al., 2013; Kamat 2000; Pascual et al., 2006; Sainz-Elipe et al., 2010). Hence, it is particularly important to establish an intensive malaria surveillance system and find imported cases as early as possible to reduce the risk of malaria transmission or outbreaks to a low level. Unfortunately, the current surveillance system is very weak in some African countries that suffer from a high malaria burden (Antonio-Nkondjio et al., 2012; Karimuribo et al., 2012; Kunene et al., 2011; Overgaard et al., 2012; Owusu-Ofori and Bates, 2012).

2.3 Malaria surveillance and response in the control and elimination stages in P.R. China

In P.R. China, the national malaria surveillance was established in the beginning of the national malaria control programme in the 1950s, and further intensified with an official document issued by Ministry of Health in 2005. At that time, the country was in the malaria control stage, with the objective to reduce malaria incidence and mortality. According to a survey in 2003 supported by the Global Fund, the malaria incidence in the whole country amounted to 740,000 of cases, and hundreds of millions of people were at risk of malaria infection in 907 counties/cities/prefectures in 18 provinces/autonomous regions/municipalities (P/A/M) (Zhou et al., 2005). Based on the epidemic situation of malaria at that time and the nationwide routine surveillance results, a total of 62 surveillance sites were set up in 2005, covering in 18 P/A/M of the country where conducting surveillance in key monitoring areas.

Since 2010, when the country declared it was entering the malaria elimination stage, the epidemiological features have changed significantly. Under the NMEP, not only the target but also the strategies and measures in the elimination stage are significantly different from those in the control stage (Cao et al., 2014; Yin et al., 2013; Xia et al., 2013; Zheng et al., 2013). The surveillance and response to individual cases and foci are the vital strategies and key interventions in the elimination stage. Consequently, the country initiated the national surveillance system for NMEP in 2012, which consists of both routine surveillance and sentinel surveillance. Sentinel surveillance sites were selected based on malaria incidence and entomological considerations, categorized into two types of sites: sites with more local cases and sites with more imported cases (Cao et al., 2014; Xia et al., 2013; Xiong et al., 2010).

3. SURVEILLANCE IN THE NATIONAL MALARIA ELIMINATION PROGRAMME

3.1 Malaria surveillance system

China's surveillance system for malaria elimination comprises routine surveillance and sentinel surveillance at three administrative levels: (1) the national level, conducting by the National Institute of Parasitic Diseases, Chinese Center for Disease Control and Prevention (CDC); (2) the provincial level performing by provincial disease control and prevention agencies; and (3) the county level, implementing through the county CDC where the surveillance site is located. The routine surveillance that covers the whole country includes daily case reporting, checkup and case investigation of malaria cases, including specific activities such as case verification, active case detection, blood examination and missing report investigations at regular intervals (China CDC, 2012). Sentinel surveillance is focused on those activities that cannot covered by the routine surveillance system in all counties, such as entomological surveillance (species, density), drug resistance monitoring and insecticide resistance surveillance.

The national malaria elimination surveillance sites (hereinafter referred to as the national surveillance sites) cover 25 P/A/M. A total of 49 national malaria surveillance sites have been set up by county/city/district as a unit and divided into two categories according to their degree of malaria prevalence and characteristics (Figure 4.1). One category is for areas with more local malaria cases, including the following eight P/A/M: Anhui

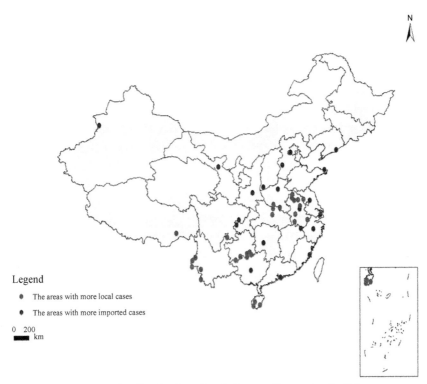

Figure 4.1 The distribution of malaria surveillance sites in China.

(Gao et al., 2012; Zhou et al., 2005), Guizhou (Li et al., 2005), Hainan (Xiao et al., 2012), Henan (Liu et al., 2006), Yunnan (Xu and Liu, 2012), Hubei (Li et al., 2013; Tian et al., 2013), Tibet (Huang et al., 2011) and Jiangsu (Cao et al., 2013). Twenty-nine counties/cities/districts with relatively higher local malaria incidence in the most recent 3 years were selected as surveillance sites. The other category is for areas with more imported malaria cases, including Beijing, Shanghai, Zhejiang (Lu et al., 2013), Fujian (Wang, 2013), Shandong (Xing et al., 2013), Liaoning (Geng et al., 2012), Guangxi (Guo et al., 2013), Guangdong, Henan, Hunan, Jiangsu, Anhui, Sichuan, Zhongqing, Hebei, Shanxi, Jiangxi, Shaanxi, Gansu and Xinjiang (Xia et al., 2013). Twenty counties/cities/districts with relatively more imported malaria cases in the most recent 3 years were selected as surveillance sites.

The nationwide routine surveillance mainly includes five activities: case reporting, diagnosis checkup, case investigation, active case detection and blood examination of unexplained fever cases. Sentinel surveillance consists of two scenarios: one scenario is focused on local transmission; the activities involved include blood examination of unexplained fever cases, serological antibody test, entomological surveillance, insecticide-resistance surveillance and drug-resistance surveillance of *P. falciparum*. The other scenario is targeted to malaria importation, in which the surveillance activities comprise sentinel hospital surveillance, returned overseas screening and drug-resistance surveillance of imported *P. falciparum*. Data collection is through an internet-based system, namely the Parasitic Diseases Management Information System Network, and an annual surveillance report is released to the public in the form of white books or online papers (Xia et al., 2013; Zhou et al., 2005).

3.2 Malaria surveillance activities
3.2.1 Routine surveillance
3.2.1.1 Case reporting and case management
Various medical institutions, CDCs, inspection and quarantine agencies, blood collection agencies at all levels as well as the village doctors fill in the infectious disease report card and make report through the network of the national disease surveillance information report management system within 24 h once they find suspected, clinically diagnosed and confirmed malaria cases. Those without direct reporting network conditions send the infectious disease report card to their local county CDC within 24 h after diagnosis, and the county CDC makes a network direct report within 2 h after receiving the infectious disease report card. When a sudden malaria epidemic occurs, a special report would be made immediately through the national public health emergency management information system, according to the relevant regulations in the malaria epidemic emergency preparedness plan (Figure 4.2).

At least one staff member in the county CDC is responsible for browsing the national disease surveillance information report management systems on a daily basis. Once a suspected, clinically diagnosed, or confirmed malaria case is reported, that staff member would immediately contact the reporting unit to check on the blood smear of the reported case. If the diagnosis of the case is changed, it would be immediately revised in the disease surveillance information report management system. Staff from county CDC would make a case investigation on each malaria case within 3 days after receiving the case reporting. The investigation includes basic information, epidemiological history, treatment history, and the diagnosis and treatment of this onset. After the completion of treatment (normally 8 days), more detailed information is collected by a well-designed questionnaire within 1 week.

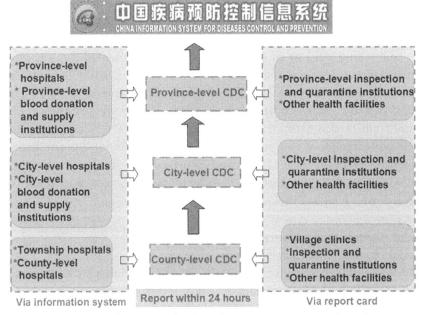

Figure 4.2 The flow chart of malaria disease reporting procedures from bottom to top levels.

At the primary medical agencies, the main diagnostic test is a microscopic examination or rapid diagnostic test (RDT). Case confirmation is conducted using microscopic examination, RDT or polymerase chain reaction in the above institution, and finally qualified assurance and qualified control (QA/QC) at provincial or national CDCs.

Taking 2013 as an example, a total of 4146 malaria cases were reported from 31 P/A/M in P.R. China. Among them, 1745 were reported by 49 surveillance sites. All cases were reported within 24 h after diagnosis. A total of 4,085 (98.58%) reported cases were checked through laboratory examination, and 4024 were laboratory confirmed as malaria with an accordance rate of 97.10%. A total of 4,065 cases (98.05%) were case investigated within 3 days.

3.2.1.2 Blood examination of fever cases

Active case detection is a key strategy in the surveillance system of NMEP as a complement to passive case detection. The specific approaches of active case detection for malaria elimination in P.R. China are as follows. Within 1 week after receiving a case report, the county CDC will organize a case screening around a passively detected case with active or inactive foci. Blood specimens also are collected from cases with a fever history in the past 2 weeks. If two or more locally acquired malaria cases or asymptomatic carriers are found, the screening expands to all the residents in the focus. Case investigations are also conducted for all screened cases and reported in the system.

Still taking 2013 as an example, 26 provinces provided data for blood examination in unexplained fever cases. A total of 4,109 of 4,662,690 examinations were tested for malaria, with a positive rate of 0.09%. In sentinel surveillance sites, 220,813 unexplained fever cases were examined; 1767 were confirmed as malaria, with a positive rate of 0.80%. Active case detection was carried out in seven provinces (including Anhui, Yunnan, Jiangsu, Zhejiang, Hebei, Jiangxi, Shanxi) and 20 surveillance sites. Among a total of 12,436 subjects, only one imported malaria case was screened by active case detection in Jiangsu Province (Table 4.1).

3.2.2 Sentinel surveillance
3.2.2.1 Entomological surveillance
Entomological surveillance and vector control not only play an important role in interrupting malaria transmission, but they also weaken the vectorial capacity to the critical level for malaria elimination through effective vector control strategies. Effective vector control strategies must be formulated on the basis of understanding the bionomics, habits, and population distribution of the vectors. Different strategies should be adopted in areas with different epidemic situations and particular vectors. In addition, personal protection and related education should be conducted in areas at risk of malaria transmission.

Entomological surveillance on *Anopheles* includes species, density, and bionomics and habits. In China, light traps are used to collect *Anopheles* specimens to identify species, and estimate the populations and density of mosquitoes in one natural village or community. Surveillance activities are mainly conducted in June through October, using a serological test twice a month in those consecutive 5 months of the year.

A total of 16 provinces carried out entomological surveillance in 2013. A total of 34,043 anopheline mosquitoes were caught, with the majority of *An. sinensis* accounting for 83.56% of cases, along with nine *An. anthropophagus* and 18 *An. minimus*. For example, entomological surveillance was carried out in June–October in Anhui Province, catching 11,282 anopheline mosquitoes, and the anopheline mosquito density was the highest in the second half of June and first half of July, reaching 12/light-night and 14.2/light-night. In Sichuan Province, a total of 1441 *An. sinensis* were caught; the peak of *An. sinensis* population and density was in late May, late July and early August, while lowest

Table 4.1 Summary of nationwide blood examination of unexplained fever cases in the malaria elimination surveillance sites in 2013

Province / Autonomous regions/ Municipality	Number of surveillance sites	Number of blood examinations of unexplained fever cases in whole province	Number of examinations of unexplained fever cases in whole province	Number of blood examinations of unexplained fever cases in surveillance sites	Number of blood examinations of unexplained fever cases in surveillance sites
Anhui	8	581,792	177	8350	0
Guizhou	6	204,350	12	4065	0
Yunnan	6	449,163	578	67,419	371
Henan	4	1,206,034	197	68,298	65
Hainan	3	120,000	16	26,705	7
Jiangsu	2	503,593	341	11,175	45
Hubei	2	407,522	129	2105	0
Tibet	1	1150	4	1105	1
Fujian	1	26,619	76	650	22
Guangdong	1	21,759	148	3909	18
Guangxi	1	218,237	1251	5708	1052
Hunan	1	62,814	196	510	7
Liaoning	1	105,692	45	1760	0
Shandong	1	202,000	131	593	0
Beijing	1	45,268	113	—	110
Shanghai	1	107,768	73	—	10
Sichuan	1	521,902	228	12,018	5
Zhejiang	1	158,217	210	—	34
Hebei	1	43,932	35	344	17
Jiangxi	1	43,429	45	50	0
Shaanxi	1	59,864	47	1059	0
Gansu	1	—	16	—	0
Xinjiang	1	19,093	5	—	0
Shanxi	1	10,817	5	238	3
Chongqing	1	45,268	31	4752	0
Total	49	4,662,690	4109	220,813	1767

in September and October. The average density of *An. sinensis* was 144/light-night. Among them, indoor density was 35.2/light-night, outdoor 35.2/light-night and cattle shed 83.8/light-night. The highest density was at cattle sheds (Figure 4.3).

When a vector population in a certain area was found at an abnormal pattern by the surveillance system, there must be a clear understanding about the insecticide resistance of vectors in this area before insecticide-based vector control strategies are employed. Surveillance of insecticide resistance of vectors in the malaria elimination stage began in 2012. The surveillance sites adopting mosquito control with insecticides were selected, and the provincial disease control and prevention agencies were responsible for conducting surveillance once every 2 years.

In 2013, two provinces submitted surveillance data of insecticide resistance for anopheline mosquitoes. Some insecticide resistance evidence was found. For example, Jiangsu province collected adult *An. sinensis* in the field of Tiangang Lake, Sihong County in July, and used the WHO-recommended adult mosquito contact tube method to determine the insecticide resistance of its filial 1 (F1) 3- to 5-day-old female mosquitoes. The results showed that the death rate of mosquitoes in 24 h after contacting 0.05%

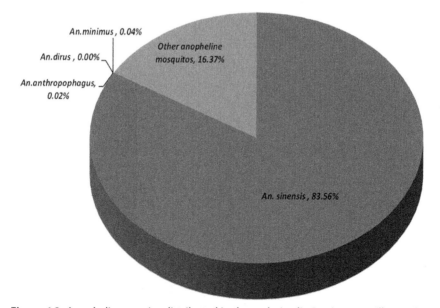

Figure 4.3 Anopheline species distributed in the malaria elimination surveillance sites in P.R. China.

deltamethrin for 1 h was 5%. The resistance degree was "R", indicating that *An. sinensis* in this area had developed relatively strong resistance to deltamethrin. In Hainan Province, resistance of *An. sinensis* to four kinds of insecticides was determined in Changjiang, Ledong and Qiongzhong cities/counties. The results showed that resistance occurred in all cities/counties with exception for Qiongzhong county, where drug resistance to deltamethrin was found but it is still sensitive to malathion.

3.2.2.2 Drug resistance surveillance

According to the requirements of the current surveillance program, in areas with relatively more number of the local malaria cases, one site was selected for conducting drug resistance surveillance in Yunnan Province. The provincial disease control and prevention agency was responsible for the surveillance and carried it out in the transmission season every year, using *in vivo* 4-week tests and *in vitro* microtechniques to determine the sensitivity of *P. falciparum* to compound artemisinin derivatives (e.g. dihydroartemisinin and piperaquine). Clear evidence has shown the resistance of *P. falciparum* parasites to artemisinin derivatives. While in areas with more number of the imported malaria cases, this surveillance was done in Jiangsu Province, using the same methods as carried out in Yunnan Proinvce. They found that all 4 cases determined by *in vivo* 4-week tests showed sensitivity. In the other 20 cases, approximately 1, 5, 3 and 2 persons had resistance to each drug of artesunate, chloroquine, mefloquine and piperaquine, respectively, as determined by *in vitro* microtechnique.

3.2.2.3 Surveillance on population at risk

The guidelines specify that in surveillance sites with more number of the imported cases, one county-level hospital with more admission of malaria cases is chosen as a sentinel hospital to carry out the routine surveillance, and special attention is paid to the imported malaria cases from other countries. The clinicians at the sentinel hospital should ask all the the febrile cases admitted about their epidemiological history abroad, collect blood samples from the febrile cases who have recent travel history in foreign countries for microscopic examination or RDT detection, and immediately report to the county CDC when the suspected malaria case found. For the imported malaria case detected, the county CDC should undertake case investigation and make timely report. At the same time, the county CDC is responsible for malaria screening of all of the companions (persons who travelled with them) of the imported malaria cases, undertaking blood test or RDT for all of the companions with fever symptoms and 10% of the companions without fever symptoms. In addition, there should be follow-up surveys of other acquaintances, and the results of investigation will enter into the questionnaire form of the national malaria surveillance site for febrile cases and asymptomatic returnees from abroad.

In 2013, sentinel surveillance was carried out in Anhui, Henan, Zhejiang and Shaanxi provinces, where there were relatively more number of the imported malaria cases, with a total of 167 individuals. Among them, a total of 23 falciparum malaria cases, three cases were mixed infection and one ovale malaria were found. At the same time, the screening of the returnees from abroad were performed in the surveillance sites of other 13 provinces, as a result, a total of 4358 persons were screened, and 737 cases were found.

4. RESPONSE STRATEGY IN THE NATIONAL MALARIA ELIMINATION PROGRAMME

4.1 1-3-7 strategy

The response strategy defined as a 1-3-7 approach were used in the NMEP (Cao et al., 2014). In the strategy, the term of "1" means case reporting, by which medical and health institutions at all levels should report cases detected including suspected cases within 1 day (24 h) through the disease surveillance information report management system of China CDC. The term of "3" means to investigate the cases and epidemiological surveys, more specifically, the county-level disease control agency should make a laboratory confirmation for all cases reported in its administrative area within 3 days, after conducting epidemiological case survey and identifying the *Plasmodium* species and sources of infection of the confirmed cases. The term of "7" refers to active surveillance and focus response, more specifically, the county CDC makes a judgement to address the questions if there is risk of transmission in places where the malaria cases reside after the active surveillance and then take appropriate response activities, such as providing preventive chemotherapy to the at-risk population and indoor residual spraying in at-risk houses (Cao et al., 2014).

Early detection and timely reporting of the cases are the most important aspects of the 1-3-7 strategy. Not only the laboratory-confirmed cases but also the clinically-diagnosed cases with laboratory-negative should be reported directly through the information management network within the specified time. The intent of the strategy is to raise the sensitivity of the surveillance system as much as possible and reduce local transmission due to a missed diagnosis or delayed diagnosis and treatment (Ministry of Health of the People's Republic of China, 2012).

All malaria cases submitted for the web-based reporting system have to be laboratory confirmed finally, since the identification of parasite species and source of infection are used as the basis for evaluation of transmission risk. Therefore, during the malaria elimination stage, not only the laboratory-confirmed cases by blood smear test and the clinically diagnosed cases need to be reconfirmed by professionals from CDCs at up level after submitted to the web-based reporting system, but also molecular biology techniques must be used for further examination of those clinically diagnosed cases with negative results in the blood smear test, as well as those with difficulty for *Plasmodium* species identification. At the same time, the source of infection of the reconfirmed cases needs epidemiological case investigation, so as to provide a basis for further conducting focus active surveillance and response against resurgence.

In the malaria elimination stage of P.R. China, "focus" is defined as the natural village or community, residential area or construction site where malaria cases occur. Therefore, interventions of NMEP including the timely report, investigation of the clue cases or the first case, focus investigation, and quick response, are the core strategies for malaria elimination surveillance and response in P.R. China (Cao et al., 2013; Yang et al., 2012; Zhou et al., 2013). By timely focus investigation of types of focus and their transmission risk as well as various appropriate disposal measures, interruption of every possible local transmission is guaranteed (Sheng et al., 2003).

Focus investigation includes the activities in three aspects, such as recoding basic information of focus, surveying vector anopheline species population and screening infections of malaria by blood smear test. Focus response includes the following three activities:

1. Health education is performed together with malaria case screening of the residents in the focus, especially educating the residents about malaria symptoms and the need to seek medical diagnosis and treatment as early as possible. In active and inactive focus, knowledge to enhance self-protection awareness by avoiding outdoor sleeping as well as promoting the use of mosquito nets, screen doors, screen windows and other mosquito-proofing facilities to reduce mosquito-biting is also disseminated.
2. Expanded treatment is performed in active focus. If missing reported malaria cases or parasite carriers are found, treatment should be expanded to the whole family of the malaria case, parasite-carrier and their neighborhood residents with the same antimalaria treatment scheme, so as to clear away the possible source of infection.

3. Vector control is undertaken in active focus. Indoor residual spraying of pyrethroid insecticides is used to reduce the density of malaria-transmitting vectors and block transmission of malaria. The range of spraying covers the patient's houses and neighbours.

If two or more missing reported cases or parasite-carriers are found through screening, the range of spraying should be expanded to the entire focus. In focus where the coverage of using long-lasting mosquito nets or insecticide-impregnated mosquito nets has reached more than 85%, insecticide spraying is not necessary.

In the first quarter of 2014, for instance, a total of 642 malaria cases were reported in the whole country. The reporting rate within 24 h after diagnosis of the cases was 100%. These cases include 631 imported cases, 10 local cases, and 1 case without detailed information. Among these cases, the laboratory examination rate was 99.69%, the laboratory confirmation rate was 97.35%, the completion rate of epidemiological case investigation within 3 days and the disposal rate of focus within 7 days were all above 99%, and no secondary case occurred. Based on these statistics, the implementation of the 1-3-7 strategy achieved successful results (Figure 4.4).

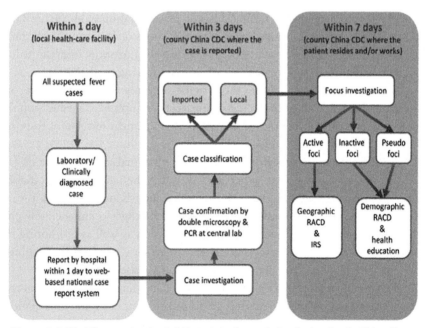

Figure 4.4 Workflow under the 1-3-7 strategy for malaria elimination in China (from Cao, et al., 2014).

4.2 Risk assessment

The surveillance system in the control stage relied on passive surveillance to collect case information in time and other monitoring data. Such data were only a temporary description of the time and spatial characteristics of some surveillance sites, without any prospective and predictive information. There are several potential risk factors affecting transmission patterns of malaria, such as changes in the distribution of vector species population (Basilua Kanza et al., 2013; Bugoro et al., 2011; Linton et al., 2005; Sahu et al., 2014), environmental changes (Confalonieri et al., 2014), global warming (Alonso et al., 2011; Aoun et al., 2010), and population movements (Osorio et al., 2007; Trung et al., 2004). If we can conduct an effective risk assessment in advance, an effective and timely prediction or warning can be achieved, the intensity of malaria transmission could be potentially reduced. A comprehensive understanding of malaria risk factors (Neuberger et al., 2010; Talisuna et al., 2012; Woyessa et al., 2013) cannot only help predict the malaria trends, but it also is very important for designing and adjustment of the surveillance and response systems in the malaria elimination stage to make them more effective and sensitive. Currently in the malaria elimination stage in China (Yin et al., 2013; Zheng et al., 2013; Zhou et al., 2013), the risks of increasing proportion of the imported cases, development of drug resistance of vectors and malaria parasites, as well as environmental changes, all bring enormous pressures and threats to the achievement of malaria elimination in P.R. China. An appropriate and effective risk assessment strategy is not only important for promoting malaria elimination in P.R. China by 2020, but also for the prevention of reemergence of malaria epidemics in the late stage of malaria elimination, particularly by providing timely warning and response strategies for some vulnerable population or areas at risk (Lin et al., 2009; Yang et al., 2010).

Malaria risk assessment can be analysed from multiple perspectives and levels. For example, international scholars regard tourists visiting family and friends as a risk factor for malaria infection (Angell and Behrens, 2005). Also, a model based on health behaviour theory that was established in the light of the geographic and cultural diversity of people provides a relatively good reference for related clinical control (Angell and Behrens, 2005; Ye et al., 2007). Furthermore, according to quantitative risk assessment based on three indicators, including vectorial capacity, vector susceptibility, and imported source of infection, it is considered that the first two indicators of entomological risk are relatively practical and closely correlated with the risk of malaria transmission (Kazembe et al., 2006; Poncon et al., 2008; Yang et al., 2010).

The risk assessments of malaria re-emergence have been extensively investigated, both in the malaria control stage and elimination stage in China (Xia et al., 2003; Xiao et al., 2010; Xiong et al., 2010; Yang et al., 2010; Zhou et al., 2010). The following are five major progresses.

Firstly, a high correlation between malaria outbreak and the rise of vectorial capacity in the central part of China was demonstrated by a survey of vectorial capacity of *An. sinensis,* which is the main risk factor affecting malaria transmission (Pan et al., 2012). The results demonstrated an increase of transmission risk in the central part of China. The reason may be the sharp drop in the number of livestock and changes in animals' behaviour, which leads to a lack of traditional biological barriers and increased risk of people and malaria vector contact.

Secondly, based on an understanding of global climate changes and biological drives, a statistic model has been developed to predict the spatiotemporal patterns of transmission of *P. vivax and P. falciparum* malaria in P.R. China, taking temperature and relative humidity as the most important environmental factors. The results of this study indicate that areas with malaria transmission year round are still limited to the southern and southeastern provinces of P.R. China, such as Yunnan, Guangdong, Guangxi and Hainan, and the endemic areas of *P. falciparum* malaria are still restricted within Yunnan and Hainan provinces (Yang et al., 2012). Control strategies for different potential endemic areas have been suggested, such as adopting appropriate surveillance and response systems in the climate-sensitive areas in the central part of the country, while strengthening surveillance and warning systems in the northeast part (Figure 4.5).

Thirdly, studies on the influence of multiple factors, including geographic, meteorological and vector, and the correlation among these complex factors and the re-emergence of malaria epidemics in the plains regions of P.R. China since 2001, indicated that the distribution of water bodies around the malaria cases, changing climate factors, the increase of vectorial capacity of *An. sinensis* and its basic reproductive rate were important risk factors causing malaria epidemics (Zhou et al., 2010).

Fourthly, taking malaria knowledge and personal protection level as an indicator to investigate the influence of factors related to the border areas of P.R. China on malaria control (Moore et al., 2008), a study found that among ethnic minorities, lack of malaria knowledge and weak personal protection consciousness are still the potential risk factors for malaria transmission in this area.

Fifthly, the imported malaria risk assessment index system was established and different risk assessment indicators were defined according to the transmission mechanism and epidemiological characteristics of imported malaria. Four risk levels were set up according to risk assessment of the

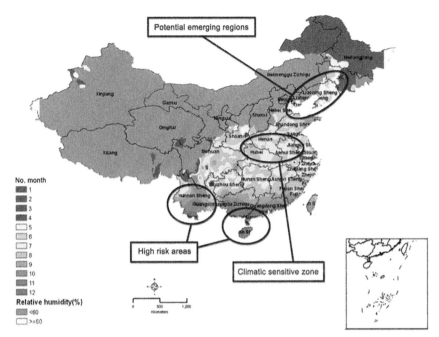

Figure 4.5 Predicted malaria risk areas highlighting the surveillance-response strategies in P.R. China (from Yang et al., 2012).

indicator system (Zeng and Yu, 2006). According to the four risk levels, recommendations were developed to implement different health and quarantine control measures in each level.

In summary, malaria risk assessment has been implemented in P.R. China for the analysis of the influence of risk factors on a single characteristic of malaria transmission, and also to integrate multiple risk factors in a mathematical model to predict overall malaria transmission trends. Different research perspectives have been used to predict malaria transmission trends by using different risk assessment methods, so as to further strengthen and improve malaria surveillance and response strategies.

4.3 Case study: imported malaria cases in Shanglin

During June 2013, several hundred imported malaria cases were reported through the surveillance system in Shanglin county, Guangxi (Xia et al., 2013). The main reason for this rapid emergence of the imported malaria cases was due to the return of thousands of Chinese gold miners from Ghana. It was declared as a public health emergency by local government, and a rapid response was carried out to effectively prevent local transmission from the imported cases, and to provide prompt treatment to patients according

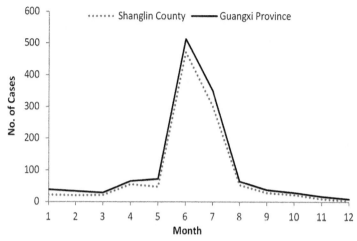

Figure 4.6 Reported malaria in Shanglin and Guangxi in 2013.

to an emergency work plan formulated by an expert group, which consisted of three steps described below (Figure 4.6).

First, a working mechanism of "one day, one report" was implemented in Shanglin county, based on the guiding principle of the national malaria surveillance and response system. As a result, detailed information on returnees' status, case diagnosis and treatment were reported from each township or town to the special coordinating office of the county government every day. The confirmed malaria cases were reported by medical institutions at various levels within 24 h. Second, the epidemiological case investigation was performed by technical personnel from local CDCs within 3 days after receiving the case report. A total of 1052 imported cases were reported in Shanglin during 2013, of which 6.59% (69 cases) were vivax malaria, 91.35% (961 cases) were falciparum malaria, 0.29% (3 cases) were quartan malaria, 1.05% (11 cases) were ovale malaria, and 0.76% (8 cases) were mixed infections. The coverage rate of case investigation was 100% within 3 days. Third, local CDC personnel completed the focus investigation and foci treatments within 7 days in collaboration with relevant departments of the local government, such as health, public security and education departments, as well as village committee in Shanglin. Insecticide spraying for mosquito control was conducted covering 26,954 m^2 in all foci identified areas and covered a total of 3939 persons. Vector surveys were carried out 32 times in six sites in three townships. In addition, a total of 751 *An. sinensis* mosquitos were caught. No other anopheline species were found, and no any sporozoite-positive mosquito was found among 31 dissected *An. sinensis*.

Thanks to the quick response, including case reporting, epidemiological investigation, vector surveillance and timely focus disposal, the occurrence of secondary malaria cases was strictly prevented. The 1-3-7 surveillance and response strategy was implemented successfully through multisectoral collaboration, resulting in zero local transmissions, no death cases and no secondary malaria case (Cao et al., 2014; Xia et al., 2013).

5. CHALLENGES

5.1 The sensitivity of surveillance systems needs to be improved

Health and medical institutions are numerous at different levels of China, including village clinics, township health centres, county hospitals, prefectural or city hospitals, CDCs of the provinces, cities and counties, individual clinics, and some private hospitals. These agencies are conducting malaria parasite examinations for all the febrile cases who seek treatment. There is broad coverage, and surveillance is performed throughout the year. In this way, the continuity of surveillance is maintained. However, the effectiveness of surveillance is affected by the number of cases seeking treatment and the malaria diagnosis abilities of the different medical institutions (Figure 4.7).

For a number of reasons, such as the decreasing local infections year by year and relatively high mobility of grassroots laboratory technicians, in some areas the registered data and the registration forms for blood examination are incomplete (Sun et al., 2012). Therefore, the task of blood examination for febrile cases has met greater pressure (Guo et al., 2013). In hypoendemic areas of malaria, a shortage of testing equipment in medical institutions seriously affects the quality and efficiency of microscopic examinations of blood smears. In addition, although malaria RDTs can be used for quick, easy and stable diagnosis of malaria parasite infection in primary health care units of remote areas, there is still reliance on further applications of molecular detection methods, particularly for the examination of asymptomatic infected persons with low density (Adhin et al., 2013; Brown et al., 2012; Ojurongbe et al., 2010; Zakeri et al., 2010).

5.2 More techniques for surveillance on the imported cases need to be developed

At present, surveillance of imported cases mainly relies on the passive surveillance system in existing medical and health systems. By the time the imported case is discovered, the infected individual may be far away, for

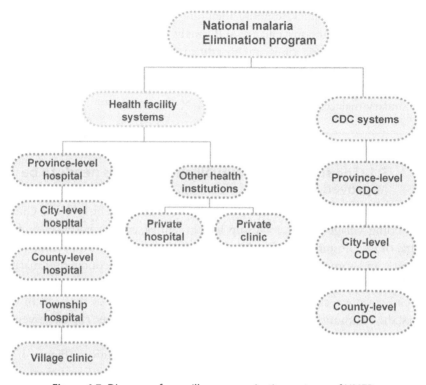

Figure 4.7 Diagram of surveillance organization systems of NMEP.

instance, from the ports of entry to his home. During this time, the threat may include: (1) deterioration in health status, which in the most serious cases may lead to death; (2) during the trip and after returning home, the infected individual maybe a mobile source of infection, and in the presence of a malaria-transmitting vector, pose a risk of secondary transmission. Therefore, the improved techniques for surveillance on the imported cases need to be developed (Smith et al., 2013; Tatem et al., 2014).

5.3 Response mechanism needs to be improved

In the malaria control stage in P.R. China, the Global Fund provided tremendous support for malaria control efforts, and malaria incidence decreased to the historically lowest level. However, when the Global Fund stopped funding programmes in the country, it had just entered the malaria elimination stage. A surveillance and response system is a large undertaking integrating manpower, equipment, organizational management, statistical analysis, and many other aspects. A well-operated and effective surveillance system certainly needs strong economic support (Doroshenko et al., 2005; Reis et al., 2007). However, the currently available surveillance funding is not sufficient for the malaria elimination stage. Furthermore, funding will be needed for necessary adjustments for new technological approaches as they become available.

6. CONCLUSIONS AND RECOMMENDATIONS

At present, China has established a nationwide malaria surveillance system for NMEP. It relies on a real-time reporting network system and is capable of submitting timely reports of various surveillance indicators. Based on the analysis in this chapter, it appears that the national system can make quick, purposeful and planned responses for different malaria epidemics, such as control future malaria outbreaks.

In the malaria elimination stage in P.R. China, the nationwide surveillance system based on the 1-3-7 strategy is designed to ensure the timely detection, monitoring and treatment of every malaria case, as well as determine its source of infection and transmission route, and prevent potential spread. The establishment of a surveillance site system offsets the deficiency of routine surveillance, recognizes the results of special surveillance in key areas, and provides a strong basis for tailoring malaria elimination strategies and epidemic responses into local settings. In addition, the application of risk assessment for predicting malaria transmission patterns also further strengthens and improves malaria surveillance and response strategies which is an essential component of NMEP.

Malaria surveillance and response is highly public, social and technical, requiring close coordination and collaboration within the whole society (Troppy et al., 2014; Velasco et al., 2014). Especially for NMEP, we must first establish a strong malaria surveillance management system and formulate practical and feasible rules and regulations, then establish a surveillance system fit for the current capacities and local seetings (Tambo et al., 2014). The surveillance capacity at different levels should be strengthened in the meantime. In addition, it is necessary to further enhance the application of surveillance and warning approaches in the fields (Gorman, 2013; Zhou et al., 2013).

ACKNOWLEDGEMENTS

This project was supported by the National S &T Major Program (grant no. 2012ZX10004220), by the National S & T Supporting Project (grant no. 2007BAC03A02) and by China UK Global Health Support Programme (grant no. GHSP-CS-OP1).

REFERENCES

Abeyasinghe, R.R., Galappaththy, G.N., Smith Gueye, C., Kahn, J.G., Feachem, R.G., 2012. Malaria control and elimination in Sri Lanka: documenting progress and success factors in a conflict setting. PLoS One 7, e43162.

Adhin, M.R., Labadie-Bracho, M., Bretas, G., 2013. Molecular surveillance as monitoring tool for drug-resistant *Plasmodium falciparum* in Suriname. Am. J. Trop. Med. Hyg. 89, 311–316.

Alonso, D., Bouma, M.J., Pascual, M., 2011. Epidemic malaria and warmer temperatures in recent decades in an East African highland. Proc. Biol. Sci. 278, 1661–1669.

Angell, S.Y., Behrens, R.H., 2005. Risk assessment and disease prevention in travelers visiting friends and relatives. Infect. Dis. Clin. North Am. 19, 49–65.

Antonio-Nkondjio, C., Defo-Talom, B., Tagne-Fotso, R., Tene-Fossog, B., Ndo, C., Lehman, L.G., et al., 2012. High mosquito burden and malaria transmission in a district of the city of Douala, Cameroon. BMC Infect. Dis. 12, 275.

Aoun, K., Siala, E., Tchibkere, D., Ben Abdallah, R., Zallagua, N., Chahed, M.K., et al., 2010. Imported malaria in Tunisia: consequences on the risk of resurgence of the disease. Med. Trop. (Mars) 70, 33–37.

Aydin, M.F., Sahin, A., 2013. Malaria epidemiology in mersin province, Turkey from 2002 to 2011. Iran. J. Parasitol. 8, 296–301.

Barboza, P., Vaillant, L., Le Strat, Y., Hartley, D.M., Nelson, N.P., Mawudeku, A., et al., 2014. Factors influencing performance of internet-based biosurveillance systems used in epidemic intelligence for early detection of infectious diseases outbreaks. PLoS One 9, e90536.

Basilua Kanza, J.P., El Fahime, E., Alaoui, S., Essassi el, M., Brooke, B., Nkebolo Malafu, A., et al., 2013. Pyrethroid, DDT and malathion resistance in the malaria vector Anopheles gambiae from the Democratic Republic of Congo. Trans. R. Soc. Trop. Med. Hyg. 107, 8–14.

Brown, T., Smith, L.S., Oo, E.K., Shawng, K., Lee, T.J., Sullivan, D., et al., 2012. Molecular surveillance for drug-resistant *Plasmodium falciparum* in clinical and subclinical populations from three border regions of Burma/Myanmar: cross-sectional data and a systematic review of resistance studies. Malar. J. 11, 333.

Bugoro, H., Iro'ofa, C., Mackenzie, D.O., Apairamo, A., Hevalao, W., Corcoran, S., et al., 2011. Changes in vector species composition and current vector biology and behaviour will favour malaria elimination in Santa Isabel Province, Solomon Islands. Malar. J. 10, 287.

Cao, J., Zhou, S.S., Zhou, H.Y., Yu, Y.B., Tang, L.H., Gao, Q., 2013. Malaria from control to elimination in China: transition of goal, strategy and interventions. Chin. J. Schisto. Control 25, 439–443 (in Chinese).

Cao, J., Sturrock, H.J., Cotter, C., Zhou, S., Zhou, H., Liu, Y., et al., 2014. Communicating and monitoring surveillance and response activities for malaria elimination: China's "1-3-7" strategy. PLoS Med. 11, e1001642.

CDC, 2013. Malaria surveillance – United States, 2011. MMWR 62 (SS-5).

China, CDC, 2012. The national malaria surveillance scheme. In: Beijing (in Chinese).

Chokejindachai, W., Conway, D.J., 2009. Case-control approach to identify *Plasmodium falciparum* polymorphisms associated with severe malaria. PLoS One 4, e5454.

Confalonieri, U.E., Margonari, C., Quintao, A.F., 2014. Environmental change and the dynamics of parasitic diseases in the Amazon. Acta Trop. 129, 33–41.

Corberan-Vallet, A., Lawson, A.B., 2014. Prospective analysis of infectious disease surveillance data using syndromic information. Stat. Methods Med. Res. 0(0), 1-19.

Dias, S., Wickramarachchi, T., Sahabandu, I., Escalante, A.A., Udagama, P.V., 2013. Population genetic structure of the *Plasmodium vivax* circumsporozoite protein (Pvcsp) in Sri Lanka. Gene 518, 381–387.

Diouf, G., Kpanyen, P.N., Tokpa, A.F., Nie, S., 2014. Changing landscape of malaria in China: progress and feasibility of malaria elimination. Asia Pac. J. Public Health 26, 93–100.

Doroshenko, A., Cooper, D., Smith, G., Gerard, E., Chinemana, F., Verlander, N., et al., 2005. Evaluation of syndromic surveillance based on National Health Service Direct derived data–England and Wales. MMWR Morb. Mortal. Wkly. Rep. (54 Suppl.), 117–122.

Duan, M.J., Duan, J.H., He, G.P., 2012. How to establish the monitoring and evaluation system in elimination stage of malaria in China: the experience from Malaria Indicator Survey system. Chin. J. Parasitol. Parasit. Dis. 30, 411–414 (in Chinese).

Galappaththy, G.N., Fernando, S.D., Abeyasinghe, R.R., 2013. Imported malaria: a possible threat to the elimination of malaria from Sri Lanka? Trop. Med. Int. Health 18, 761–768.

Gao, H.W., Wang, L.P., Liang, S., Liu, Y.X., Tong, S.L., Wang, J.J., et al., 2012. Change in rainfall drives malaria re-emergence in Anhui Province, China. PLoS One 7, e43686.

Geng, Y.Z., Mao, L.L., Teng, C., An, C.L., Wang, B., Chen, J.Y., et al., 2012. Allele genotype analysis of *Plasmodium vivax* merozoite surface protein 1 (PvMSP-1) from Dangdong, Liaoning Province. Chin. J. Parasitol. Parasit. Dis. 30, 242–244 (in Chinese).

Gorman, S., 2013. How can we improve global infectious disease surveillance and prevent the next outbreak? Scand. J. Infect. Dis. 45, 944–947.

Guan, Y.Y., Tang, L.H., Hu, L., Feng, X.P., Liu, D.Q., 2005. The point mutations in Pfcrt and Pfmdr1 genes in *Plasmodium falciparum* isolated from Hainan Province. Chin. J. Parasitol. Parasit. Dis. 23, 135–139 (in Chinese).

Gunawardena, S., Ferreira, M.U., Kapilananda, G.M., Wirth, D.F., Karunaweera, N.D., 2014. The Sri Lankan paradox: high genetic diversity in *Plasmodium vivax* populations despite decreasing levels of malaria transmission. Parasitology 1–11.

Guo, C.K., Li, J., Li, J.H., Huang, Y.M., Mao, W., Lin, K.M., 2013. Status of malaria epidemic and feasibility of malaria elimination in Guangxi Zhuang Autonomous Region. Chin. J. Parasitol. Parasit. Dis. 25, 36–39 (in Chinese).

Huang, F., Zhou, S., Zhang, S., Wang, H., Tang, L., 2011. Temporal correlation analysis between malaria and meteorological factors in Motuo County, Tibet. Malar. J. 10, 54.

Huang, F., Tang, L., Yang, H., Zhou, S., Liu, H., Li, J., et al., 2012a. Molecular epidemiology of drug resistance markers of *Plasmodium falciparum* in Yunnan Province, China. Malar. J. 11, 243.

Huang, F., Tang, L., Yang, H., Zhou, S., Sun, X., Liu, H., 2012b. Therapeutic efficacy of artesunate in the treatment of uncomplicated *Plasmodium falciparum* malaria and anti-malarial, drug-resistance marker polymorphisms in populations near the China-Myanmar border. Malar. J. 11, 278.

Kamat, V., 2000. Resurgence of malaria in Bombay (Mumbai) in the 1990s: a historical perspective. Parassitologia 42, 135–148.

Karimuribo, E.D., Sayalel, K., Beda, E., Short, N., Wambura, P., Mboera, L.G., et al., 2012. Towards one health disease surveillance: the Southern African Centre for Infectious Disease Surveillance approach. Onderstepoort J. Vet. Res. 79, 454.

Kazembe, L.N., Kleinschmidt, I., Holtz, T.H., Sharp, B.L., 2006. Spatial analysis and mapping of malaria risk in Malawi using point-referenced prevalence of infection data. Int. J. Health Geogr. 5, 41.

Kelly, G.C., Hale, E., Donald, W., Batarii, W., Bugoro, H., Nausien, J., et al., 2013. A high-resolution geospatial surveillance-response system for malaria elimination in Solomon Islands and Vanuatu. Malar. J. 12, 108.

Kunene, S., Phillips, A.A., Gosling, R.D., Kandula, D., Novotny, J.M., 2011. A national policy for malaria elimination in Swaziland: a first for sub-Saharan Africa. Malar. J. 10, 313.

Li, K.J., Huang, G.Q., Zhang, H.X., Lin, W., Dong, X.R., Pi, Q., et al., 2013. Epidemic situation and control strategy of imported malaria in Hubei Province from 2006 to 2011. Chin. J. Schisto. Control 25, 259–262 (in Chinese).

Li, X.M., Guo, C.K., Li, J.H., Huang, Y.M., Du, J.F., Fu, W.Z., 2005. A discussion on the CSP genotyping of *Plasmodium vivax* and malaria control in five southern provinces of China. Chin. J. Parasitol. Parasit. Dis. 23, 274–276 282 (in Chinese).

Lin, H., Lu, L., Tian, L., Zhou, S., Wu, H., Bi, Y., et al., 2009. Spatial and temporal distribution of falciparum malaria in China. Malar. J. 8, 130.

Linton, Y.M., Dusfour, I., Howard, T.M., Ruiz, L.F., Duc Manh, N., Ho Dinh, T., et al., 2005. Anopheles (Cellia) epiroticus (Diptera: Culicidae), a new malaria vector species in the Southeast Asian Sundaicus Complex. Bull. Entomol. Res. 95, 329–339.

Liu, X.Z., Xu, B.L., 2006. Malaria situation and evaluation on the control effect in Henan Province during 1990–2005. Chin. J. Parasitol. Parasit. Dis. 24, 226–229 (in Chinese).

Lu, Q.B., Xu, X.Q., Lin, J.F., Wang, Z., Zhang, H.L., Lai, S.J., et al., 2013. The application of China infectious diseases Automated-alert and response system in Zhejiang Province, 2012. Chin. J. Epi. 34, 594–597 (in Chinese).

MacLean, J.D., Demers, A.M., Ndao, M., Kokoskin, E., Ward, B.J., Gyorkos, T.W., 2004. Malaria epidemics and surveillance systems in Canada. Emerg. Infect. Dis. 10, 1195–1201.

Maharaj, R., Morris, N., Seocharan, I., Kruger, P., Moonasar, D., Mabuza, A., et al., 2012. The feasibility of malaria elimination in South Africa. Malar. J. 11, 423.

Ministry of Health, 2010. Action plan of China malaria elimination (2010–2020). In: Government Document Beijing, P.R. China.

Ministry of Health of the People's Republic of China, 2012. Action Plan of China Malaria Elimination (2010–2020). http://www.gov.cn/gzdt/att/att/site1/20100526/001e3741a2cc0d67233801.doc (accessed 15.07.14.) (in Chinese).

Moore, S.J., Min, X., Hill, N., Jones, C., Zaixing, Z., Cameron, M.M., 2008. Border malaria in China: knowledge and use of personal protection by minority populations and implications for malaria control: a questionnaire-based survey. BMC Public Health 8, 344.

Neuberger, A., Klement, E., Reyes, C.M., Stamler, A., 2010. A cohort study of risk factors for malaria among healthcare workers in equatorial Guinea: stay away from the ground floor. J. Travel Med. 17, 339–345.

Ojurongbe, O., Oyedeji, S.I., Oyibo, W.A., Fagbenro-Beyioku, A.F., Kun, J.F., 2010. Molecular surveillance of drug-resistant *Plasmodium falciparum* in two distinct geographical areas of Nigeria. Wien Klin. Wochenschr 122, 681–685.

Osorio, L., Todd, J., Pearce, R., Bradley, D.J., 2007. The role of imported cases in the epidemiology of urban *Plasmodium falciparum* malaria in Quibdo, Colombia. Trop. Med. Int. Health 12, 331–341.

Overgaard, H.J., Reddy, V.P., Abaga, S., Matias, A., Reddy, M.R., Kulkarni, V., et al., 2012. Malaria transmission after five years of vector control on Bioko Island, Equatorial Guinea. Parasit. Vectors 5, 253.

Owusu-Ofori, A.K., Bates, I., 2012. Impact of inconsistent policies for transfusion-transmitted malaria on clinical practice in Ghana. PLoS One 7, e34201.

Pan, J.Y., Zhou, S.S., Zheng, X., Huang, F., Wang, D.Q., Shen, Y.Z., et al., 2012. Vector capacity of *Anopheles sinensis* in malaria outbreak areas of central China. Parasit. Vectors 5, 136.

Pascual, M., Ahumada, J.A., Chaves, L.F., Rodo, X., Bouma, M., 2006. Malaria resurgence in the East African highlands: temperature trends revisited. Proc. Natl. Acad. Sci. U.S.A. 103, 5829–5834.

Poncon, N., Tran, A., Toty, C., Luty, A.J., Fontenille, D., 2008. A quantitative risk assessment approach for mosquito-borne diseases: malaria re-emergence in southern France. Malar. J. 7, 147.

Reis, B.Y., Kohane, I.S., Mandl, K.D., 2007. An epidemiological network model for disease outbreak detection. PLoS Med. 4, e210.

Richard, G.A., Feachem, A.A.P., Geoffrey, 2009. Shrinking the Malaria Map: A Prospectus on Malaria Elimination. Global Health Science. San Francisco, USA.

Sahu, S.S., Gunasekaran, K., Raju, H.K., Vanamail, P., Pradhan, M.M., Jambulingam, P., 2014. Response of malaria vectors to conventional insecticides in the southern districts of Odisha State, India. Indian J. Med. Res. 139, 294–300.

Sainz-Elipe, S., Latorre, J.M., Escosa, R., Masia, M., Fuentes, M.V., Mas-Coma, S., et al., 2010. Malaria resurgence risk in southern Europe: climate assessment in an historically endemic area of rice fields at the Mediterranean shore of Spain. Malar. J. 9, 221.

Sanders, K.C., Rundi, C., Jelip, J., Rashman, Y., Smith Gueye, C., Gosling, R.D., 2014. Eliminating malaria in Malaysia: the role of partnerships between the public and commercial sectors in Sabah. Malar. J. 13, 24.

Sheng, H.F., Zhou, S.S., Gu, Z.C., Zheng, X., 2003. Malaria situation in the People's Republic of China in 2002. Chin. J. Parasitol. Parasit. Dis. 21, 193–196 (in Chinese).

Simon, C., Moakofhi, K., Mosweunyane, T., Jibril, H.B., Nkomo, B., Motlaleng, M., et al., 2013. Malaria control in Botswana, 2008-2012: the path towards elimination. Malar. J. 12, 458.

Smith, D.L., Cohen, J.M., Chiyaka, C., Johnston, G., Gething, P.W., Gosling, R., et al., 2013. A sticky situation: the unexpected stability of malaria elimination. Philos. Trans. R. Soc. Lond. Ser. B Biol. Sci. 368, 20120145.

Sun, B.C., Zeng, Y.L., Xia, M.Y., 2012. Epidemic situation and control strategy of malaria in Yancheng City, Jiangsu Province. Chin. J. Parasitol. Parasit. Dis. 24, 672–675 (in Chinese).

Talisuna, A.O., Karema, C., Ogutu, B., Juma, E., Logedi, J., Nyandigisi, A., et al., 2012. Mitigating the threat of artemisinin resistance in Africa: improvement of drug-resistance surveillance and response systems. Lancet Infect. Dis. 12, 888–896.

Tambo, E., Ai, L., Zhou, X., Chen, J.H., Hu, W., Bergquist, R., Guo, J.G., Utzinger, J., Tanner, M., Zhou, X.N., 2014. Surveillance-response systems: the key to elimination of tropical diseases. Infect. Dis. Poverty. 3, 17.

Tatem, A.J., Huang, Z., Narib, C., Kumar, U., Kandula, D., Pindolia, D.K., et al., 2014. Integrating rapid risk mapping and mobile phone call record data for strategic malaria elimination planning. Malar. J. 13, 52.

Tian, L.H., Tan, L., Fan, Y.Z., Wang, Y., Zhang, J., Cheng, L.W., et al., 2013. The application of integrated surveillance system for symptoms in surveillance of influenza among children. Chin. J. Prevent. Med. 47, 1095–1099 (in Chinese).

Troppy, S., Haney, G., Cocoros, N., Cranston, K., DeMaria Jr., A., 2014. Infectious disease surveillance in the 21st century: an integrated web-based surveillance and case management system. Public Health Rep. 129, 132–138.

Trung, H.D., Van Bortel, W., Sochantha, T., Keokenchanh, K., Quang, N.T., Cong, L.D., et al., 2004. Malaria transmission and major malaria vectors in different geographical areas of Southeast Asia. Trop. Med. Int. Health 9, 230–237.

van Bortel, W., Trung, H.D., Hoi le, X., Van Ham, N., Van Chut, N., Luu, N.D., et al., 2010. Malaria transmission and vector behaviour in a forested malaria focus in central Vietnam and the implications for vector control. Malar. J. 9, 373.

Velasco, E., Agheneza, T., Denecke, K., Kirchner, G., Eckmanns, T., 2014. Social media and internet-based data in global systems for public health surveillance: a systematic review. Milbank Q. 92, 7–33.

Wang, X.Y., 2013. Imported malaria and control strategies in Quanzhou City. Chin. J. Schisto. Control. 25, 96–97 (in Chinese).

WHO, 2012a. Disease Surveillance for Malaria Elimination: An Operational Manual. WHO Press, Geneva, Switzerland.

WHO, 2012b. Disease Surveillance for Malaria Control: An Operational Manual. WHO Press, Geneva, Switzerland.

WHO, 2013a. Malaria Policy Advisory Committee to the WHO: conclusions and recommendations of March 2013 meeting. Malar. J. 12, 213.

WHO, 2013b. World Malaria Report 2012. WHO, Geneva.

Woyessa, A., Deressa, W., Ali, A., Lindtjorn, B., 2013. Malaria risk factors in Butajira area, south-central Ethiopia: a multilevel analysis. Malar. J. 12, 273.

Xia, Z.G., Tang, L.H., Gu, Z.C., Huang, G.Q., Zheng, X., Wang, Y., et al., 2003. Study on the thresholds of malaria transmission by Anopheles anthropophagus in Hubei Province. Chin. J. Parasitol. Parasit. Dis. 21, 224–226 (in Chinese).

Xia, Z.G., Feng, J., Zhou, S.S., 2013. Malaria situation in the People's Republic of China in 2012. Chin. J. Parasitol. Parasit. Dis. 31, 413–418 (in Chinese).

Xiao, D., Long, Y., Wang, S., Fang, L., Xu, D., Wang, G., et al., 2010. Spatiotemporal distribution of malaria and the association between its epidemic and climate factors in Hainan, China. Malar. J. 9, 185.

Xiao, D., Long, Y., Wang, S., Wu, K., Xu, D., Li, H., et al., 2012. Epidemic distribution and variation of *Plasmodium falciparum* and *Plasmodium vivax* malaria in hainan, China during 1995-2008. Am. J. Trop. Med. Hyg. 87, 646–654.

Xing, L.L., Wang, Y.B., Cui, X.F., Bu, X.Q., Kong, X.L., Zhang, B.G., et al., 2013. Analysis of malaria endemic situation in Shanxian County from 2002 to 2011. Chin. J. Schisto. Control 25, 408–410 (in Chinese).

Xiong, W., Lv, J., Li, L., 2010. A survey of core and support activities of communicable disease surveillance systems at operating-level CDCs in China. BMC Public Health 10, 704.

Xu, J., Liu, H., 2012. The challenges of malaria elimination in Yunnan Province, People's Republic of China. Southeast Asian J. Trop. Med. Public Health 43, 819–824.

Yang, G.J., Gao, Q., Zhou, S.S., Malone, J.B., McCarroll, J.C., Tanner, M., et al., 2010. Mapping and predicting malaria transmission in the People's Republic of China, using integrated biology-driven and statistical models. Geospat. Health 5, 11–22.

Yang, G.J., Tanner, M., Utzinger, J., Malone, J.B., Bergquist, R., Chan, E.Y., et al., 2012. Malaria surveillance-response strategies in different transmission zones of the People's Republic of China: preparing for climate change. Malar. J. 11, 426.

Yang, Z., Zhang, Z., Sun, X., Wan, W., Cui, L., Zhang, X., et al., 2007. Molecular analysis of chloroquine resistance in *Plasmodium falciparum* in Yunnan Province, China. Trop. Med. Int. Health 12, 1051–1060.

Ye, Y., Louis, V.R., Simboro, S., Sauerborn, R., 2007. Effect of meteorological factors on clinical malaria risk among children: an assessment using village-based meteorological stations and community-based parasitological survey. BMC Public Health 7, 101.

Yin, J.H., Yang, M.N., Zhou, S.S., Wang, Y., Feng, J., Xia, Z.G., 2013. Changing malaria transmission and implications in China towards national malaria elimination programme between 2010 and 2012. PLoS One 8, e74228.

Zakeri, S., Afsharpad, M., Ghasemi, F., Raeisi, A., Safi, N., Butt, W., et al., 2010. Molecular surveillance of *Plasmodium vivax* dhfr and dhps mutations in isolates from Afghanistan. Malar. J. 9, 75.

Zeng, Z.F., Yu, Y., 2006. Constitution and application of health quarantine risk assessment system for malaria. Chin. J. of Front. Health and Quar. 29, 23–26.

Zheng, Q., Vanderslott, S., Jiang, B., Xu, L.L., Liu, C.S., Huo, L.L., et al., 2013. Research gaps for three main tropical diseases in the People's Republic of China. Infect. Dis. Poverty 2, 15.

Zhou, S.S., Tang, L.H., Sheng, H.F., 2005. Malaria situation in the People's Republic of China in 2003. Chin. J. Parasitol. Parasit. Dis. 385–387 (in Chinese).

Zhou, S.S., Huang, F., Wang, J.J., Zhang, S.S., Su, Y.P., Tang, L.H., 2010. Geographical, meteorological and vectorial factors related to malaria re-emergence in Huang-Huai River of central China. Malar. J. 9, 337.

Zhou, X.N., Cai, L., Zhang, X.P., Sheng, H.F., Ma, X.B., Jin, Y.J., et al., 2007. Potential risks for transmission of schistosomiasis caused by mobile population in Shanghai. Chin. J. Parasitol. Parasit. Dis. 25, 180–184 (in Chinese).

Zhou, X.N., Bergquist, R., Tanner, M., 2013. Elimination of tropical disease through surveillance and response. Infect. Dis. Poverty 2, 1.

CHAPTER FIVE

Operational Research Needs Toward Malaria Elimination in China

Shen-Bo Chen[1], Chuan Ju[1], Jun-Hu Chen[1,*], Bin Zheng[1,*], Fang Huang[1], Ning Xiao[1], Xia Zhou[1], Tambo Ernest[2], Xiao-Nong Zhou[1,*]

[1]National Institute of Parasitic Diseases, Chinese Center for Disease Control and Prevention; Key Laboratory of Parasite and Vector Biology of the Chinese Ministry of Health; WHO Collaborating Centre for Malaria, Schistosomiasis and Filariasis, Shanghai, People's Republic of China
[2]Center for Sustainable Malaria Control, Faculty of Natural and Environmental Science; Center for Sustainable Malaria Control, Biochemistry Department, Faculty of Natural and Agricultural Sciences, University of Pretoria, Pretoria, South Africa
*Corresponding authors: E-mails: junhuchen@hotmail.com, cdcipdzhengbin@126.com, zhouxn1@chinacdc.cn

Contents

1. Introduction	110
2. Methods	111
2.1 Search strategy and selection criteria	111
2.2 Statistical analysis	111
3. Research Challenges in the Stage of Malaria Elimination	112
3.1 General information	112
3.2 Diagnostics and detection	114
3.3 Drugs and drug resistance	115
3.4 Epidemiology and disease control	116
3.5 Entomology and insecticides	116
3.6 Immunology and vaccines	117
3.7 Research gap analysis	118
4. Research Priority toward Malaria Elimination in China	120
4.1 Development of detection techniques for *Plasmodium* with low parasitemia	120
4.2 Tracing the original source of *Plasmodium* parasites	121
4.3 Border malaria in P.R. China	122
4.4 Drug resistance monitoring	123
4.5 Detection of G6PD deficiency	124
4.6 Active malaria screening methods	124
4.7 Effect of the environment and climate variation on vector distribution	125
5. Conclusions and Recommendations	126
Acknowledgements	127
References	127

Abstract

Owing to the implementation of a national malaria elimination programme from 2010 to 2020, we performed a systematic review to assess research challenges in the People's Republic of China (P.R. China) and define research priorities in the next few years. A systematic search was conducted for articles published from January 2000 to December 2012 in international journals from PubMed and Chinese journals from the China National Knowledge Infrastructure (CNKI). In total, 2532 articles from CNKI and 308 articles from PubMed published between 2010 and 2012 related to malaria after unrelated references and review or comment were further excluded, and a set of research gaps have been identified that could hinder progress toward malaria elimination in P.R. China. For example, there is a lack of sensitive and specific tests for the diagnosis of malaria cases with low parasitemia, and there is a need for surveillance tools that can evaluate the epidemic status for guiding the elimination strategy. Hence, we argue that malaria elimination will be accelerated in P.R. China through the development of new tests, such as detection of parasite or drug resistance, monitoring glucose-6-phosphate dehydrogenase (G6PD) deficiency, active malaria screening methods, and understanding the effects of the environment and climate variation on vector distribution.

1. INTRODUCTION

With the completion of genome sequences and stage-specific transcriptomes of the intraerythrocytic developmental cycle of *Plasmodium falciparum* and *P. vivax*, a postgenomic era has begun and has opened the eyes of scientists to design new approaches for the prevention and control of malaria infections in the world (Gardner et al., 2002; Bozdech et al., 2003, 2008; Carlton et al., 2008; Neafsey et al., 2012). Additionally, much technical advancement was achieved in the discovery of diagnostic tools, vaccines and drug targets, as well as the biomarkers for antimalarial drug resistance (Chen et al., 2010a; Crompton et al., 2010; Hu et al., 2010; Fan et al., 2013; Miotto et al., 2013; Ariey et al., 2014). However, there were still an estimated 216 million episodes of malaria and 655,000 malaria deaths in 2010, which was less than 50% of morbidity and mortality levels in 2000 (Murray et al., 2012). Moreover, the efficacy of RTS,S/AS01E vaccine over the four-year period was only 16.8%, and the efficacy of the vaccine declined over time and with increasing malaria exposure (Olotu et al., 2013). Thus, the interruption of malaria transmission worldwide is one of the greatest challenges for global health and development communities (Alonso et al., 2011).

Much effort has been devoted to prevent and control malaria in the People's Republic of China (P.R. China), as well as basic and operational research, e.g.

the first discovery of artemisinin in the world and its use in the treatment of malaria – a medical advance that has saved millions of lives across the world, especially in the developing world. As a result, the 2011 Lasker DeBakey Clinical Research Award went to Youyou Tu, a Chinese Professor, for the discovery of artemisinin (Andersen et al., 2011; Miller et al., 2011). Despite the significant reduction of malaria over the last decade in P.R. China, considerable effort will be needed to prevent a resurgence, after launching the National Malaria Eradication Programme (NMEP). Control and eventual elimination of human parasitic disease in P.R. China requires novel, innovative approaches, particularly in areas of diagnostics, mathematical modelling, monitoring and evaluation, surveillance and public health response (Chen et al., 2012).

2. METHODS

2.1 Search strategy and selection criteria

A systematic search was conducted for articles published from January 2000 to December 2012 in international journals from PubMed and the Chinese journals from the China National Knowledge Infrastructure (CNKI), respectively. We used the following search criteria, *i.e.* (malaria or plasmodium[Title/Abstract]) AND China AND English[Language] AND ("2000/01/01"[Date - Create]: "2012/12/31"[Date - Create]), to search the MEDLINE database through PubMed. The references were exported from PubMed into Endnote X1 (Thompson Reuters, San Francisco, USA), duplicates were removed and the file was transferred to Excel 2007 (Microsoft Corp., Seattle, USA). Articles published in Chinese journals were searched using similar criteria (Figure 5.1).

Each article was assigned to at least one of the following subjects based on the keywords included in the reference: drug/drug resistance, immunology/vaccines, basic sciences, epidemiology/control, entomology/insecticides, diagnostics/detection and clinical. We excluded the following articles: review, education, software, comment, book, health promotion and commercial-related topics.

2.2 Statistical analysis

Data were processed using Excel (Microsoft, WA, USA) filters, and Excel was also used to compute polynomial regression for references published in CNKI and index analysis for references published in PubMed.

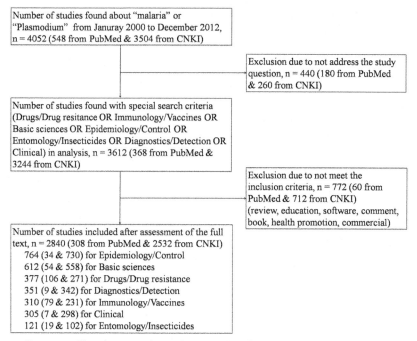

Figure 5.1 Flowchart visualizing the procedure for identifying relevant articles.

3. RESEARCH CHALLENGES IN THE STAGE OF MALARIA ELIMINATION

3.1 General information

There were far fewer references located from PubMed than from CNKI (Figure 5.2). In contrast to a binomial tendency of references published in CNKI ($R^2 = 0.8621$), an exponential increase of references published in PubMed ($R^2 = 0.8918$) was observed from 2000 to 2012.

In total, 308 references were found for malaria research in P.R. China from PubMed after unrelated references and reviews or comments were excluded from the search results. The remaining references were categorized according to subject, including drugs or drug resistance (34.4%), immunology or vaccines (25.6%), basic sciences (17.5%), epidemiology or control (11.0%), entomology or insecticides (6.2%), diagnostics or detection (2.9%) and clinical (2.3%) (Figure 5.3). The distribution of references in seven categories was much different from the distribution in a previous report about malaria references from the Asia Pacific Malaria Elimination Network (APMEN) (Andersen et al., 2011).

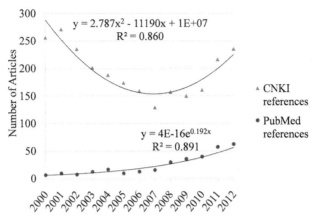

Figure 5.2 Distribution of references for malaria research in P.R. China searched from PubMed and CNKI.

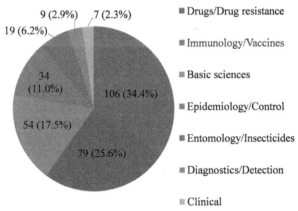

Figure 5.3 Percentage distribution of references for the malaria research in P.R. China searched from PubMed.

In total, 2532 references were found for malaria research in P.R. China from CNKI after unrelated references and reviews or comments were excluded from the search results. The remaining references were categorized according to subject, including epidemiology or control (28.8%), basic sciences (22.0%), diagnostics or detection (13.5%), clinical (11.8%), drugs or drug resistance (10.7%), immunology or vaccines (9.1%) and entomology or insecticides (4.0%) (Figure 5.4). The distribution of references in seven categories was much different from the distribution of references for malaria research in P.R. China from PubMed. In articles from the CNKI database, there was less research on

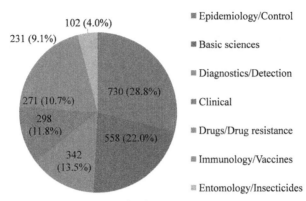

Figure 5.4 Percentage distribution of references for the malaria research in P.R. China searched from CNKI.

epidemiology/control, as well as on diagnostics/detection, and more research on immunology/vaccines.

Research references in the basic sciences and clinical sciences of malaria were excluded in the following analysis because they were not in close connection with the prevention and control of malaria.

3.2 Diagnostics and detection

Diagnostics and detection accounts for nine references from PubMed and 342 from CNKI, respectively (Figures 5.3 and 5.4). When the composition of research topics was analysed for diagnostics and detection, it was shown that most references focused on diagnostic tools, such as polymerase chain reaction (PCR), rapid diagnostic tests (RDTs), microscopic examination (eight from PubMed and 245 from CNKI), and imported and border malaria (one from PubMed and 97 from CNKI), while no reference concentrated on detection of low parasitemia and case tracing (Figure 5.5).

In the last few years, innovative tools to detect *Plasmodium* infection were developed and improved (Chen et al., 2012). The loop-mediated isothermal amplification (LAMP) test is a high-performance method for detecting DNA, which holds promise for use in the first-line battle against malaria. LAMP uses a set of primers that initiate large-scale nucleic acid synthesis by *Bst* DNA polymerase at isothermal conditions. It has been claimed that the LAMP method can detect as few as 100 copies of DNA template in blood samples (equal to roughly five parasites/µl of blood). This sensitivity is notably higher than any currently known immunochromatography-based malaria rapid diagnostic test (RDT) as recommended by the World Health Organization (WHO) as

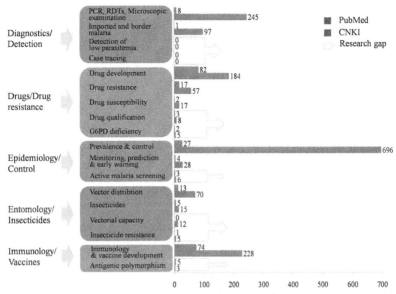

Figure 5.5 Distribution of malaria-related articles in P.R. China by topic searched from PubMed and CNKI.

part of the global malaria control strategy (Moody, 2002). A visualized LAMP method was established by the addition of a microcrystalline wax dye capsule containing the highly sensitive DNA fluorescence dye SYBR Green I to a normal LAMP reaction prior to the initiation of the reaction (Tao et al., 2011). Although further validation is needed and indeed ongoing, we can conclude that this novel, cheap and quick visualized LAMP method is feasible for malaria diagnosis under resource-constrained field settings in rural parts of P.R. China.

3.3 Drugs and drug resistance

In total, 106 references from PubMed and 271 references from CNKI were identified as associated with drug development and drug resistance (Figures 5.3 and 5.4). The composition of papers on drug and drug resistance are shown in Figure 5.5. Most articles were associated with drug development (82 from PubMed and 184 from CNKI) and drug resistance (17 from PubMed and 57 from CNKI), followed by drug susceptibility (two from PubMed and 17 from CNKI), drug quantification (three from PubMed and eight from CNKI) and Glucose-6-phosphate dehydrogenase (G6PD) deficiency (two from PubMed and five from CNKI).

The research on antimalarial drugs has a long history in P.R. China, with notable achievements. Due to the first discovery of artemisinin in the world, Chinese scientists have paid close attention to the antimalarial studies using artemisinin and artemisinin derivatives (Li et al., 2000, 2010; Wang et al., 2010; Liu et al., 2011; Zheng et al., 2013). However, *P. falciparum* parasites with reduced in vivo susceptibility to artemisinin derivatives have emerged in western Cambodia (Noedl et al., 2008; Dondorp et al., 2009, 2010; Amaratunga et al., 2012; Phyo et al., 2012).

3.4 Epidemiology and disease control

Epidemiology and disease control accounts for 34 references from PubMed and 730 references from CNKI, respectively (Figures 5.3 and 5.4). Of these papers (Figure 5.5), most references report disease prevention and control (27 from PubMed and 696 from CNKI). The remaining references were found to be related to monitoring, prediction and early warning (four from PubMed and 28 from CNKI) and active malaria screening (three from PubMed and six from CNKI).

In P.R. China, continued effort has restrained the transmission and re-emergence of malaria (Zhou et al., 2013b). The NMEP needs to establish effective surveillance response systems tailored to local transmission patterns. Currently, the malaria transmission in P.R. China can be divided into three major strata in terms of intensity. The first stratum focuses on the southern and southwestern regions. The second stratum focuses on the central part of the country. The remaining areas, where malaria transmission is very low or might have been interrupted, belong to the third stratum (Yang et al., 2012).

Malaria transmission is influenced by various factors including climatic and nonclimatic factors. The spatial and seasonal distribution of malaria is largely determined by climate, and climatic factors including rainfall, temperature and humidity have been widely used and recognized in the malaria early warning system (MEWS). However, climatic factors are not enough for MEWS, which requires comprehensive and integrated indicators. To predict the timing and severity of malaria epidemics in MEWS, epidemiological surveillance indicators (e.g. slide positivity rates) should be considered (Bi et al., 2012).

3.5 Entomology and insecticides

In all references, 19 papers from PubMed and 102 papers from CNKI were identified as associated with entomology and insecticides (Figures 5.3 and 5.4). When the composition of research topics was analysed for entomology and insecticides (Figure 5.5), it was shown that the highest number of reported literatures was on the topic of distribution of vectors (13 from PubMed and

70 from CNKI), followed by insecticides (five from PubMed and 15 from CNKI) and vectorial capacity (zero from PubMed and 12 from CNKI).

In the NMEP of P.R. China vector control is one of the important components in rapid response to the malaria transmission in outbreak foci as well as to improve the efficiency of case management, which has been promoted by both the WHO and Roll Back Malaria Partnership (RBM) for reduction of malaria transmission (Pan et al., 2012). One of the most difficult issues in the elimination process is to have real-time surveillance and response systems to monitor the changes of transmission patterns in order to guide the elimination efforts in the high risk areas (Yang et al., 2010). In order to better understand the role of the vector in the transmission of malaria during outbreaks, the vector capacity of *Anopheles sinensis* in the Huanghuai valley of central P.R. China was investigated. The study suggested that *P. vivax* malaria outbreaks in Huanhuai valley is highly related to the enhancement in vector capacity of *An. sinensis* for *P. vivax*, which was attributed to the local residents' habits and the remarkable drop in the number of large livestock leading to the disappearance of traditional biological barriers (Pan et al., 2012).

3.6 Immunology and vaccines

Immunology and vaccines account for 79 references from PubMed and 231 references from CNKI, respectively (Figures 5.3 and 5.4). The composition of papers reported on immunology and vaccines are shown in Figure 5.5. It was shown that most references focused on immunology and vaccine development (74 from PubMed and 228 from CNKI) rather than antigenic polymorphism (five from PubMed and three from CNKI).

Malaria vaccine and related immunological studies are essential and important in the global malaria communities. Until now, protective efficacy of the RTS,S/AS01 *P. falciparum* malaria vaccine was 55%, and the overall reduction in severe malaria was 35% (Agnandji et al., 2011, 2012; White, 2011; Agnandji et al., 2012; Daily, 2012). The most effective malaria vaccination, up to date, is on the immunization of human volunteers by means of repeated exposure to live *P. falciparum* sporozoites through bites of sporozoite-infected *An. stephensi* mosquitoes; parasitemia developed in none of the 10 immunized volunteers but did develop in all five nonimmunized volunteers (Campbell, 2009; Roestenberg et al., 2009). Recently, *P. falciparum* sporozoites (PfSPZ) vaccine, composed of attenuated, aseptic, purified, cryopreserved PfSPZ, was administered four to six times intravenously to adults. Zero of six subjects who received five doses of 1.35 x

10^5 PfSPZ vaccine developed into malaria episode after controlled human malaria infection, a standard process in malaria vaccine trials that the vaccinee was immunized by the bites of mosquitoes carrying the PfSPZ Vaccine. These data indicate that high-level protection against malaria can be achieved with intravenously administration of the vaccine that is safe and meets regulatory standards (Good, 2013; Seder et al., 2013). Malaria scientists at Second Military Medical University and Tongji University School of Medicine, Shanghai, P.R. China have developed PfCP2.9, a fusion protein containing domain III of AMA1 strain 3D7 and the 19 kDa c-terminal portion of MSP1 strain K1/FVO, expressed in *Pichia pastoris* and adjuvanted with Montanide ISA 720. PfCP2.9 was the only vaccine candidate from P.R. China listed on the WHO malaria vaccine rainbow table (Qian et al., 2002; Pan et al., 2004; Zhang et al., 2005; Hu et al., 2008; Xue et al., 2010; Schwartz et al., 2012).

Vaccines based on polymorphic malaria proteins may not elicit responses against all variants of the target antigen circulating in the parasite population, thus it is important to understand the natural variation in the frequency of polymorphisms in a malaria vaccine antigen (Takala et al., 2007). Genetic diversity was only analysed in several vaccine candidates in P.R. China, such as MSP1 and GLURP of *P. falciparum* (Zhu et al., 2002; Pan et al., 2010) and AMA1, MSP1, Pvs25 and Pvs28 of *P. vivax* (Figtree et al., 2000; Feng et al., 2011).

Currently, the most promising vaccine, RTS,S, is in phase III clinical trials in various African countries; however, both the efficacy (30–50%) and the duration of protection (a few months) are limited (Good, 2013), and the vaccines are far from the usage in the prevention and control of malaria, thus it is excluded in the further analysis.

3.7 Research gap analysis

A series of research gaps have been identified that will hinder progress toward malaria elimination in P.R. China (Figures 5.5 and 5.6). First, in the control and pre-elimination stage, PCR, RDTs and the microscopic examination of a slide were commonly used in the field, but there is a lack of sensitive and specific tests for the diagnosis of malaria cases with low parasitemia and tracing the original source of *Plasmodium* parasites. Besides that, most references focused on detection of imported malaria, while only one reference discussed transmission patterns of malaria in the border areas where more mobile population crossed the border frequently.

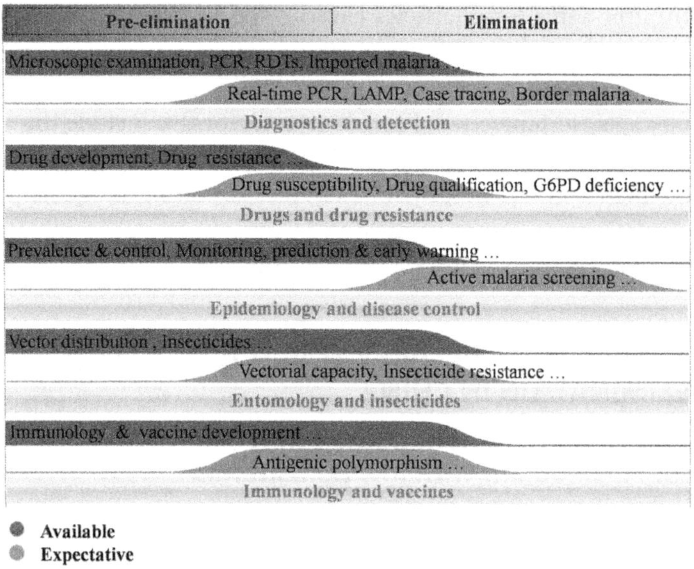

Pre-elimination	Elimination
Microscopic examination, PCR, RDTs, Imported malaria ...	Real-time PCR, LAMP, Case tracing, Border malaria ...
Diagnostics and detection	
Drug development, Drug resistance ...	Drug susceptibility, Drug qualification, G6PD deficiency ...
Drugs and drug resistance	
Prevalence & control, Monitoring, prediction & early warning ...	Active malaria screening ...
Epidemiology and disease control	
Vector distribution, insecticides ...	Vectorial capacity, Insecticide resistance ...
Entomology and insecticides	
Immunology & vaccine development ...	Antigenic polymorphism ...
Immunology and vaccines	

● Available
● Expectative

Figure 5.6 Gap analysis of research references toward malaria elimination in P.R. China, searched from CNKI and PubMed.

Second, the research output of drug development and drug susceptibility accounted for the highest number of publications in the field of drugs and drug resistance. However, it is short of innovative and efficient techniques for screening G6PD deficiency and monitoring the drug qualification. In the past few years, researchers have paid great attention to drug resistance, and it is still the research focus during the pre-elimination/elimination stage in China.

Third, a great number of articles focused on prevalence and control, monitoring, prediction and early warning of malaria, which suggests that monitoring and controlling the transmission of malaria is a key priority during the pre-elimination stage in P.R. China. Furthermore, there is also an urgent need for surveillance tools that can evaluate the epidemic status for guiding the elimination strategy.

Fourth, further study is warranted to investigate the effect of the environment and climate variation on vector distribution during the pre-elimination/elimination stage despite efforts by researchers concentrated simply on vector distribution in the past. Meanwhile, it is an essential to pay more attention to the assessment of insecticidal impact on the malaria mosquito's vectorial capacity and keep a close eye on insecticide resistance in the future investigation.

Fifth, although a lot of articles have been focused on immunology and vaccines development rather than antigenic polymorphism in malaria, *Plasmodium* parasite infections do not readily evoke an effective protective immunity against reinfection. Possible reasons for this include the ability of the parasites to interfere with the host's immune response and to evade the response in an immune host by, for example, exploiting antigenic polymorphism. Hence, there is a strong need to investigate antigenic polymorphism in malaria and its implication for immune evasion.

4. RESEARCH PRIORITY TOWARD MALARIA ELIMINATION IN CHINA

4.1 Development of detection techniques for *Plasmodium* with low parasitemia

Early diagnosis followed by treatment is an essential component to the national malaria elimination programme. Microscopy is regarded as the gold standard for malaria diagnosis (WHO, 1999) and is used as the principal means for diagnosis and surveillance of malaria in P.R. China as well. However, the lack of skilled technologists in medical facilities in endemic areas often leads to poor interpretation of data. Furthermore, microscopy is time-consuming and labour-intensive, cannot detect sequestered *P. falciparum* parasites (Leke et al., 1999) and is less reliable at low-density parasitaemia, such as parasitaemia < 50 parasites $(\mu l\ blood)^{-1}$ (Kilian et al., 2000; Bell et al., 2005). Most rapid diagnostic tests (RDTs) are made based on antigen capture, and do not work with low parasitemias (Zhou et al., 2013a). By now, there are several *Plasmodium* antigens for diagnosis targets, such as histidine-rich protein-2 (HRP-2) and parasite-specific lactate dehydrogenase (pLDH). Of these, HRP-2 is able to distinguish from other *Plasmodium* parasites because of its specific expression in *P. falciparum*. Hence, recombinant antigen of the China strain is recommended as coating antigen for detection of *P. falciparum* and *P. vivax* by colloidal gold immunochromatographic assay (GICA) strips which has been used in the elimination stage of the NMEP. Moreover, real-time PCR has been developed and applied for the diagnosis of malaria. Microscopic examination, GICA and real-time PCR were employed to identify parasitemia patients and people without parasitemia (Khairnar et al., 2009). There were no statistical differences in specificity among three methods, but sensitivity of real-time PCR was better than others, especially for low parasitemias.

The recently developed LAMP assay is a relatively simple and field-applicable technique to detect parasite infections that can overcome the shortage of the time-consuming and expensive purification of DNA prior to amplification (Han et al., 2007; Chen et al., 2010b; Han, 2013). LAMP tests have recently been evaluated in detecting the infections of *P. falciparum* (Yi et al., 2010 and *P. vivax* (Zhu et al., 2010; Lu et al., 2012; Wang et al., 2012) in a China reference laboratory and a rural clinic in field, with promising results. LAMP test was evaluated for samples, from 272 outpatients at a rural Ugandan clinic. For samples, with a *P. falciparum* qPCR titer of ≥2 parasites/μL, LAMP sensitivity was 97.8%, similar to that of single well-nested PCR in a United Kingdom reference laboratory. LAMP dramatically lowers the detection threshold achievable in malaria-endemic settings, providing a new tool for diagnosis in elimination strategies (Hopkins et al., 2013).

4.2 Tracing the original source of *Plasmodium* parasites

In P.R. China, as a result of active implementation of malaria control measures for more than 40 years, considerable success to control and eliminate malaria transmission has achieved. No local infection cases within three years is one of important indexes for malaria elimination, so the tracing origin technique based on molecular biology is necessary for malaria elimination. In P.R. China, MAD20-type MSP1 allele and 3D7-type MSP2 allele were dominant in *P. falciparum* population in the Hainan Province. The mixed infection rate of different types of MSP1 or MSP2 alleles was low (Jiang et al., 2003). Moreover, MSP2 gene polymorphisms of different geographical strains of *P. falciparum* in Yunnan is evident through random amplified DNA polymorphism- PCR techniques (Yang et al., 2004).

As genetic makers, microsatellites are abundant and distributed throughout the eukaryotic genomes. Their advantages include high polymorphism and abundance, codominance, selective neutrality and high reliability. Microsatellite has been applied for the investigation of source tracing and genetic diversity (Koepfli et al., 2011; Iwagami et al., 2012, 2013). In the 1990s, microsatellites (MS) were introduced to map the chloroquine (CQ)-resistant gene (Su et al., 1996, 1997). Analysis of the population structure with microsatellites demonstrated that *P. vivax* isolates from different areas of China showed high genetic diversity and regionally centered difference (Guo et al., 2012).

The completion of many malaria parasite genomes provides great opportunities for genome-wide characterization of gene expression and high-throughput genotyping (Carlton et al., 2008; Winzeler, 2008). Substantial progress in malaria genomics and genotyping has been made recently, particularly the development of various microarray platforms for large-scale characterization of the *P. falciparum* genome. Microarray has been used for gene expression analysis, detection of single nucleotide polymorphism (SNP) and copy number variation (CNV), characterization of chromatin modifications and other applications. In the 2000s, SNP is becoming the marker of choice because of the development of high-throughput SNP genotyping methods (Su et al., 2007; Maresso et al., 2008). SNP genotyping is actually based on the same polymorphism as polymerase chain reaction-restriction fragment length polymorphism (PCR-RFLP); the difference is in the method for detecting the polymorphism. Large numbers of MS and SNP have been developed for *P. falciparum* (Su et al., 1999; Jeffares et al., 2007; Mu et al., 2007; Volkman et al., 2007) and genetic markers for other parasites such as *P. vivax* and rodent malaria parasites are also available or being developed (Grech et al., 2002; Feng et al., 2003; Martinelli et al., 2004; Li et al., 2007; Karunaweera et al., 2008). High-throughput genotyping methods are now available for typing DNA from *P. falciparum* and for mapping parasite traits, and many more typing methods are under development, including those for other malaria species. Unfortunately, the malaria parasite is a single-cell organism, and it is thus challenging to detect or measure reproducible phenotypic variation between individual parasites. The future direction for mapping malaria traits should focus on developing methods to accurately characterize and measure phenotypic variation among individual parasites.

4.3 Border malaria in P.R. China

There are more than 35 minorities distributed in Yunnan Province, and most of them are living in remote areas, including the border with other countries, such as Myanmar, Vietnam, Lao PDR, etc. Malaria prevalence is at highest level among ethnic minorities, migrants and forest workers, with the most vulnerable population being pregnant women, the very poor and the malnourished (ADB, 1998; Xia et al., 2013). The previous study investigated knowledge of malaria prevention and use of personal protection measures among ethnic minority populations in the Yunnan Province in Southwest China. Most Chinese are Han-Chinese, but the populations interviewed during this study are recognized as ethnic nationalities by the state: having culture, language and lifestyle unique to their groups (Xu et al., 2003).

Along the border areas, it is very common that people from one ethnic group may reside in several countries. They frequently cross international borders to visit family members, make cross-border marriages, and conduct trades (Xu et al., 2003), where they have exposed to mosquito biting while travelling, and finally import the disease when they go back to their villages (Xu et al., 1997). The Yunnan Province is one of the two remaining epidemic areas of China with high annual transmission of both *P. vivax* and *P. falciparum*; the other being Hainan Island before 2010. In 2004, the annual reported malaria incidence was 3.09/10,000 (Zhou et al., 2006), although the estimated number of actual cases is at least 18 times greater (Sheng et al., 2003). Malaria is a particularly severe social and health problem along the border with Myanmar, where mobile workers move back and forth across the border and malaria control is weak. One-third of malaria cases in P.R. China came from the Yunnan Province in 2005, and about a quarter of these were actually infected in Myanmar during trips to visit relatives and conduct business (Fund, 2007).

The use of personal protection must be increased, particularly among outdoor workers that have higher risk of malaria infection. However, personal protection is widely used and widely accepted to prevent nuisance-biting mosquitoes, with the major barrier to use being affordability. Therefore, social marketing campaigns aimed at women and those that work outdoors that provide highly subsidized products, especially insecticide impregnation kits for bed nets and hammock nets are most likely to succeed in lowering malaria morbidity among non-Han-Chinese groups in rural China (Moore et al., 2008).

4.4 Drug resistance monitoring

The discovery in the 1940s that the synthetic drug CQ could effectively treat individuals safely and cheaply helped spur malaria eradication efforts in the 1950s. However, the emergence of CQ resistance diminished its therapeutic efficacy and doomed initial efforts to eradicate the disease. This is of particular importance because in recent years artemisinin class drugs, the current recommended first-line treatment for uncomplicated and severe malaria (WHO, 2006), may lose their effectiveness. However, a delayed-clearance phenotype has already been reported both in western Cambodia (Dondorp et al., 2010) and Thailand (Phyo et al., 2012). This delayed-clearance phenotype, whilst not of clinical significance yet (Phyo et al., 2012), is the first indication that resistance to artemisinin may emerge soon. This has important implications for global eradication efforts, as it will likely be at least a decade before a new compound is capable of replacing the artemisinins (Olliaro et al., 2009). The sensitivity surveillance of *Plasmodium* parasites to drug contains in vivo and in vitro. Culture system in vitro of *P. falciparum* has been well developed. However culture method in vitro of *P. vivax* is not mature yet. Two monoclonal strains of dihydroarthemisinin-resistant *P. falciparum* were obtained from chloroquine-resistant *P. falciparum* originating in Yunnan, China (Yang et al., 2013). *P. falciparum* isolates in China–Myanmar showed resistance to chloroquine and pyronaridine, and most isolates were still sensitive to piperaquine (Zhang et al., 2012).

4.5 Detection of G6PD deficiency

Primaquine is an essential tool for malaria control and elimination since it is the only available drug preventing multiple clinical attacks by relapses of *P. vivax*. It is also the only antimalarial agent against the sexual stages of *P. falciparum* infectious to mosquitoes, and is thus useful in preventing malaria transmission (Baird et al., 2011; Bousema et al., 2011). However, the difficulties of identifying glucose-6-phosphate dehydrogenase deficiency (G6PDd) greatly hinder primaquine's widespread use, as this common genetic disorder makes patients susceptible to potentially severe and fatal primaquine-induced haemolysis. The risk of such an outcome varies widely among G6PD gene variants (Howes et al., 2013b).

G6PD is a potentially pathogenic inherited enzyme abnormality and, similar to other human red blood cell polymorphisms, is particularly prevalent in historically malaria-endemic countries (Cappellini et al., 2008; Howes et al., 2012). The spatial extent of *P. vivax* malaria overlaps widely with that of G6PD deficiency; unfortunately, the only drug licensed for the radical cure and relapse prevention of *P. vivax*, primaquine, can trigger severe haemolytic anaemia in G6PD deficient individuals. According to the past and current data on this unique pharmacogenetic association, G6PDd is becoming increasingly important as several nations now consider strategies to eliminate malaria transmission rather than control its clinical burden (Wells et al., 2010). G6PD deficiency is a highly variable disorder, in terms of spatial heterogeneity in prevalence and molecular variants, as well as its interactions with *P. vivax* and primaquine. Consideration of factors including aspects of basic physiology, diagnosis and clinical triggers of primaquine-induced haemolysis is required to assess the risks and benefits of applying primaquine in various geographic and demographic settings. Given that haemolytically toxic antirelapse drugs will likely be the only therapeutic options for the coming decade, it is clear that we need to understand G6PD deficiency and primaquine-induced haemolysis in depth to determine safe and effective therapeutic strategies to overcome this hurdle and achieve malaria elimination (Howes et al., 2013a).

4.6 Active malaria screening methods

Many of malaria's signs and symptoms are indistinguishable from those of other febrile diseases. Detection of the presence of *Plasmodium* parasites is essential, therefore, to guide case management in NMEP (Bojang et al., 2000). Improved diagnostic tools are required to enable targeted treatment of infected individuals. In addition, field-ready diagnostic tools for mass screening and surveillance that can detect asymptomatic infections of very low parasite densities are needed to monitor transmission reduction and ensure elimination. Antibody-based tests for infection and novel methods based on biomarkers need further development and validation, as do methods for the detection and treatment of *P. vivax*. Current rapid diagnostic tests (RDTs) targeting *P. vivax* are generally less effective than those targeting *P. falciparum* (malERA Consultative Group on Diagnoses and Diagnostics, 2011). Identification of parasitemia in febrile patients is essential in all of the programmatic phases of the continuum, from malaria control to elimination, although the challenges for health systems in maintaining this activity in areas where malaria has become rare will be more prominent, as will the importance of detecting asymptomatic infections of low parasite density. Analyses of past experiences and operations research are required to guide decisions on when these changes in emphasis should take place as control progresses (malERA Consultative Group on Health Systems and Operational Research, 2011; malERA Consultative Group on Modeling, 2011).

Current antigen-detecting RDTs are likely to miss a significant proportion of asymptomatic cases in low-transmission settings (Roper et al., 1996; Kidson et al., 1998; Collins et al., 1999). Thus, although the current generation of RDTs can indicate the presence of malaria in a community, they cannot determine the true prevalence of parasite carriage. Research aimed towards increasing the sensitivity of existing RDTs may not change this situation because of the limitations of the currently available technology. Some antigen or antibody detecting enzyme-linked immunosorbent assay (ELISA) are more sensitive than RDTs (Chen et al., 2011). Furthermore, because ELISAs can also be used to quantify antigen, they have been used to monitor drug efficacy, and may also facilitate high-throughput testing. However, their use is currently limited by laboratory and training requirements. Detection of antisporozoite antibodies (so-called anti-CSP antibodies) alone or in combination with antibodies to blood-stage parasites has also been suggested as a surrogate for detecting individuals with a high likelihood of carrying *P. vivax* hypnozoites to provide evidence of infection (Cho et al., 2001; Kim et al., 2003; Lee et al., 2003; Park et al., 2003; Suh et al., 2004).

4.7 Effect of the environment and climate variation on vector distribution

Malaria is transmitted to humans by mosquitoes of the genus *Anopheles*. Improved vector control is essential for the elimination or eradication of malaria in P.R. China. In regions where transmission rates are at low or moderate level, existing tools may be sufficient to achieve elimination, but in many malaria-endemic regions, new vector control interventions, including new insecticides and formulations, are needed. Better understanding of vector biology is an essential prerequisite for the development of new control interventions (malERA Consultative Group on Vector Control, 2011). The overarching goal of malaria vector control is to reduce the vectorial capacity of local vector populations below the critical threshold needed to achieve a malaria reproduction rate ($R0$, the expected number of human cases that arise from each human case in a population) of less than one. Because of the long extrinsic incubation time of *Plasmodium* in its *Anopheles* vectors, the most effective vector control strategies in use today rely on insecticide interventions like indoor residual insecticide sprays (IRSs) and long-lasting insecticide-treated nets (LLINs) that reduce vector daily survival rates (Enayati et al., 2010). For many malaria-endemic regions, these tools can make substantial contributions to malaria control and may be sufficient for local malaria elimination.

5. CONCLUSIONS AND RECOMMENDATIONS

A systematic search was conducted for literatures published from January 2000 to December 2012 in the international journals from PubMed and the Chinese journals from CNKI. The number of studies found from PubMed were far fewer than those searched from CNKI. In contrast to a binomial tendency of references published in CNKI, an exponential increasing number of references published in PubMed was observed from 2000 to 2012.

In total, 308 and 2532 references were found on the malaria research carried out in P.R. China from PubMed and CNKI, respectively. Articles on drugs or drug resistance were the largest proportion of publications located from PubMed; in contrast, articles on epidemiology or control were the largest proportion of publications from CNKI.

A series of research gaps have been identified that will hinder progress toward malaria elimination in P.R. China. For example, there is a lack of sensitive and specific tests for the diagnosis of malaria cases with low parasitemia, and there is a need for surveillance tools that can evaluate the epidemic status for guiding the elimination strategy. Hence, we expected that malaria elimination could be accelerated in China through development of new tests both used in the detection of parasite or drug resistance and monitoring G6PD deficiency, active malaria screening methods, and understanding the effect of the environment and climate variation on vector distribution.

Moreover, combined with improved tools for diagnosis and surveillance and connected to effective response packages will lead to integrated, multipronged strategies for the control and elimination of malaria in P.R. China. Experiences reviewed here will be important for other middle and low income countries that are moving from morbidity control towards transmission interruption and eventual elimination of malaria.

ACKNOWLEDGEMENTS
This work was supported in part by the Foundation of National Science and Technology Major Programme (Grant No. 2012ZX10004-220), by the Special Fund for Health Research in the Public Interest (Grant No. 201202019) and by China UK Global Health Support Programme (grant no. GHSP-CS-OP1).

REFERENCES
ADB, 1998. Technical assistance for the study of the health and educational needs of ethnic minorities in the Greater Mekong Sub Region. Asian Development Bank, Manila.
Agnandji, S.T., Lell, B., Fernandes, J.F., Abossolo, B.P., Methogo, B.G., Kabwende, A.L., et al., 2012. A phase 3 trial of RTS,S/AS01 malaria vaccine in African infants. N. Engl. J. Med. 367, 2284–2295.
Agnandji, S.T., Lell, B., Soulanoudjingar, S.S., Fernandes, J.F., Abossolo, B.P., Conzelmann, C., et al., 2011. First results of phase 3 trial of RTS,S/AS01 malaria vaccine in African children. N. Engl. J. Med. 365, 1863–1875.
Alonso, P.L., Brown, G., Arevalo-Herrera, M., Binka, F., Chitnis, C., Collins, F., et al., 2011. A research agenda to underpin malaria eradication. PLoS Med. 8, e1000406.
Amaratunga, C., Sreng, S., Suon, S., Phelps, E.S., Stepniewska, K., Lim, P., et al., 2012. Artemisinin-resistant *Plasmodium falciparum* in Pursat province, western Cambodia: a parasite clearance rate study. Lancet Infect. Dis. 12, 851–858.
Andersen, F., Douglas, N.M., Bustos, D., Galappaththy, G., Qi, G., Hsiang, M.S., et al., 2011. Trends in malaria research in 11 Asian Pacific countries: an analysis of peer-reviewed publications over two decades. Malar. J. 10, 131.
Ariey, F., Witkowski, B., Amaratunga, C., Beghain, J., Langlois, A.C., Khim, N., et al., 2014. A molecular marker of artemisinin-resistant *Plasmodium falciparum* malaria. Nature 505, 50–55.
Baird, J.K., Surjadjaja, C., 2011. Consideration of ethics in primaquine therapy against malaria transmission. Trends. Parasitol 12, 11–16.
Bell, D.R., Wilson, D.W., Martin, L.B., 2005. False-positive results of a *Plasmodium falciparum* histidine-rich protein 2-detecting malaria rapid diagnostic test due to high sensitivity in a community with fluctuating low parasite density. Am. J. Trop. Med. Hyg. 73, 199–203.
Bi, Y., Hu, W., Liu, H., Xiao, Y., Guo, Y., Chen, S., et al., 2012. Can slide positivity rates predict malaria transmission? Malar. J. 11, 117.
Bojang, K.A., Obaro, S., Morison, L.A., Greenwood, B.M., 2000. A prospective evaluation of a clinical algorithm for the diagnosis of malaria in Gambian children. Trop. Med. Int. Health 5, 231–236.
Bousema, T., Drakeley, C., 2011. Epidemiology and infectivity of *Plasmodium falciparum* and *Plasmodium vivax* gametocytes in relation to malaria control and elimination. Clin. Microbiol. Rev. 12, 377–410.
Bozdech, Z., Llinas, M., Pulliam, B.L., Wong, E.D., Zhu, J., DeRisi, J.L., 2003. The transcriptome of the intraerythrocytic developmental cycle of *Plasmodium falciparum*. PLoS Biol. 1, E5.
Bozdech, Z., Mok, S., Hu, G., Imwong, M., Jaidee, A., Russell, B., et al., 2008. The transcriptome of *Plasmodium vivax* reveals divergence and diversity of transcriptional regulation in malaria parasites. Proc. Natl. Acad. Sci. U.S.A. 105, 16290–16295.
Campbell, C.C., 2009. Malaria control–addressing challenges to ambitious goals. N. Engl. J. Med. 361, 522–523.
Cappellini, M.D., Fiorelli, G., 2008. Glucose-6-phosphate dehydrogenase deficiency. Lancet 371, 64–74.
Carlton, J.M., Adams, J.H., Silva, J.C., Bidwell, S.L., Lorenzi, H., Caler, E., et al., 2008. Comparative genomics of the neglected human malaria parasite *Plasmodium vivax*. Nature 455, 757–763.

Chen, J.H., Jung, J.W., Wang, Y., Ha, K.S., Lu, F., Lim, C.S., et al., 2010a. Immunoproteomics profiling of blood stage *Plasmodium vivax* infection by high-throughput screening assays. J. Proteome Res. 9, 6479–6489.

Chen, J.H., Lu, F., Lim, C.S., Kim, J.Y., Ahn, H.J., Suh, I.B., et al., 2010b. Detection of *Plasmodium vivax* infection in the Republic of Korea by loop-mediated isothermal amplification (LAMP). Acta Trop. 113, 61–65.

Chen, J.H., Wang, H., Chen, J.X., Bergquist, R., Tanner, M., Utzinger, J., Zhou, X.N., 2012. Frontiers of parasitology research in the People's Republic of China: infection, diagnosis, protection and surveillance. Parasit. Vectors 5, 221.

Chen, J.H., Wang, Y., Ha, K.S., Lu, F., Suh, I.B., Lim, C.S., et al., 2011. Measurement of naturally acquired humoral immune responses against the C-terminal region of the *Plasmodium vivax* MSP1 protein using protein arrays. Parasitol. Res. 109, 1259–1266.

Cho, D., Kim, K.H., Park, S.C., Kim, Y.K., Lee, K.N., Lim, C.S., 2001. Evaluation of rapid immunocapture assays for diagnosis of *Plasmodium vivax* in Korea. Parasitol. Res. 87, 445–448.

Collins, W.E., Jeffery, G.M., 1999. A retrospective examination of the patterns of recrudescence in patients infected with *Plasmodium falciparum*. Am. J. Trop. Med. Hyg. 61, 44–48.

Crompton, P.D., Kayala, M.A., Traore, B., Kayentao, K., Ongoiba, A., Weiss, G.E., et al., 2010. A prospective analysis of the Ab response to *Plasmodium falciparum* before and after a malaria season by protein microarray. Proc. Natl. Acad. Sci. U.S.A. 107, 6958–6963.

Daily, J.P., 2012. Malaria vaccine trials–beyond efficacy end points. N. Engl. J. Med. 367, 2349–2351.

Dondorp, A.M., Nosten, F., Yi, P., Das, D., Phyo, A.P., Tarning, J., et al., 2009. Artemisinin resistance in *Plasmodium falciparum* malaria. N. Engl. J. Med. 361, 455–467.

Dondorp, A.M., Yeung, S., White, L., Nguon, C., Day, N.P., Socheat, D., von Seidlein, L., 2010. Artemisinin resistance: current status and scenarios for containment. Nat. Rev. Microbiol. 8, 272–280.

Enayati, A., Hemingway, J., 2010. Malaria management: past, present, and future. Annu. Rev. Entomol. 55, 569–591.

Fan, Y.T., Wang, Y., Ju, C., Zhang, T., Xu, B., Hu, W., Chen, J.H., 2013. Systematic analysis of natural antibody responses to *P. falciparum* merozoite antigens by protein arrays. J. Proteomics 78, 148–158.

Feng, H., Zheng, L., Zhu, X., Wang, G., Pan, Y., Li, Y., et al., 2011. Genetic diversity of transmission-blocking vaccine candidates Pvs25 and Pvs28 in *Plasmodium vivax* isolates from Yunnan Province, China. Parasit. Vectors 4, 224.

Feng, X., Carlton, J.M., Joy, D.A., Mu, J., Furuya, T., Suh, B.B., et al., 2003. Single-nucleotide polymorphisms and genome diversity in *Plasmodium vivax*. Proc. Natl. Acad. Sci. U.S.A. 100, 8502–8507.

Figtree, M., Pasay, C.J., Slade, R., Cheng, Q., Cloonan, N., Walker, J., Saul, A., 2000. *Plasmodium vivax* synonymous substitution frequencies, evolution and population structure deduced from diversity in AMA 1 and MSP 1 genes. Mol. Biochem. Parasitol. 108, 53–66.

Fund, G., 2007. East Asia and Pacific Regional Overview: Successes, Challenges and Achivements to Date. The Global Fund to Fight AIDS, Tuberculosis and Malaria, Geneva.

Gardner, M.J., Hall, N., Fung, E., White, O., Berriman, M., Hyman, R.W., et al., 2002. Genome sequence of the human malaria parasite *Plasmodium falciparum*. Nature 419, 498–511.

Good, M.F., 2013. Immunology. Pasteur approach to a malaria vaccine may take the lead. Science 341, 1352–1353.

Grech, K., Martinelli, A., Pathirana, S., Walliker, D., Hunt, P., Carter, R., 2002. Numerous, robust genetic markers for *Plasmodium chabaudi* by the method of amplified fragment length polymorphism. Mol. Biochem. Parasitol. 123, 95–104.

Guo, X., Zhang, D., Wang, J., Zhang, C., Pan, W., 2012. Analysis of the population structure of *Plasmodium vivax* isolates from different areas in China using microsatellites. Int. J. Med. Parasit. Dis. 39, 197–201 (in Chinese).

Han, E.T., 2013. Loop-mediated isothermal amplification test for the molecular diagnosis of malaria. Expert Rev. Mol. Diagn. 13, 205–218.

Han, E.T., Watanabe, R., Sattabongkot, J., Khuntirat, B., Sirichaisinthop, J., Iriko, H., et al., 2007. Detection of four *Plasmodium* species by genus- and species-specific loop-mediated isothermal amplification for clinical diagnosis. J. Clin. Microbiol. 45, 2521–2528.

Hopkins, H., Gonzalez, I.J., Polley, S.D., Angutoko, P., Ategeka, J., Asiimwe, C., et al., 2013. Highly sensitive detection of malaria parasitemia in a malaria-endemic setting: performance of a new loop-mediated isothermal amplification kit in a remote clinic in Uganda. J. Infect. Dis. 208, 645–652.

Howes, R.E., Battle, K.E., Satyagraha, A.W., Baird, J.K., Hay, S.I., 2013a. G6PD deficiency: global distribution, genetic variants and primaquine therapy. Adv. Parasitol. 81, 133–201.

Howes, R.E., Dewi, M., Piel, F.B., Monteiro, W.M., Battle, K.E., Messina, J.P., et al., 2013b. Spatial distribution of G6PD deficiency variants across malaria-endemic regions. Malar. J. 12, 418.

Howes, R.E., Piel, F.B., Patil, A.P., Nyanqiri, O.A., Gething, P.W., Dewi, M., et al., 2012. G6PD deficiency prevalence and estimates of affected populations in malaria endemic countries. a geostatistical model-based map PLoS Med. 9, e1001339.

Hu, G., Cabrera, A., Kono, M., Mok, S., Chaal, B.K., Haase, S., et al., 2010. Transcriptional profiling of growth perturbations of the human malaria parasite *Plasmodium falciparum*. Nat. Biotechnol. 28, 91–98.

Hu, J., Chen, Z., Gu, J., Wan, M., Shen, Q., Kieny, M.P., et al., 2008. Safety and immunogenicity of a malaria vaccine, *Plasmodium falciparum* AMA-1/MSP-1 chimeric protein formulated in montanide ISA 720 in healthy adults. PLoS One 3, e1952.

Iwagami, M., Fukumoto, M., Hwang, S.Y., Kim, S.H., Kho, W.G., Kano, S., 2012. Population structure and transmission dynamics of *Plasmodium vivax* in the Republic of Korea based on microsatellite DNA analysis. PLoS Negl. Trop. Dis. 6, e1592.

Iwagami, M., Hwang, S.Y., Kim, S.H., Park, S.J., Lee, G.Y., Matsumoto-Takahashi, E.L., et al., 2013. Microsatellite DNA analysis revealed a drastic genetic change of *Plasmodium vivax* population in the Republic of Korea during 2002 and 2003. PLoS Negl. Trop. Dis. 7, e2522.

Jeffares, D.C., Pain, A., Berry, A., Cox, A.V., Stalker, J., Ingle, C.E., et al., 2007. Genome variation and evolution of the malaria parasite *Plasmodium falciparum*. Nat. Genet. 39, 120–125.

Jiang, G., Song, J., Chen, P., Wang, S., 2003. Study on the genotypes of MSP1 and MSP2 genes of *Plasmodium falciparum* isolates from Hainan Province. South China J. Prev. Med. 29, 9–11 (in Chinese).

Karunaweera, N.D., Ferreira, M.U., Munasinghe, A., Barnwell, J.W., Collins, W.E., King, C.L., et al., 2008. Extensive microsatellite diversity in the human malaria parasite *Plasmodium vivax*. Gene 410, 105–112.

Khairnar, K., Martin, D., Lau, R., Ralevski, F., Pillai, D.R., 2009. Multiplex real-time quantitative PCR, microscopy and rapid diagnostic immuno-chromatographic tests for the detection of *Plasmodium* spp: performance, limit of detection analysis and quality assurance. Malar. J. 8, 284.

Kidson, C., Indaratna, K., 1998. Ecology, economics and political will: the vicissitudes of malaria strategies in Asia. Parassitologia 40, 39–46.

Kilian, A.H., Metzger, W.G., Mutschelknauss, E.J., Kabagambe, G., Langi, P., Korte, R., von Sonnenburg, F., 2000. Reliability of malaria microscopy in epidemiological studies: results of quality control. Trop. Med. Int. Health 5, 3–8.

Kim, S., AHN, H.J., Kim, T.S., NAM, H.W., 2003. ELISA detection of vivax malaria with recombinant multiple stage-sepcific antigens and its application to survey of residents in endemic areas. Korea J. Parasitol. 41, 203–207.

Koepfli, C., Ross, A., Kiniboro, B., Smith, T.A., Zimmerman, P.A., Siba, P., et al., 2011. Multiplicity and diversity of *Plasmodium vivax* infections in a highly endemic region in Papua New Guinea. PLoS Negl. Trop. Dis. 5, e1424.

Lee, K.N., Suh, I.B., Chang, E.A., Kim, S.D., Cho, N.S., Park, P.W., et al., 2003. Prevalence of antibodies to the circumsporozite protein of *Plasmodium vivax* in five different regions of Korea. Trop. Med. Int. Health 8, 1062–1067.

Leke, R.F., Djokam, R.R., Mbu, R., Leke, R.J., Fogako, J., Megnekou, R., et al., 1999. Detection of the *Plasmodium falciparum* antigen histidine-rich protein 2 in blood of pregnant women: implications for diagnosing placental malaria. J. Clin. Microbiol. 37, 2992–2996.

Li, J., Zhang, Y., Sullivan, M., Hong, L., Huang, L., Lu, F., et al., 2007. Typing *Plasmodium yoelii* microsatellites using a simple and affordable fluorescent labeling method. Mol. Biochem. Parasitol. 155, 94–102.

Li, J., Zhou, B., 2010. Biological actions of artemisinin: insights from medicinal chemistry studies. Molecules 15, 1378–1397.

Li, Y., Zhu, Y.M., Jiang, H.J., Pan, J.P., Wu, G.S., Wu, J.M., et al., 2000. Synthesis and antimalarial activity of artemisinin derivatives containing an amino group. J. Med. Chem. 43, 1635–1640.

Liu, Y., Cui, K., Lu, W., Luo, W., Wang, J., Huang, J., Guo, C., 2011. Synthesis and antimalarial activity of novel dihydro-artemisinin derivatives. Molecules 16, 4527–4538.

Lu, F., Gao, Q., Zhou, H., Cao, J., Wang, W., Lim, C.S., et al., 2012. Molecular test for vivax malaria with loop-mediated isothermal amplification method in central China. Parasitol. Res. 110, 2439–2444.

malERA Consultative Group on Diagnoses and Diagnostics, 2011. A research agenda for malaria eradication: diagnoses and diagnostics. PLoS Med. 8, e1000396.

malERA Consultative Group on Health Systems and Operational Research, 2011. A research agenda for malaria eradication: health systems and operational research. PLoS Med. 8, e1000397.

malERA Consultative Group on Modeling, 2011. A research agenda for malaria eradication: modeling. PLoS Med. 8, e1000403.

malERA Consultative Group on Vector Control, 2011. A research agenda for malaria eradication: vector control. PLoS Med. 8, e1000401.

Maresso, K., Broeckel, U., 2008. Genotyping platforms for mass-throughput genotyping with SNPs, including human genome-wide scans. Adv. Genet. 60, 107–139.

Martinelli, A., Hunt, P., Cheesman, S.J., Carter, R., 2004. Amplified fragment length polymorphism measures proportions of malaria parasites carrying specific alleles in complex genetic mixtures. Mol. Biochem. Parasitol. 136, 117–122.

Miller, L.H., Su, X., 2011. Artemisinin: discovery from the Chinese herbal garden. Cell 146, 855–858.

Miotto, O., Almagro-Garcia, J., Manske, M., Macinnis, B., Campino, S., Rockett, K.A., et al., 2013. Multiple populations of artemisinin-resistant *Plasmodium falciparum* in Cambodia. Nat. Genet. 45, 648–655.

Moody, A., 2002. Rapid diagnostic tests for malaria parasites. Clin. Microbiol. Rev. 15, 66–78.

Moore, S.J., Min, X., Hill, N., Jones, C., Zaixing, Z., Cameron, M.M., 2008. Border malaria in China: knowledge and use of personal protection by minority populations and implications for malaria control: a questionnaire-based survey. BMC Public Health 8, 344.

Mu, J., Awadalla, P., Duan, J., McGee, K.M., Keebler, J., Seydel, K., et al., 2007. Genome-wide variation and identification of vaccine targets in the *Plasmodium falciparum* genome. Nat. Genet. 39, 126–130.

Murray, C.J., Rosenfeld, L.C., Lim, S.S., Andrews, K.G., Foreman, K.J., Haring, D., et al., 2012. Global malaria mortality between 1980 and 2010: a systematic analysis. Lancet 379, 413–431.

Neafsey, D.E., Galinsky, K., Jiang, R.H., Young, L., Sykes, S.M., Saif, S., et al., 2012. The malaria parasite *Plasmodium vivax* exhibits greater genetic diversity than *Plasmodium falciparum*. Nat. Genet. 44, 1046–1050.

Noedl, H., Se, Y., Schaecher, K., Smith, B.L., Socheat, D., Fukuda, M.M., 2008. Evidence of artemisinin-resistant malaria in western Cambodia. N. Engl. J. Med. 359, 2619–2620.

Olliaro, P., Wells, T.N., 2009. The global portfolio of new antimalarial medicines under development. Clin. Pharmacol. Ther. 85, 584–595.

Olotu, A., Fegan, G., Wambua, J., Nyangweso, G., Awuondo, K.O., Leach, A., et al., 2013. Four-year efficacy of RTS,S/AS01E and its interaction with malaria exposure. N. Engl. J. Med. 368, 1111–1120.

Pan, D., Hu, J., Ma, Q., Pan, W., Li, M., 2010. Diversity and prevalence of the C-terminal region of *Plasmodium falciparum* merozoite surface protein 1 in China. Acta Trop. 116, 200–205.

Pan, J.Y., Zhou, S.S., Zheng, X., Huang, F., Wang, D.Q., Shen, Y.Z., et al., 2012. Vector capacity of *Anopheles sinensis* in malaria outbreak areas of central China. Parasit. Vectors 5, 136.

Pan, W., Huang, D., Zhang, Q., Qu, L., Zhang, D., Zhang, X., et al., 2004. Fusion of two malaria vaccine candidate antigens enhances product yield, immunogenicity, and antibody-mediated inhibition of parasite growth in vitro. J. Immunol. 172, 6167–6174.

Park, S.K., Lee, K.W., Hong, S.H., Kim, D.S., Lee, J.H., Jeon, B.H., et al., 2003. Development and evaluation of an immunochromatographic kit for the detection of antibody to *Plasmodium vivax* infection in South Korea. Yonsei Med. J. 44, 747–750.

Phyo, A.P., Nkhoma, S., Stepniewska, K., Ashley, E.A., Nair, S., McGready, R., et al., 2012. Emergence of artemisinin-resistant malaria on the western border of Thailand: a longitudinal study. Lancet 379, 1960–1966.

Qian, F., Pan, W., 2002. Construction of a tetR-integrated *Salmonella enterica* serovar Typhi CVD908 strain that tightly controls expression of the major merozoite surface protein of *Plasmodium falciparum* for applications in human Vaccine production. Infect. Immun. 70, 2029–2038.

Roestenberg, M., McCall, M., Hopman, J., Wiersma, J., Luty, A.J., van Gemert, G.J., et al., 2009. Protection against a malaria challenge by sporozoite inoculation. N. Engl. J. Med. 361, 468–477.

Roper, C., Elhassan, I.M., Hviid, L., Giha, H., Richardson, W., Babiker, H., et al., 1996. Detection of very low level *Plasmodium falciparum* infections using the nested polymerase chain reaction and a reassessment of the epidemiology of unstable malaria in Sudan. Am. J. Trop. Med. Hyg. 54, 325–331.

Schwartz, L., Brown, G.V., Genton, B., Moorthy, V.S., 2012. A review of malaria vaccine clinical projects based on the WHO rainbow table. Malar. J. 11, 11.

Seder, R.A., Chang, L.J., Enama, M.E., Zephir, K.L., Sarwar, U.N., Gordon, I.J., et al., 2013. Protection against malaria by intravenous immunization with a nonreplicating sporozoite vaccine. Science 341, 1359–1365.

Sheng, H.F., Zhou, S.S., Gu, Z.C., Zheng, X., 2003. Malaria situation in the People's Republic of China in 2002. Chin. J. Parasitol. Parasit. Dis. 21, 193–196 (in Chinese).

Su, X., Wellems, T.E., 1996. Toward a high-resolution *Plasmodium falciparum* linkage map: polymorphic markers from hundreds of simple sequence repeats. Genomics 33, 430–444.

Su, X.Z., Ferdig, M.T., Huang, Y., Huynh, C.Q., Liu, A., You, J., et al., 1999. A genetic map and recombination parameters of the human malaria parasite *Plasmodium falciparum*. Science 286, 1351–1353.

Su, X.Z., Hayton, K., Wellems, T.E., 2007. Genetic linkage and associattion analyses for trait mapping in *Plasmodium falciparum*. Nat. Rev. Genet. 8, 497–506.

Su, X.Z., Wellems, T.E., 1997. *Plasmodium falciparum*: a rapid DNA fingerprinting method using microsatellite sequences within var clusters. Exp. Parasitol. 86, 235–236.

Suh, I.B., Lee, K.H., Kim, Y.R., Woo, S.K., Kang, H.Y., Won, Y.D., et al., 2004. Comparison of immunological responses to the various types circumsporozoite proteins of *Plasmodium vivax* in malaria patients of Korea. Microbiol. Immunol. 48, 119–123.

Takala, S.L., Coulibaly, D., Thera, M.A., Dicko, A., Smith, D.L., Guindo, A.B., et al., 2007. Dynamics of polymorphism in a malaria vaccine antigen at a vaccine-testing site in Mali. PLoS Med. 4, e93.

Tao, Z.Y., Zhou, H.Y., Xia, H., Xu, S., Zhu, H.W., Culleton, R.L., et al., 2011. Adaptation of a visualized loop-mediated isothermal amplification technique for field detection of *Plasmodium vivax* infection. Parasit. Vectors 4, 115.

Volkman, S.K., Sabeti, P.C., DeCaprio, D., Neafsey, D.E., Schaffner, S.F., Milner Jr., D.A., et al., 2007. A genome-wide map of diversity in *Plasmodium falciparum*. Nat. Genet. 39, 113–119.

Wang, J., Huang, L., Li, J., Fan, Q., Long, Y., Li, Y., Zhou, B., 2010. Artemisinin directly targets malarial mitochondria through its specific mitochondrial activation. PLoS One 5, e9582.

Wang, Z.Y., Jiang, L., Cai, L., Wang, W.J., Zhang, Y.G., Hong, G.B., et al., 2012. Analysis and establishment of loop-mediated isothermal amplication for the diagnosis of *Plasmodium vivax*. J. Trop. Med 12, 157–161 (in Chinese).

Wells, T.N., Burrows, J.N., Baird, J.K., 2010. Targeting the hypnozoite reservoir of *Plasmodium vivax*: the hidden obstacle to malaria elimination. Trends. Parasitol 26, 145–151.

White, N.J., 2011. A vaccine for malaria. N. Engl. J. Med. 365, 1926–1927.

WHO, 1999. New Perspectives: Malaria Diagnosis. WHO, Geneva, Switzerland.

WHO, 2006. Guidelines for the Treatment of Malaria. WHO, Geneva, Switzerland.

Winzeler, E.A., 2008. Malaria research in the post-genomic era. Nature 455, 751–756.

Xia, S., Allotey, P., Reidpath, D.D., Yang, P., Sheng, H.F., Zhou, X.N., 2013. Combating infectious diseases of poverty: a year on. Infect. Dis. Poverty 2, 27.

Xu, J., Liu, H., 1997. Border malaria in yunnan, China. Southeast Asian J. Trop. Med. Public Health 28, 456–459.

Xu, J., Salas, M., 2003. Moving the Periphery to the Centre: Indigenous People, Culture and Knowledge in Changing Yunnan. Rockefeller Foundation, Bangkok.

Xue, X., Ding, F., Zhang, Q., Pan, X., Qu, L., Pan, W., 2010. Stability and potency of the *Plasmodium falciparum* MSP1-19/AMA-1(III) chimeric vaccine candidate with Montanide ISA720 adjuvant. Vaccine 28, 3152–3158.

Yang, G.J., Gao, Q., Zhou, S.S., Malone, J.B., McCarroll, J.C., Tanner, M., et al., 2010. Mapping and predicting malaria transmission in the People's Republic of China, using integrated biology-driven and statistical models. Geospat. Health 5, 11–22.

Yang, G.J., Tanner, M., Utzinger, J., Malone, J.B., Bergquist, R., Chan, E.Y., et al., 2012. Malaria surveillance-response strategies in different transmission zones of the People's Republic of China: preparing for climate change. Malar. J. 11, 426.

Yang, Y., Zhang, G., Liu, H., Zhang, Z., Gao, B., 2004. Study on MSP-2 gene polymorphism of different geographical strains of *Plasmodium falciparum* in Yunnan. Chin. Trop. Med. 4, 901–906 (in Chinese).

Yang, Y.M., Li, X.P., Zhang, C.L., Li, B.F., Liu, H., Li, L., 2013. Creation of cloned strains of chloroquine-resistant *Plasmodium falciparum* from Yunnan, China with resistance to dihdroartemisinin. J. Path. Biol. 8, 714–717 (in Chinese).

Yi, H.H., Xu, B., Fang, C., Song, Y.W., Wu, P.L., Wang, Y.F., et al., 2010. Study on the detection of *Plasmodium falciparum* by fluorescent quantitative loop-mediated isothermal amplication. Chin. Trop. Med 10, 1178–1180 (in Chinese).

Zhang, C., Zhou, H., Wang, J., Liu, H., 2012. In vitro sensitivity of *Plasmodium falciparum* isolates from China-Myanmar border region to chloroquine piperaquine and pyronaridine. Chin. J. Parasitol. Parasit. Dis. 30, 41–44 (in Chinese).

Zhang, D., Pan, W., 2005. Evaluation of three Pichia pastoris-expressed *Plasmodium falciparum* merozoite proteins as a combination vaccine against infection with blood-stage parasites. Infect. Immun. 73, 6530–6536.

Zheng, Q., Vanderslott, S., Jiang, B., Xu, L.L., Liu, C.S., Huo, L.L., et al., 2013. Research gaps for three main tropical diseases in the People's Republic of China. Infect. Dis. Poverty 2, 15.

Zhou, S.S., Tang, L.H., Sheng, H.F., Wang, Y., 2006. Malaria situation in the People's Republic of China in 2004. Chin. J. Parasitol. Parasit. Dis. 24, 1–3 (in Chinese).

Zhou, X., Li, S.G., Chen, S.B., Wang, J.Z., Xu, B., Zhou, H.J., et al., 2013a. Co-infections with Babesia microti and *Plasmodium* parasites along the China-Myanmar border. Infect. Dis. Poverty 2, 24.

Zhou, X.N., Bergquist, R., Tanner, M., 2013b. Elimination of tropical disease through surveillance and response. Infect. Dis. Poverty 2, 1.

Zhu, H.W., Cao, J., Zhou, H.Y., Li, J.L., Zhu, G.D., Gu, Y.P., et al., 2010. Detection of *Plasmodium vivax* sporozoites-carrying mosquitoes using loop-mediated isothermal amplication (LAMP). Chin. J. Schisto. Control 22, 158–163 (in Chinese).

Zhu, X.P., Zhang, X.M., Zhou, L., Yang, Y.P., Gao, X., 2002. Sequence analysis and genotypes of glutamate rich protein of *Plasmodium falciparum* isolates from different malaria endemic areas in China. Biomed. Environ. Sci. 15, 1–7.

CHAPTER SIX

Approaches to the Evaluation of Malaria Elimination at County Level: Case Study in the Yangtze River Delta Region

Min Zhu[1,§], Wei Ruan[2,§], Sheng-Jun Fei[3], Jian-Qiang Song[4], Yu Zhang[5], Xiao-Gang Mou[6], Qi-Chao Pan[1], Ling-Ling Zhang[2], Xiao-Qin Guo[3], Jun-Hua Xu[4], Tian-Ming Chen[5], Bin Zhou[6], Peiling Yap[7], Li-Nong Yao[2,*], Li Cai[1,*]

[1]Shanghai Municipal Center for Disease Control & Prevention, Shanghai, People's Republic of China
[2]Zhejiang Provincial Center for Disease Control & Prevention, Hangzhou, People's Republic of China
[3]Songjiang District Center for Disease Control & Prevention, Shanghai, People's Republic of China
[4]Zhabei District Center for Disease Control & Prevention, Shanghai, People's Republic of China
[5]Haiyan County Center for Disease Control & Prevention, Haiyan, People's Republic of China
[6]Anji County Center for Disease Control & Prevention, Anji, People's Republic of China
[7]Department of Epidemiology and Public Health, Swiss Tropical and Public Health Institute, Basel, Switzerland; University of Basel, Basel, Switzerland
*Corresponding authors: E-mail: ylinong@163.com; Caili@scdc.sh.cn

Contents

1. Background	137
2. Introduction	138
2.1 Introduction of Shanghai and Zhejiang	138
2.2 History of epidemic situation	140
2.2.1 Parasite species	140
2.2.2 Transmission vector	141
2.2.3 Transmission trends	141
2.3 Process of control activities	142
3. Pilot Project of the National Malaria Elimination Programme	143
3.1 General information of the pilot counties/districts	143
3.1.1 Zhabei district, Shanghai	143
3.1.2 Songjiang district, Shanghai	143
3.1.3 Haiyan County, Zhejiang	145
3.1.4 Anji County, Zhejiang Province	145
3.2 Goal of the pilot project	146
3.3 Measures	146
3.4 Targets	147
3.5 Time schedule	147

§Min Zhu and Wei Ruan contributed equally

- **3.6** Baseline survey — 148
  - *3.6.1 Sampling method* — *148*
  - *3.6.2 Detailed content of survey* — *148*
  - *3.6.3 Quality control* — *149*
- **3.7** Assessment and evaluation — 149
  - *3.7.1 Self-evaluation and application of county/district* — *149*
  - *3.7.2 Method of assessment and evaluation* — *150*
- **3.8** Organization and guarantee — 151
  - *3.8.1 Policy guarantee* — *151*
  - *3.8.2 Setup of particular sections* — *156*
  - *3.8.3 Technical guarantee* — *156*
  - *3.8.4 Financial guarantee* — *157*
- **4.** Results and Achievements — 158
  - **4.1** Main indicators for evaluation of the NMEP — 158
    - *4.1.1 Malaria epidemic situation* — *158*
    - *4.1.2 Case detection* — *158*
    - *4.1.3 Residents' awareness of malaria control knowledge and utilization of mosquito-proof facilities* — *159*
    - *4.1.4 Malaria vector* — *161*
  - **4.2** Improving health services and raising control capacity — 161
    - *4.2.1 Establishing work mechanism and network platform* — *161*
    - *4.2.2 Equipment, supplies and fund guarantee* — *163*
    - *4.2.3 Professional skill improvement of personal of disease control and prevention* — *163*
  - **4.3** Implementation of main measures — 164
    - *4.3.1 Finding the malaria cases* — *164*
    - *4.3.2 Management of malaria cases* — *164*
    - *4.3.3 Survey of case and focus* — *164*
    - *4.3.4 Professional training* — *168*
    - *4.3.5 Management of mobile populations* — *168*
    - *4.3.6 Health education* — *170*
    - *4.3.7 Quality control of blood examination* — *170*
    - *4.3.8 Supervision of the process of the NMEP* — *170*
  - **4.4** Evaluation and certification — 171
    - *4.4.1 Assessment score* — *171*
    - *4.4.2 Summary of assessment and evaluation* — *171*
  - **4.5** Experiences — 173
    - *4.5.1 Disease control network* — *173*
    - *4.5.2 Sector cooperation* — *174*
    - *4.5.3 Finding malaria cases* — *174*
    - *4.5.4 Funding mechanism* — *175*
  - **4.6** Lessons — 175
    - *4.6.1 Number of blood examinations* — *175*
    - *4.6.2 Selection of blood examination methods* — *177*
    - *4.6.3 People's awareness of malaria control knowledge* — *178*
    - *4.6.4 Malaria vector surveillance* — *179*
- **5.** Conclusions — 180
- Acknowledgements — 180
- References — 180

Abstract

As the progress on transition from malaria control to malaria elimination in the People's Republic of China (P.R. China), four counties/districts, namely Zhabei District and Songjiang District of Shanghai municipality, and Anji County and Haiyan County of Zhejiang Province, representatives of the Yangtze River Delta region, were included in the pilot project of the national malaria elimination programme in P.R. China. A baseline survey was conducted first. The main measures performed were blood examination of febrile cases, improving the information management system of malaria cases, providing standard diagnosis and treatment, standardized disposal of epidemic focus, and health education and health promotion, strengthening the management of mobile population, etc. All the measures were assessed and evaluated through data examination and on-site investigation. In the whole process of the pilot project, quality control was especially emphasized. During the implementation of pilot project, the three-level control system was improved, professional staff was enriched and the working fund was ensured (a total fund of RMB 2,923,600). Thirty-nine training courses were conducted. Among 102,451 febrile cases receiving blood examination, all of the 23 malaria cases were confirmed as imported from other provinces or foreign countries. All the epidemic foci were surveyed and some control measures were carried out. Various health education and promotion activities were carried out including publicizing malaria control knowledge through news media, newspapers and periodicals and networks. Assessment and evaluation of the project was done by the Zhejiang and Shanghai Government, comprehensive score was >95 points under the evaluation system which indicated all four pilot counties/districts had first achieved the goal of elimination of malaria in P.R. China. Experiences and lessons about the measures carried out in the project were discussed.

1. BACKGROUND

Malaria has been prevalent in China for a long time. The character of 'malaria' was found in the oracle and bronze inscriptions in 2003–1006 BC (Tang et al., 2012). In the following 4000 years, a large amount of information about the epidemics, damage, treatment and pathology of malaria or malaria-like diseases in China has been recorded in a variety of literature. Before the founding of the People's Republic of China (P.R. China), malaria was seriously prevalent in different provinces; the annual national incidence was over 30 million, and about three-fourths of the Chinese people were at the threat of malaria (Shang et al., 2012). After the founding of P.R. China, through parasitological treatment, vector control and other effective measures, the human incidence of malaria has been brought under control, and the percentage of malaria cases in the total number of infectious diseases has declined (Liu, 2014); the number of counties without cases or with low incidence have increased, the areas with falciparum malaria transmission have decreased remarkably, and more and more counties have reached the criteria for basic elimination of malaria (annual incidence less than 10 per 100,000). Shanghai and Zhejiang reached the criteria for basic elimination of malaria at the provincial level in 1986 and 1991, respectively. In 2009, 12 counties/districts of six provinces/municipalities including Shanghai and Zhejiang, started the pilot project under the national malaria elimination programme (NMEP). This chapter summarizes the pilot work of Shanghai and Zhejiang as representatives of the Yangtze River Delta region.

2. INTRODUCTION

2.1 Introduction of Shanghai and Zhejiang

Shanghai is located in the lower reaches of the Yangtze River, a flood plain of the Yangtze River Estuary, in the east edge of the Yangtze River Delta, at 30°40′–31°41′ north latitude and 120°58′–121°44′ east longitude. The city has a land area of 6340.5 km^2, accounting for 0.06% of the country's total area; water covers 12% of the total area, which is a typical plain region with a network of watercourses (Figure 6.1). The average annual temperature was about 17.8 °C in 2009, the highest monthly average temperature was in June (26.3 °C), and the lowest monthly average temperature was in January (4.3 °C). The annual precipitation was 1457.9 mm, mostly in June–September. At present, the municipality has 16 districts and 1 county. At the end of 2009, it had a total permanent residence of 19,213,200, of which 14,007,000 were registered permanent residence. Shanghai is one of the largest centers of economy, finance, trade and shipping in P.R. China. From the industrial structure, the city's gross domestic product (GDP) in 2009 was 1504.65 billion Renminbi (RMB) yuan. It has a grain cropping area of 193,300 ha and paddy cultivation area of 169,800 ha, and the per capita GDP was 78,989 RBM yuan (Shanghai Municipal Statistics Bureau, 2010).

Zhejiang Province is located in the south wing of the Yangtze River Delta, facing the East China Sea in the east, bordering Fujian Province in the south, Jiangxi Province and Anhui Province in the west, and Shanghai Municipality and Jiangsu Province in the north, at 27°01′–31°10′ north latitude and 118°01′–123°08′ east longitude. It has 101,800 km^2 land area (Figure 6.1). Zhejiang Province has a subtropical monsoon climate, warm and humid with four distinct seasons. Its annual average temperature is 15–19 °C, and its isotherm is roughly parallel to the longitudinal axis, with a frost-free period that lasts 230–270 days. The average annual precipitation was 1640 mm, mainly in April–September. Zhejiang Province has 11 cities and

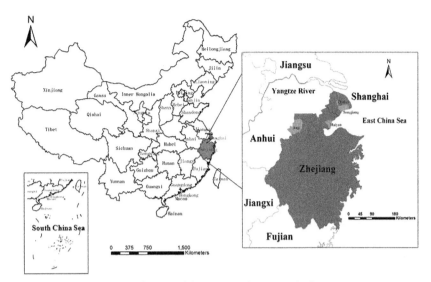

Figure 6.1 Sketch map of the geographic area of pilot counties.

90 counties. At the end of 2009, its permanent residence was 51,800,000; among them 47,161,800 were registered population. Its agricultural production is dominated by grain planting, and the northern part of Zhejiang is known as a granary. In the 1950s, a double cropping system mainly with single harvest rice was adopted; then in the late 1960s, a triple-cropping system was practiced, single harvest rice was changed to double harvest rice, and dry farmland was changed to paddy field, then returned to single harvest rice in the late 1990s. In 2009, the provincial GDP was 2299.04 billion RBM yuan, and per capita GDP was 44,641 RBM yuan. It has a grain cropping area of 1,290,100 ha (Investigation Term of Zhejiang, 2010).

Both Shanghai and Zhejiang have an excellent public health insurance system and strong financial support. They have established medical institutions covering the entire administrative area, and a disease control system covering all administrative villages. In 2009, there were 296 and 652 general hospitals, 284 and 2006 health centers or community health service centers, 1447 and 13,922 village clinics, 164,774 and 255,611 all kinds of medical and health personnel, 9.48 and 5.65 health technical workers per 1000 people, 21 and 101 centers for disease control and prevention, respectively, in the two places. In 2009, the financial expenditure was 13.285 billion and 17.705 billion RBM yuan, respectively, in the two places, and per capita health input was 391.30 and 431.80 RBM yuan, respectively. Since both places have excellent medical and disease control systems, the incidence of relevant infectious diseases has been controlled at the historically lowest levels. In 2009, the incidence of Class A and B notifiable infectious diseases was 221.26 per 100,000 and 324.60 per 100,000, respectively, and malaria incidence was only 0.31 per 100,000 and 0.43 per 100,000 respectively (MOH, 2010b).

2.2 History of epidemic situation

Malaria has been prevalent for a long time in China, covering a vast area, and epidemics occurred frequently, greatly endangering the population's health. In the 1930s, a survey of residents revealed that the malaria prevalence rate was as high as 32% in the residence of Gaoqiao Town, Shanghai, where malaria cases accounted for 28% of the total outpatients, spleen rate of the healthy primary school students was 18.3% and parasite rate was 5.2%, including *Plasmodium vivax* (52.4%), *P. malaria* (40.1%) and *P. falciparum* (7.5%) (Cai et al., 2012; Zheng, et al., 2013).

2.2.1 Parasite species
2.2.1.1 Vivax malaria
The vivax malaria is the most widely distributed with the longest transmission. During last 40 years after the founding of P.R. China, vivax malaria occurred every year, usually was endemic, and the outbreak occurred under optimal conditions (Cai et al., 2012; Yao and Yu 2012).

2.2.1.2 Falciparum malaria
The falciparum malaria was only prevalent in the early days after the founding of P.R. China. In the 1950s, it was endemic in 11 prefectures/cities of Zhejiang Province, but its distribution range has diminished greatly since 1964; the last case was found in 1971. In Shanghai, there were only a few case reports in the middle and late 1960s, and all were imported cases (Cai et al., 2012; Yao and Yu 2012).

2.2.1.3 Quartan malaria
The quartan malaria mainly occurred before the founding of P.R. China and became very rare afterwards. In 1935, a survey of the residents in Gaoqiao Town of Shanghai revealed that quartan malaria cases accounted for 40.1%. Since 1949, no quartan malaria has been found in Shanghai. In Zhejiang Province, it nearly disappeared in the late 1950s and early 1960s, and the last two cases were found in Anji County in 1961 and 1963 respectively (Cai et al., 2012; Yao and Yu 2012).

2.2.1.4 Ovale malaria
Ovale malaria cases were recognized in south area of China sporadicaly, but no ovale malaria patients was reported in Shanghai and Zhejiang. (Cai et al., 2012; Yao and Yu 2012).

2.2.2 Transmission vector
Based on historical data of the survey, there were two species of mosquito transmitting malaria in Shanghai and Zhejiang, namely, *Anopheles sinensis* and *An. anthropophagus* (Cai et al., 2012; Yao and Yu 2012).

2.2.2.1 Anopheles sinensis
Anopheline mosquito is widely distributed in large numbers, a semi-domestic species, mainly feeding on domestic animals, such as cows, pigs, etc. and also on humans. Generally, it had two peaks of night activity, at 19:30–21:30 and 0:30–2:30. Human blood proportion was 0.011–0.082, and its vectorial capacity in transmission season was 0.014–0.316 (Cai et al., 2012; Yao and Yu 2012).

2.2.2.2 Anopheles anthropophagus
In Shanghai, *An. anthropophagus* was found in Jinshan District adjacent to the north of Zhejiang, but no more was found after the mid-1960s; it had been also found in Songjiang District. The distribution of *An. anthropophagus* in North Zhejiang Plain joined with Anhui and Jiangsu, and the local density of *An. anthropophagus* was higher than that of *An. sinensis*; it was in dotted distribution in southeast part of Zhejiang. Since the early 1970s, the distribution of *An. anthropophagus* in Zhejiang gradually reduced, and only a few of the *An. anthropophagus* species were found in the survey in north Zhejiang in 1974; none have been found since 1984 (Shang, 2008; Liu, 1992; Cai et al., 2012; Yao and Yu 2012).

2.2.3 Transmission trends
Historically, the region of Shanghai and Zhejiang belonged to unstable mesoendemic and hypoendemic areas of malaria, and had two transmission patterns, namely endemic and epidemic outbreaks. The malaria transmission season was in May–October, especially in July–September. Three big epidemic outbreaks occurred in the two places in the 1950s, 1960s and 1970s, and each outbreak continued for 2–6 years at an interval of 8–10 years. The first outbreak occurred in 1952–1956, mainly with vivax malaria, and the malaria incidence was more than 50,000 per 100,000 in some individual natural villages. The second outbreak occurred in 1961–1966, and in

Figure 6.2 Malaria incidence in Shanghai and Zhejiang in 1950–2008.

addition to vivax malaria, falciparum malaria was also found in some areas, and some cases even died. The highest annual incidence rate was 3231.2 per 100,000 in Shanghai and 3300.9 per 100,000 in Zhejiang. The incidence in some counties was as high as 5631.2 per 100,000. The third outbreak occurred in the early 1970s with short duration and weak prevalence, and was effectively control after three years. During that period of time, 30,441 malaria cases (Shanghai, 1971–1973) and 613,807 malaria cases (Zhejiang, 1972–1974) were reported; the highest annual incidence rate was 98.20 per 100,000 (Shanghai, 1972) and 747.18 per 100,000 (Zhejiang, 1973). After 1974, malaria prevalence declined in both places, and remained low endemic. The malaria incidence reached <1 per 100,000 by county as a unit, respectively, in 1983 (Shanghai) and 1988 (Zhejiang) (Figure 6.2) (Cai et al., 2012; Yao and Yu 2012).

2.3 Process of control activities

Based on the principle of classified guidance for malaria control, Shanghai and Zhejiang adopted the integrated control measures mainly to eliminate the source of infection, combined with the patriotic health campaign to kill mosquitoes and improve mosquito-proof facilities; strengthen the management of moving population; and implement corresponding control measures according to malaria prevalence and process control programmes to promote malaria control work gradually (Yao et al., 2013; Zheng et al., 2013; Ning et al., 1999; Liu et al., 2014).

Malaria control in Shanghai and Zhejiang has basically experienced four stages; through over 60 years of control activities, malaria was brought under effective control, and passed the assessment and evaluation of expert panel of Ministry of Health, respectively, in 1986 and 1993, and then was certified as reaching the criteria for basic elimination of malaria. Since then, both places adhered to malaria control with emphasis on malaria surveillance; malaria incidence has remained stable under 1 per 100,000, with local malaria cases reducing gradually, but imported cases increasing year by year (Table 6.1) (Cai et al., 2012;Yao andYu 2012).

3. PILOT PROJECT OF THE NATIONAL MALARIA ELIMINATION PROGRAMME

In order to explore the feasibility in the implementation of NMEP, accumulate experience in treatment of the imported cases, establish assessment and evaluation system of the NMEP and promote the nationwide NMEP work, based on the epidemic situation of malaria at present in China, the Ministry of Health (MOH) initiated the pilot project of the NMEP in China (MOH, 2009) in November 2009, implementing it first in 12 counties/districts from Shanghai and Zhejiang and in another four provinces. Zhabei District, Songjiang District of Shanghai, and Anji County and Haiyan County of Zhejiang were included in the pilot project representing the Yangtze River Delta region.

3.1 General information of the pilot counties/districts

3.1.1 Zhabei district, Shanghai

It is located in the north of the central area of Shanghai. In 2009, the total number of permanent residents was 760,300, medical and health expenditure was 245.31 million RMB yuan, and community health prevention funding was 500,000 RMB yuan per 10,000 permanent residents (Government of Zhabei District). Malaria reports of Zhabei District date back to 1952; its malaria incidence reached the highest level 1364 per 100,000 in 1962 (8883 cases), and declined rapidly afterwards, kept less than 1 per 100,000 since 1987. In 2001–2009, 15 imported malaria cases were managed (Figure 6.3).

3.1.2 Songjiang district, Shanghai

It is located in the southwest of Shanghai. In 2009, the total number of permanent residents was 849,430, among them 559,442 were registered population (Government of Songjiang District). The epidemic situation of malaria in Songjiang District was unstable, a malaria outbreak occurred in the early 1960s, and the incidence was as high as 5018.31 per 100,000; afterwards it turned stable, and the annual malaria incidence fluctuated at low levels. Songjiang District reached the criteria for basic elimination of malaria (annual incidence leas than 10 per 100,000) in 1985; and since 2000,

Table 6.1 Comparison of malaria control of different periods in Shanghai and Zhejiang

Region	Period	Time	Strategy	Main measures
Shanghai	Control of transmission	1952–1966	Integrated control	Establish organization, formulate control programme, personnel training, case detection, anti-relapse treatment, vector control, health education
	Consolidation and elimination of transmission	1967–1981	Control with case management	Management of source of infection, intermittent period treatment, prophylactic medication on target population
	Transmission elimination	1982–1986	Assessment and evaluation	Improve management system, eliminate source of infection, manage malaria in moving population
	Surveillance and maintenance	1987–present	Intensified surveillance	Comprehensive surveillance, timely detection and treatment of cases, explore key work in surveillance period
Zhejiang	Early stage of control programme	1950–1957	Drug treatment of patients and mosquito control	Set up institutions and organize teams, carry out pilot project, sum up experience
	Control of transmission	1958–1978	Integrated control measures with emphasis on control and elimination of source of infection	Formulate work plan, detect and treat patients in timely manner, intensify epidemic report, provide prophylactic drug administration, control vector density
	Transmission elimination	1978–1991	Consolidate the achievements and prevent recurrence, assessment and evaluation	Strengthen surveillance network, set up county level microscopic station, manage malaria in moving population, assess basic elimination of malaria
	Surveillance and maintaince	1992–present	Comprehensive surveillance	Standardize process, detect patients, implement surveillance measures

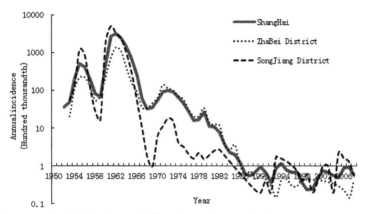

Figure 6.3 Malaria incidence in Shanghai and two pilot districts in 1950–2008.

the annual malaria incidence has declined to less than 2 per 100,000. In 2000–2009, a total of 58 malaria cases were reported, among them 38 were vivax malaria, 19 were unidentified and one was falciparum malaria. The epidemiological investigation revealed that 44 cases got infections in other provinces and onset in local places, eight cases were locally infected, and six cases were unknown about their infection place (Figure 6.3).

3.1.3 Haiyan County, Zhejiang

It is affiliated with Jiaxing City of Zhejiang Province. In 2009, it had a total registered population of 370,702 (Government of Haiyan County, 2009). It was a mesoendemic area of malaria, two incidence peaks appeared in 1954 and 1962, respectively, and the incidence was 3009.89 per 100,000 and 4998.46/100,000, respectively. Through parasitological treatment and large-scale patriotic health campaign and mosquito control, the prevalence was under effective control. Since 1980, malaria surveillance has been carried out in the whole county, and malaria incidence declined to 0.3 per 100,000 in 1985. In 1987, it reached the criteria for basic elimination of malaria. By the end of 2011, no local infected malaria case had been found for continuous 23 years in the whole county, and no case died of malaria (Figure 6.4).

3.1.4 Anji County, Zhejiang Province

It is affiliated with Huzhou City of Zhejiang Province. In 2009, it had a total registered population of 457,700 (Government of Anji). It was a hyperendemic area of malaria; three malaria pandemics occurred, respectively, in 1951–1956, 1962–1963 and 1971–1974. The largest outbreak occurred in 1962–1963. In that year there were 74,908 malaria cases, and among

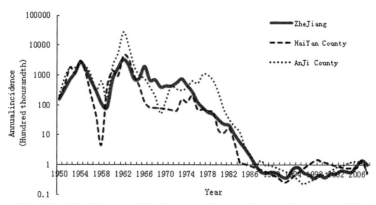

Figure 6.4 Malaria incidence in Zhejiang Province and two pilot county/districts in 1950–2008.

them, 8416 were falciparum malaria and 336 died; malaria vectors included *An. sinensis* and *An. anthropophagus*. In 1989, Anji County reached the criteria for basic elimination of malaria, and the malaria prevalence has remained stable; only 28 cases were found in the last 10 years, among them, 16 were local infected cases and 12 were imported cases. Of the 28 cases, 27 were vivax malaria and one was falciparum malaria. Effective control measures have been adopted and no new secondary infected cases due to imported malaria were found. At the same time, the surveillance results showed that in Anji County *An. sinensis* is a unique malaria vector, and no *An. anthropophagus* was found (Figure 6.4).

3.2 Goal of the pilot project

By the end of 2011, each pilot county/district (PCD) is tying to achieve or confirm that there would be no local infected malaria case, establish and improve the assessment system of the NMEP, and conduct assessment and evaluation of the entire work of the pilot project in 2012.

3.3 Measures

Measures would be implemented in the following nine aspects:
- Conduct blood examination of febrile cases to find as many malaria cases as possible;
- Improve the malaria epidemic information management system to ensure no underreporting of malaria cases;
- Provide standard medical services to ensure timely detection and standardized treatment of malaria cases;

- Conduct standardized disposal of epidemic focus to prevent spread of epidemic situation;
- Carry out health education and health promotion to raise people's awareness of malaria protection;
- Strengthen the moving population management to prevent the import and spread of malaria cases;
- Conduct a baseline data survey to facilitate the assessment of the results and sum up experiences;
- Strengthen data management to ensure the basis for assessment;
- Establish collaborating network of the NMEP to implement source and information sharing.

3.4 Targets

By the end of 2011, each PCD will complete the following indicators:
- Annual blood examination of febrile cases not less than 2% of the total population in the administrative area, and the number of blood examinations in May–October not less than 80% of the total annual blood examinations;
- Laboratory examination rate of malaria cases up to 100%;
- Report rate of malaria cases within 24 h after diagnosis up to 100%;
- Standard treatment rate of malaria cases up to 100%;
- Percentage of reported malaria cases receiving epidemiological investigation up to 100%;
- Awareness rate of the residents in epidemic focus on malaria control up to 90%;
- Residents' protection rate by mosquito-proof facilities up to 90%.

3.5 Time schedule

The pilot project is divided into three stages, namely, preparatory stage, implementation stage and assessment and evaluation stage. In preparatory stage (September–December 2009), the work plan was formulated and improved, PCD was determined, a kick-off meeting on the pilot project of the NMEP was held, the pilot work was arranged, and concentrated training of relevant personnel of the PCD was conducted. In implementation stage (January 2010–December 2011), baseline survey was carried out, various items of the work plan were organized and implemented, and the awareness of malaria control knowledge was investigated. In assessment and evaluation stage (January–December 2012), evaluation and summing up of the pilot project of malaria elimination (ME) was conducted, experience and lessons were discussed (Figure 6.5).

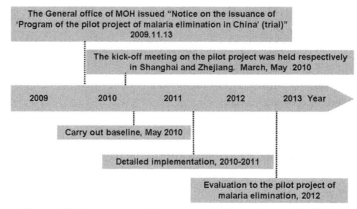

Figure 6.5 Sketch map of the process of pilot project of the NMEP.

3.6 Baseline survey

In order to get a comprehensive understanding of malaria control work done before, and investigate the ability of each PCD to conduct the NMEP and make preparations for the pilot project, baseline survey was carried out before implementation of the pilot project of the NMEP.

3.6.1 Sampling method

Cluster sampling was employed. The PCD was divided into five areas (east, south, west, north and central), then one township/community from each area was selected as a sampling spot. Based on different items of survey, one administrative village/neighborhood committee from each sampling spot was selected as an observation spot.

3.6.2 Detailed content of survey

3.6.2.1 Basic information

A unified questionnaire was used for investigating the basic information of the whole area and sampling spot in 2009.

3.6.2.2 Malaria incidence

In the observation spot, person's febrile condition within two weeks, malaria infection in 2009, and mosquito-proof facilities and their using status were investigated through household-entering surveys; blood smears and/or filter paper blood were collected from febrile cases to detect the malaria parasite. At the same time, one township/community primary school was selected for collecting filter paper blood, of more than 300 pupils, to test the antibody titer by performing indirect fluorescent antibody test (IFA) (antigen slide was provided by the National Institute of Parasitic Diseases, Chinese Centers for Disease Control and Prevention (CDC)).

3.6.2.3 Vector survey
In the observation spot, five households were selected from the direction of east, south, west, north and central areas, one room or corridor in the ground floor was selected from each household for catching mosquitos with an electric mosquito collector for 15 min, then the mosquitos were identified according to the classification key.

3.6.2.4 Investigation of residents' awareness about malaria control knowledge
From each observation spot, one natural village/resident's quarter was randomly drawn for carrying out the survey of residents' awareness of malaria control knowledge; more than 200 people were investigated.

3.6.3 Quality control
The provincial CDC would formulate a detailed implementation programme of baseline survey and unified questionnaire, conduct training at different levels, specify the sampling methods, and clearly define the sampling methods, survey data and collection methods, methods for various testing items, etc. A designated person was responsible for data collection, entry and checkup in every PCD.

3.7 Assessment and evaluation
3.7.1 Self-evaluation and application of county/district
Each PCD collated and summarized its data materials of the NMEP, and made conscientious summaries and analyses in the following aspects: natural, social, economic and health conditions in its administrative area; historical and recent 10 years' epidemic situation; malaria vector distribution and the last local infected malaria case; setup of institutions, personnel construction and financial support for guaranteeing the pilot work and capacity building; blood examination of febrile cases, management of malaria epidemic situation, standardized treatment of patients, investigation and treatment of epidemic focus, health education and publicity, surveillance of moving population and quality control, etc. Finally it was certified that all indicators had been completed through more than two years' pilot work, and all the data indicated that there was no local infected malaria case for three continuous years, and had achieved the goal of the NMEP. The administrative health

departments of PCD officially submitted the application to their provincial health department for assessment and evaluation of NMEP.

3.7.2 Method of assessment and evaluation

According to the 'Detailed Rules for ME Certification in Pilot Counties' (MOH, 2012), Shanghai Municipal Health Bureau and Zhejiang Provincial Health Department organized an evaluation team consisting of six experts on disease control, health administration, epidemiology, laboratory examination and vector control from MOH, other provinces and their own province. The evaluation team conducted assessment and evaluation in October–November 2012, after application of assessment and evaluation for four PCDs.

3.7.2.1 Briefing and discussion

After reading the self-evaluation report and hearing the briefing introduction of the work by different PCD, the evaluation team held discussions with the people from health administrative departments, the CDC, hospitals and community health service centers (CHSCs) on the implementation and the problems encountered in the process of NMEP.

3.7.2.2 Data assessment

The evaluation team came to interview with the authorities in the center for disease control and prevention, peoples' hospital and two randomly drawn CHSCs in the PCD, and checked up the data materials of the pilot project in the most recent three years, mainly including documents on relevant policy, setup of institutions, control staff, sources and use of working fund, working plan and summary, personnel training, blood examination and quality control, case detection and treatment, supervision and inspection, health education, etc.

3.7.2.3 On-site inspection

Four aspects were evaluated for the on-site inspection to each of PCD. First, it was assessment of the microscopist's skill. Five microscopists from county/district CDC, hospital and two CHSCs were selected to assesses their skill on examination of malaria parasites; each professional was asked to examine six blood slides in 1 h, and identify the parasite species on a positive blood smear provided by provincial CDC. Second, it was assessment of blood smears. All the positive and 30 negative blood smears kept since the beginning of pilot project of the NMEP were rechecked. Third, it was assessment of knowledge on diagnosis and treatment of malaria. Ten people were evaluated on their knowledge about malaria diagnosis and treatment; six were clinicians from county/district hospital and four were general practitioners from two CHSCs (two for each center); 10 examination questions randomly drawn from the question bank by the examination team were used for the examination. Fourth, it was investigation of residents' knowledge about malaria control and mosquito-proof measures. Fifty households in the village with the last reported malaria case in the PCD were visited to answer a questionnaire survey on the awareness of malaria control knowledge, and the use of mosquito-proof measures.

3.7.2.4 Assessment score

The evaluation team referred to the 'score form of the data materials of pilot project of the NMEP in Shanghai and Zhejiang' and ' score form of the on-site assessment of pilot project of NMEP in Shanghai and Zhejiang' (Tables 6.2 and 6.3) to score the working data and on-site assessment, and finally to calculate the total score of the assessment of the NMEP, by taking the score of data material and the score of on-site assessment as 50% of the total, respectively. When the comprehensive score ≥90 it would be regarded as qualified.

3.8 Organization and guarantee

Based on the characteristics of China's national conditions, for carrying out and implementing the various measures of the NMEP in the PCDs smoothly, the health administrative departments in Shanghai and Zhejiang gave strong support in policy making, institutions, personnel, funding and technical programmes, and many other supports.

3.8.1 Policy guarantee

All the governments and responsible departments of four PCDs paid attention to the NMEP, a leading group of the NMEP headed by the administrative department of public health and a technical steering group headed by the professional institutions were established at provincial and county levels. The programme for the pilot project and its implementation were formulated and issued based on the comprehensive evaluation of the control process and various conditions in different areas.

All of the PCDs practiced the working mechanism of 'government leadership, department cooperation and the masses participation'. 'Liaison meetings of public health' participated in by departments of health, education, business, tourism, entry and exit inspection and quarantine, public security and finance were established by the provincial and county governments,

Table 6.2 Score form for the data materials of pilot project of the NMEP in Shanghai and Zhejiang

Item	Details	Requirement	Rating criteria	Full score
Policy guarantee (10 point)	Policy	Formulate working plan for the pilot county of the NMEP	2 points for already formulated; 0 for not formulated	2
	Setup of institution	Set up sections responsible for malaria control	Must have clear division of responsibilities, if not no score	3
	Personnel	Establish leading group, technical team (or expert panel) and full-time and part-time personnel	2.5 points for having document, deduct 0.5 points if missing 1item	2.5
	Funds	Arrange special funds for the NMEP	2.5 points for having funding allocation vouchers for continuous 3years, deduct 1 point if missing 1year	2.5
Management of data (15 points)	Working plan and summary	Formulate working plan and summary every year since 2010	Deduct 2 points if missing 1 year, deduct 1 point if missing1 item	6
	Report form	Keep complete working report forms every year since 2010	Deduct 1 point if missing 1 year	6
	Filing archives	Filing by data type	Check up the filing archives, deduct points where appropriate	3
Training (15 points)	Training materials	At least one training course on malaria control and microscopic examination skill every year (including notice of training, check-in, training materials, examination, photo, summary)	Deduct 4 points if missing 1 training, deduct 0.5 point for each incomplete material	12
	Coverage rate of training staff	Medical institutions in the administrative area (township health center or community health service center, county level general hospital) coverage rate up to 95%	No deduction for coverage ≥90 every year, deduct 0.2 point for coverage 80~90%, deduct 0.4 for coverage 70~80%, and so on	3

Blood examination (20 points)	Rate of institutions which conducting blood examination	Medical institutions conducting blood examination (township health center or community health service center, county level general hospital) coverage rate up to 95%	No deduction for coverage ≥90 every year, deduct 1 point for coverage 80~90%, deduct 2 points for coverage 70~80%, and so on	4
	Number of blood examination	Annual number of blood examination of febrile cases ≥2% of the total population of the administrative area	No deduction for annual blood examination rate ≥2%, deduct 0.5 point for 1.5~2%, deduct 1 for 1.0~1.5%, and so on	10
	Number of blood examination in transmission season	Number of blood examination in May–October ≥80% of annual number of blood examinations	No deduction for annual percentage of blood examination in transmission season ≥80, deduct 0.5 point for 75~80%, deduct 1 for coverage 70~75%, and so on	3
	Quality control of blood examination	Annual number of blood slides rechecked ≥20% of annual number of blood examinations	No deduction for annual rechecked blood slides ≥20, deduct 1 point for 15~20%, deduct 1 point for coverage 10~15%, and so on	3
Disposal of malaria cases (20 points)	Timely report	≤24 h	Check the infectious diseases report network, calculate the time for report card generation and diagnostic time. Deduct 0.2 point for 1 overtime case	2
	Laboratory examination rate	100%	Deduct 1 point for missing 1 examination result, no score for no examination	3

Continued

Table 6.2 Score form for the data materials of pilot project of the NMEP in Shanghai and Zhejiang—cont'd

Item	Details	Requirement	Rating criteria	Full score
	Case survey	Completed in 4 workdays after report	Check the case questionnaire, deduction would be made according to the percentage of score points of each cases, deduct 25% of the score points of each case if not completed in time.	4
	Standard treatment	Refer to 'Principle of Antimalarial Medication'	Check the case questionnaire, deduct the points according to the percentage of score points of each case if no standard treatment	2
	Record and report of radical cure of vivax malaria in intermittent period	Radical cure in intermittent period on the vivax malaria cases managed in last year	2 points for keeping the records of radical cure in intermittent season for all cases or the records for missing follow-up cases since 2010, deduct the points according to percentage of score points of each case if no relevant records	2
	Epidemic focus survey	Including population survey and vector survey	4 points for keeping the survey records of all epidemic foci since 2010, deduction would be made according to the percentage of score points of each survey if the record is not available	4
	Vector control	Focal vector control in transmission season	3 points for keeping the records of vector control in all epidemic focus in transmission season since 2010, deduction would be made according to the percentage of score points of each focus if there is no relevant record	3
Supervision (10 points)	Supervision	Complete records, including notice, plan, check-in, record sheet, photos	At least 2 supervisions every year, deduct 3 points if missing 1 year's materials, deduct 0.5 point if missing 1 item every year	10
Health education (10 points)	Planning and summary	Corresponding plan and summary	Deduct 1 point if missing 1 year's materials, deduct 0.5 point if missing 1 item every year	3
	Health education	Complete work record, including check-in, publicity materials, photos and summary	At least 7 propaganda activities since 2010, deduct 1 point if missing 1 activity, deduct 0.2 point if missing 1 item of activity	7
Total				100

Table 6.3 Score form for the on-site assessment of pilot project of the NMEP in Shanghai and Zhejiang

Details	Evaluation method	Score calculation criteria	Full score
Assessment of microscopic examination skill	Randomly select five microscopists from county hospital, county CDC and two CHSCs for conducting microscopic examination skill on malaria parasite, each person would examine 6 blood slides in 1 h	30 points if all correct, deduct 10 points for 1 qualitative judgment error, and deduct 5 points for 1 error on parasite species	30
Knowledge on malaria control	Randomly select 10 clinicians from county hospital and CHSC to examine their knowledge about malaria diagnosis and treatment, 6 examination questions were randomly drawn from the question bank	20 points if all correct, deduct 0.3 point for 1 error/1 person	20
Assessment of blood smear	Recheck the positive blood smear kept since the beginning of the pilot project of NMEP (no more than 20 slides), and select 30 negative blood slides for recheck	30 points if all correct, deduct 5 points for 1 error	30
Investigation of mosquito-proof measure	Visit 50 households in the village (community) with the last reported malaria case in the county to investigate the use of mosquito-proof measures, including mosquito net, mosquito coils, repellent, screen door, screen window; it would be regarded as protection if any one of the measures was adopted	10 points for protection rate ≥90%, deduct 1 point for 85–90%, and deduct 2 points for 80–85%, and so on	10
Investigation of residents' awareness of NMEP knowledge	Visit 50 households in the village (community) with the last reported malaria case to conduct questionnaire survey (5 core knowledge questions)	It would be regarded as qualified if 3 or more questions were correctly answered, 10 points for qualification rate ≥90%, deduct 1 point for 85–90%, and deduct 2 points for 80–85%, and so on	10
Total			100

respectively; the responsibilities of different departments were well defined, meetings of leaders of membership units and liaisons were held at irregular intervals every year for coordination and organization of the implementation of relevant works. A multisectoral coordination mechanism was established. Administrative departments of public health at different levels included the NMEP as an important task, held kick-off meetings for the pilot project, signed the letter of responsibility of target management of the NMEP, worked out implementation programmes for the pilot project, and ensured accountability to the people.

3.8.2 Setup of particular sections

The CDCs at provincial and county levels were responsible for the specific work of pilot projects, and the measures of the NMEP were implemented through the three-level (province, county, township) network of disease control, and a particular section of parasitic disease control was established, respectively. There were 25 personnel of parasitic diseases control and 10 full-time malaria control personnel, at provincial level in Shanghai and Zhejiang, and 12 professional malaria control personnel in four pilot counties. A technical steering group was established at provincial and county levels to be responsible for the technical guidance and training.

In the PCD, various medical institutions at different levels were responsible for detecting, diagnosing and reporting of the malaria cases.

3.8.3 Technical guarantee

To facilitate implementation of unified measures of the NMEP, CDCs at different levels were assigned special responsibilities. The provincial CDC would help the administrative department of health formulate various plans, conducting technical guidance, training, supervision and evaluation, analysis of surveillance data, and quality control, checking and verifying every case, organizing monitoring, investigation of unreported cases, guide health education activities on malaria control and elimination, etc. The CDC at the county level would organize the implementation of various measures, including conducting epidemical surveys and disposal of case focus, checking and verifying the case, training, collecting and analyzing, conducting the quality control, etc. CHSCs would find malaria cases and help the staff of county CDC conduct the investigation of case and disposal of case focus, conduct health education on malaria control and prevention. Hospitals would find and treat the malaria case, and help the staff of county CDC to conduct the investigation of case.

County-level microscopy stations of malaria were established in every PCD for ensuring quality control of blood examination for the febrile cases. Blood slides of all the malaria cases found by medical institutions must be made and sent to CDCs at county and province/city levels for recheck; the negative blood slides of febrile cases were randomly rechecked at county/district and province/city levels. The county microscopic examination station randomly drew 20% of the negative blood films from each different medical institution for microscopic recheck, and blood film production was also scored every month; and again provincial and city microscopic examination stations would randomly draw 1% of the blood examinations for making a second microscopic recheck and scoring blood film production every month. The leading groups of the NMEP of each PCD would carry out supervision and inspection on the implementation of the work plan of the NMEP, so as to find and solve problems in time. The CDC in PCD would conduct one technical guidance and assessment on the relevant works of various health institutions in the administrative area.

3.8.4 Financial guarantee

During the pilot project, a financial guarantee mechanism of 'government investment as the main share, burden supporting by different level and multi-channel financing' was established gradually in Shanghai and Zhejiang. The governments in pilot provinces and counties/districts provided working funds based on their own real conditions, and at the same time, 'Chinese Central Government Subsidy for Malaria Control Project' and 'Global Fund to Fight AIDS, Tuberculosis and Malaria (GFATM) on Malaria Project' also gave financial support. During the 3 years of implementation of the pilot project (2010–2012), four PCDs received financial support of 2,923,600 RMB yuan. This fund was used for purchasing antimalarials, insecticides, related equipment, test reagents and consumables, personal training, blood examination, etc. (Table 6.4).

Table 6.4 Finance of pilot county of the NMEP in Shanghai and Zhejiang in 2010–2012

Name of PCD	Local finance (RMB)	Central subsidy (RMB)	Global fund (RMB)	Total (RMB)
Zhabei, Shanghai	471,300	152,300	0	623,600
Songjiang, Shanghai	1,037,600	110,200	170,500	1,318,300
Haiyan, Zhejiang	320,000	160,000	0	480,000
Anji, Zhejiang	176,000	202,000	123,700	501,700
Total	2,004,900	624,500	294,100	2,923,600

4. RESULTS AND ACHIEVEMENTS
4.1 Main indicators for evaluation of the NMEP
4.1.1 Malaria epidemic situation
The baseline survey in 2009 showed that the annual malaria incidence in Shanghai and Zhejiang was 0.34 per 100,000 and 0.49 per 100,000, respectively, but it declined year by year within 3 years, to 0.09 per 100,000 and 0.30 per 100,000, respectively, in 2012. In the four PCDs, the number of malaria cases was 0–4 every year, and all those malaria cases were not locally infected by epidemic and tracer survey (Table 6.5).

4.1.2 Case detection
Active case detection: In Shanghai and Zhejiang, a total of 27,658 people of 12,124 households in 20 observation spots were investigated by household-entering survey in 2009, and it was found that 87 of them had fever within 2 weeks, and the incidence rate was 0.31%. Blood examination was performed on the 87 people, but no malaria parasite was found (Table 6.6). During 2009–2012, a total of 1711 person-times were investigated around the malaria focus in PCDs, and no malaria infection was found (Tables 6.7).

In the PCDs, a total of 3631 blood samples from the pupils of 12 primary schools were collected in 2009 to detect malaria antibody by IFA. The results showed that only six were found IFA positive; they were all from Haiyan of Zhejiang. Through follow-up investigation, it was found that among them, five were migrant children, respectively, from Anhui Province (2), Henan Province (2) and Sichuan Province (1), and only one was a local registered resident. In 2012, malaria antibody detection was performed on

Table 6.5 Malaria incidence situation of the pilot counties/districts in Shanghai and Zhejiang in 2009–2012

	No. malaria cases				Annual incidence (1/100,000)			
Pilot	2009	2010	2011	2012	2009	2010	2011	2012
Shanghai	64	42	43	21	0.34	0.22	0.19	0.09
Zhabei	0	0	1	1	0.00	0.00	0.16	0.14
Songjiang	4	6	3	1	0.54	0.79	0.19	0.06
Zhejiang	225	122	124	141	0.49	0.26	0.26	0.30
Haiyan	2	0	1	1	0.54	0.00	0.27	0.27
Anji	1	0	1	1	0.22	0.00	0.22	0.22

Table 6.6 Active case detection of malaria infection in pilot counties/districts (PCDs) in 2009

Name of PCD	No. persons investigated	No. persons with fever	No persons with malaria history	No. blood examination	No. positive in blood examination
Zhabei, Shanghai	7589	3	0	3	0
Songjiang, Shanghai	9374	42	0	42	0
Haiyan, Zhejiang	2741	12	0	12	0
Anji, Zhejiang	7954	30	0	30	0
Total	27,658	87	0	87	0

Table 6.7 Active case detection of malaria in malaria focus in pilot counties/districts (PCDs) in 2009–2012

Pilot	2009	2010	2011	2012	Total
Zhabei, Shanghai	0	0	539	222	761
Songjiang, Shanghai	4	379	378	104	865
Haiyan, Zhejiang	36	0	4	20	60
Anji, Zhejiang	19	0	1	5	25
Total	59	379	922	351	1711

450 pupils and 150 residents in the two PCDs of Shanghai, and all were negative results (Table 6.8).

4.1.3 Residents' awareness of malaria control knowledge and utilization of mosquito-proof facilities

In 2009, 4277 residents of four PCDs were asked on 25,662 question-times; among them, 15,412 questions were correctly answered, the average resident's awareness rate was 60.06%. The awareness rate was higher in Anji County, but it was relatively low in the other three PCDs. At the end of pilot project of the NMEP in 2012, residents' awareness rate (all above 90%) was increased by 60% in comparison with the baseline survey, and the difference was statistically significant ($x^2 = 5558$, $P < 0.01$) (Table 6.9).

A total of 9063 households of PCDs were investigated by household-entering survey about their utilization of mosquito-proof facilities in 2009, and the results showed that 8374 households had mosquito-proof facilities, accounting for 92.40% of the total households investigated. Among them, the residents' utilization rate of mosquito-proof facilities in Zhabei District

Table 6.8 The result of malaria antibody detection on the pupils and residents in four pilot counties/districts (PCDs) in 2009 and 2012

Pilot	2009		2012		
	No. pupils tested	No. persons antibody-positive	No. pupils investigated	No. residents investigated	No. persons antibody-positive
Zhabei, Shanghai	1524	0	300	0	0
Songjiang, Shanghai	1503	0	150	150	0
Haiyan, Zhejiang	300	6	–	–	–
Anji, Zhejiang	304	0	–	–	–
Total	3631	6	450	150	0

Table 6.9 Residents' awareness of malaria control knowledge in pilot counties/districts (PCDs) of the NMEP in Shanghai and Zhejiang

Pilot	2009			2012		
	No. persons investigated	Questions asked	Awareness rate (%)	No. persons investigated	Questions asked	Awareness rate (%)
Zhabei, Shanghai	1172	7032	49.19	452	2712	95.00
Songjiang, Shanghai	1090	6540	57.23	1812	10,872	93.60
Haiyan, Zhejiang	1005	6030	40.43	50	300	90.00
Anji, Zhejiang	1010	6060	95.25	304	1824	91.00
Total	4277	25,662	60.06	2618	15,708	93.66

was relatively low, being 78.40% (Table 6.10). At the end of the pilot project in 2012, among the 452 households in Zhabei District investigated again about their utilization of mosquito-proof facilities, 449 households had mosquito-proof facilities, the utilization rate was 99.34%, being significantly increased ($x^2 = 112.58$, $P < 0.01$).

4.1.4 Malaria vector
A malaria vector survey was carried out in 20 observation spots in the PCDs in 2009 and a total of 3173 mosquitoes were caught. Among them, 251 were *An. sinensis*, 1792 were *Culex pipiens pallens*, 428 were *Culex tritaeniorhynchus* and 18 were *Aedes albopictus*. In Zhabei District, 40 *Cu. pipiens pallens* were caught, and no malaria-transmitting *An. sinensis* were found (Table 6.11). The results indicated that *An. sinensis* existed in all pilot counties except Zhabei District.

4.2 Improving health services and raising control capacity
4.2.1 Establishing work mechanism and network platform
4.2.1.1 Establish a work mechanism of government leadership and multi-sector participation
The governments and responsible departments in PCDs paid great attention to the NMEP. A leading group of the NMEP headed by the health administrative department and a technical steering group headed by a professional institution were established, respectively, at province, city and county levels; and a 'liaison meeting of public health' participated in by departments of health, education, business, tourism, entry and exit inspection and quarantine, public security and finance was established, respectively; the responsibilities of different departments were well defined, meetings of leaders of membership units and liaisons were held at irregular intervals every year for coordination and organization of the implementation of relevant works; a leading group of the NMEP was established, and multisectoral coordination mechanisms were established; the administrative department of health included the NMEP as an important task, held kick-off meetings for the pilot project of the NMEP, signed the memorandum of responsibility of target management of the NMEP with the authorities at different levels, worked out an implementation programme for the pilot project of the NMEP, to ensure accountability to the people, make task in place and put the target clear.

4.2.1.2 Make full use of the three-level disease prevention network to implement technical measures to eliminate malaria
Three-dimensional networks of disease prevention 'province/city-district/county-township/town' and 'medical institutions-disease control-community

Table 6.10 Utilization of mosquito-proof facilities in the pilot of the NMEP in Shanghai and Zhejiang in 2009

Pilot	No. households investigated	No. households using mosquito nets	No. households using screened doors and windows	No. households using repellents	No. households having mosquito-proof facilities	Utilization rate of mosquito-proof facilities (%)
Zhabei, Shanghai	3051	49	1886	986	2392	78.40
Songjiang, Shanghai	2656	508	2508	2392	2628	98.95
Haiyan, Zhejiang	982	666	2592	2633	980	99.80
Anji, Zhejiang	2374	961	2196	2331	2374	100.00
Total	9063	2184	9185	3642	8374	92.40

Table 6.11 The mosquito survey in pilot counties/districts (PCDs) in 2009

Pilot	No. mosquitoes caught	No. Anopheles Sinensis	No. Culex pipiens pallens	No. Culex tritaenio-rhynchus	No. Aedes albopictus	No. Armigeres subalbatus
Zhabei, Shanghai	48	0	40	0	8	0
Songjiang, Shanghai	52	4	41	7	0	0
Haiyan, Zhejiang	2193	96	1645	0	0	452
Anji, Zhejiang	880	151	66	421	10	232
Total	3173	251	1792	428	18	684

health service center' were all established in Shanghai and Zhejiang, and various measures of the NMEP were effectively implemented through this network. A technical expert group was established, respectively, at province and county levels, responsible for technical guidance and professional training on the NMEP. Various medical institutions at different levels in the PCDs were responsible for diagnosis, reports, treatment, rescue of critical patients and treatment of heavy antimalarial toxicity and adverse reaction, and took part in case detection, antimalarial drug administration, mosquito control and other measures in the administrative area.

A particular section of parasitic disease control was established in the center for disease control and prevention at provincial and county levels. There were 25 personnel of parasitic diseases control and 10 full-time malaria control personnel at the provincial level in Shanghai and Zhejiang, and 12 professional malaria control personnel in four pilot counties, respectively.

4.2.2 Equipment, supplies and fund guarantee

During the pilot project, a funding mechanism by government investment as the main share, grading burden and multichannel financing was established gradually. The governments in pilot provinces and counties provided working funds based on their own real conditions, and at the same time, 'Chinese Central Government Subsidies for Malaria Control Project' and 'Global Fund on Malaria Project' also gave financial support (Table 6.4).

4.2.3 Professional skill improvement of personal of disease control and prevention

During the assessment of, different pilot counties/districts in 2012, six clinicians from county-level medical institutions and four general practitioners from two community health service centers (two for each center) were randomly drawn for examination of their knowledge about malaria diagnosis and treatment. Ten examination questions were randomly drawn from the question bank by the examination team. The results showed that the awareness rate of the NMEP knowledge among selected 40 doctors was 100%. Five microscopists were selected, respectively, from one county general hospital, one county center for disease control and prevention and two township health centers (community health service centers) for assessing their skill on examination of malaria parasites; each person was asked to examine six blood slides in 60 min, and the results of all blood slides examined by 20 microscopists were all correct.

4.3 Implementation of main measures

4.3.1 Finding the malaria cases
Blood examination was performed on febrile patients who clinically diagnosed as malaria cases or as suspected malaria cases, and who had fever with no significant reasons, by the medical institutions at different levels. Blood examination of 34,990 and 42,581 person-times were performed, respectively, in 2010 and 2011, accounting for 2% of the annual population in four PCDs, being significantly higher than the number of blood examinations before the beginning of the pilot project of the NMEP in 2009 and after the end of the pilot project in 2012 (Table 6.12). In 2010 and 2011, a total of 12 malaria cases were found in the PCDs, respectively, with 7 cases in 2009 before the beginning of the pilot project, and 3 cases in 2012 after the end of the pilot project.

4.3.2 Management of malaria cases
All the blood slides of malaria cases were sent to CDC in each PCD by hospitals where the malaria patients were identified and rechecked by microscopists. Then those slides were sent to provincial CDC to recheck again and which species of parasite the patient was infected with was identified. Results showed that among 23 malaria cases, 10 were vivax malaria, 12 were falciparum malaria, and 1 was quartan malaria. All the malaria cases were reported within 24 h. The 24-h report rate was 100%.

After the diagnosis was confirmed, all the cases also received standardized treatment according to the principle of drug administration of malaria in hospital or as outpatient. For the patients who were treated as outpatients, CDC was responsible for the supervision of drug administration. Critically ill patients would be transferred to designated hospitals for hospitalization. Vivax malaria cases would be followed and treated with primaquine 8-day therapy for radical cure in intermittent periods in May of the following year. A total of 5 out 10 vivax malaria cases were traced and given the primaquine therapy, and another 5 cases were lost at follow-up investigation (Table 6.13).

4.3.3 Survey of case and focus
All the patients with malaria were investigated by CDC staff in each PCD, to find out when the malaria happened, where they were possibly infected etc. The results showed that all the malaria cases were imported from abroad (Zheng et al., 1989; Zhou et al., 2011). No case was locally infected.

At the same time, an epidemiological survey was conducted within three workdays after the case was reported, and the results showed that all cases were imported from other provinces or foreign countries. Health education

Table 6.12 Blood examination of febrile cases and detection of malaria cases in pilot counties/districts (PCDs)

Pilot	No. blood examinations of febrile cases					Blood examination rate of febrile cases					No. malaria cases detected				
	2009	2010	2011	2012	Subtotal	2009	2010	2011	2012	Average	2009	2010	2011	2012	Total
Zhabei, Shanghai	1004	7134	12,358	5336	25,832	0.16	1.11	1.96	0.77	1.00	0	0	1	1	2
Songjiang, Shanghai	2455	10,760	12,749	7805	33,769	0.44	1.92	2.20	1.29	1.46	4	6	3	1	14
Haiyan, Zhejiang	952	7856	7971	973	17,752	0.26	2.12	2.14	0.26	1.19	2	0	1	1	4
Anji, Zhejiang	1388	9240	9503	4967	25,098	0.30	2.02	2.07	1.08	1.37	1	0	1	1	3
Total	5799	34,990	42,581	19,081	102,451	0.29	1.73	2.09	0.89	1.25	7	6	6	4	23

Table 6.13 List of malaria case found in pilot counties/districts, in 2009–2012

Patient	Sex	Age	Occupation	Name of county(1)	Name of county(2)	Date of on-set	Species	Date of report	Drugs for treatment	Radical treatment Treated or not	The reasons if not
Case 1	F	17	Student	Minghang	Songjiang	2009-2-23	*P. vivax*	2009-3-4	Chloroquine + primaquine	Yes	
Case 2	M	35	Worker	Haiyan	Haiyan	2009-3-8	*P. falciparum*	2009-4-4	Artesunate	—	
Case 3	M	16	Student	Songjiang	Songjiang	2009-7-4	*P. vivax*	2009-7-10	Chloroquine + primaquine	Yes	
Case 4	M	21	Worker	Haiyan	Haiyan	2009-7-19	*P. vivax*	2009-7-26	Chloroquine + primaquine	No	Lost of follow up
Case 5	M	18	Businessman	Minghang	Songjiang	2009-7-29	*P. vivax*	2009-7-29	Chloroquine + primaquine	Yes	
Case 6	M	22	Worker	Minghang	Songjiang	2009-8-17	*P. vivax*	2009-8-22	Chloroquine + primaquine	Yes	
Case 7	M	44	Worker	Anji	Anji	2009-8-19	*P. vivax*	2009-8-21	Chloroquine + primaquine	Yes	
Case 8	M	28	Other	Jinshan	Songjiang	2010-2-1	*P. falciparum*	2010-2-6	Artesunate	—	
Case 9	F	29	Other	Jinshan	Songjiang	2010-5-16	*P. falciparum*	2010-6-18	Artesunate	—	
Case 10	M	20	Businessman	Songjiang	Songjiang	2010-6-17	*P. vivax*	2010-6-17	No (3)	Go back to homeland	
Case 11	M	55	Worker	Songjiang	Songjiang	2010-8-6	*P. vivax*	2010-8-18	Chloroquine + primaquine	No	Back to homeland

Continued

Table 6.13 List of malaria case found in pilot county/districts, in 2009-2012—cont'd

Patient	Sex	Age	Occupation	Name of county(1)	Name of county(2)	Date of outbreak	Species(3)	Date of report	Drugs for treatment	Radical treatment Treated or not	The reasons if not
Case 12	M	40	Other	Jinshan	Songjiang	2010-8-26	P.falciparum	2010-9-18	Artesunate	—	
Case 13	M	50	Businessman	Jinshan	Songjiang	2010-11-30	P.falciparum	2010-12-6	Artesunate	—	
Case 14	M	30	Officer	Hongzhoui	Anji	2010-12-14	P.falciparum	2011-1-1	Artesunate	—	0
Case 15	M	62	Worker	Anji	Anji	2012-12-23	P.falciparum	2012-12-30	Artesunate + promaquine	—	0
Case 16	M	47	Other	Jinshan	Songjiang	2011-6-19	P.falciparum	2011-6-24	Artesunate	—	
Case 17	M	33	Other	Jinan	Songjiang	2011-7-20	P. quartan	2011-7-21	Artesunate	—	
Case 18	M	55	Officer	Huangpu	Zhabei	2011-8-15	P.falciparum	2011-8-15	Artemether	—	
Case 19	M	50	Worker	Xuhui	Songjiang	2011-9-9	P. vivax	2011-9-19	Chloroquine + primaquine	No	Lost of follow up
Case 20	M	26	Worker	Haiyan	Haiyan	2011-10-2	P. vivax	2011-11-1	Chloroquine + primaquine	No	Back to homeland
Case 21	M	29	Businessman	Jiaxin	Haiyan	2012-5-1	P.falciparum	2012-5-8	Artesunate	—	
Case 22	M	26	Worker	Jinshan	Zhabei	2012-11-12	P.falciparum	2012-11-13	Artesunate	—	
Case 23	M	30	Other	Pudong	Songjiang	2012-11-23	P.falciparum	2012-11-28	Artesunate	—	

Note: 1. The name of county means where the malaria case was found or diagnosed.
2. The name of county means where the malaria cases lived as the residents, so the survey and disposition of focus would be carried out.
3. After Case 8 was diagnosed as malaria, he returned to his homeland to get treatment.

was addressed to the people around the malaria case, 2281 pieces of publicity materials were distributed, and 11,275 person-times were benefitted; at the same time, 1710 person-times with malaria history and fever within recent two weeks were investigated, 161 person-times with malaria history, fever within recent 2 weeks and close contact with malaria cases were examined for malaria parasites; no parasite positive case was found. Among them, 27 received prophylactic medication. Mosquito density and mosquito breeding places within 300 m around the patient's home were investigated, and indoor residual spraying in the patient's home and its neighborhood. Insecticide was sprayed for mosquito control for 3 days continuously within 100 m around the patient's home, covering 158,330 m^2. After that, the mosquito density was lower than 4/man-hour, indicating that malaria transmission had been effectively controlled. No second generation of malaria case was found finally (Table 6.14).

4.3.4 Professional training

In 2009–2012, 30 courses of professional training were conducted at the provincial level in Shanghai and Zhejiang, and a total of 2403 person-times were trained. In the four PCDs, 39 courses were held, and 996 person-times were trained (Table 6.14). For different trainees, such as malaria control staff, clinicians, laboratory technicians, etc. they were trained in malaria control knowledge, microscopic examination skills, clinical diagnosis, case survey and disposal, malaria surveillance, vector surveillance ad control, etc. (Zofou et al., 2014) Different training methods were used, such as, lectures, skills operations, field exercises and competitions and contests (Table 6.15).

4.3.5 Management of mobile populations

Both Shanghai and Zhejiang are located in the economically developed Yangtze River Delta region with frequent population movement. In order to discover the imported malaria cases in a timely manner, Shanghai and Zhejiang Entry-Exit Inspection and Quarantine Bureau signed a 'memorandum of cooperation in joint prevention and control of infectious diseases', and specified each port counterpart contact a designated hospital, further deepened bilateral cooperation in information communication, case surveillance, health education, research and other aspects; relying on the platform of joint meetings, the health department worked jointly with business departments, the tourism sector, the National Development and Reform Commission, the Entry-Exit Inspection and Quarantine Bureau and other departments, performing their respective duties and working in close coordination to strengthen the publicity about malaria control

Approaches to the Evaluation of Malaria Elimination at County Level 169

Table 6.14 Investigation and disposal of case foci in pilot counties/districts (PCDs) of the NMEP in 2009–2012

Pilot	No. case focus	No. case survey	No. population investigated	No. person screened by blood examination	No. person with prophylactic medication	No. person benefitted	No. publicity materials distributed	Mosquito control area m²
Zhabei, Shanghai	2	2	761	0	0	761	300	1000
Songjiang, Shanghai	14	14	865	78	6	10,350	1817	15,6130
Haiyan, Zhejiang	4	4	59	59	2	25	25	900
Anji, Zhejiang	3	3	25	24	19	139	139	300
Total	23	23	1710	161	27	11,275	2281	158,330

Table 6.15 Professional training of the NMEP in Shanghai and Zhejiang

Category of training	Name of Area	2009		2010		2011		2012		Total	
		Course-times	Person-times	Course-times	Person-times	Course-times	Person-times	Course-times	Person-times	Course-times	Person-times
Province level	Shanghai	4	282	5	266	6	307	5	368	20	1223
	Zhejiang	1	100	2	300	5	550	2	230	10	1180
	Subtotal	5	382	7	566	11	857	7	598	30	2403
County level	Zhabei	2	39	4	74	4	93	5	98	15	304
	Songjiang	2	46	4	88	2	46	5	166	13	346
	Haiyan	1	40	2	38	1	44	1	36	5	158
	Anji	1	33	2	52	2	69	1	34	6	188
	Subtotal	6	158	12	252	9	252	12	334	39	996
Total		11	540	19	818	20	1109	19	932	69	3399

knowledge among mobile populations and entry-exit people, training and malaria control and surveillance (Bi and Tong, 2014; Liu et al., 2014). The PCDs strengthened the surveillance of imported malaria cases from hyperendemic areas of malaria, for instance, Haiyan County issued 'Haiyan county programme of monitoring the people from hyper-endemic countries (regions)', and established an exchange mechanism of key monitoring windows of imported malaria (Tambo et al., 2014; Zhang et al., 2013).

4.3.6 Health education
During the 'National Malaria Day' on 26th April each year, various health education and promotion activities were carried out in pilot counties, publicizing extensively malaria control knowledge through news media, newspapers and periodicals and networks, and went to the schools, communities, construction sites, agricultural product markets and other key places to conduct on-site Malaria Day propaganda activities. Some of their activities included hanging banners, putting out display panels, distributing informational folders, providing consultations with experts, producing propaganda performances, and award-winning questions and answers, to raise the awareness of the citizens and transient population about disease prevention and to enhance the knowledge of malaria control and prevention, so as to create an atmosphere of comprehensive public participation and implementation of the NMEP goal. In 2009–2012, the PCDs printed and produced 55,830 pieces of propaganda materials.

4.3.7 Quality control of blood examination
In 2009–2012, all blood slides of 23 malaria cases found by the medical institutions in the administrative region were rechecked by CDC in each PCD, and the results were consistent. In the same period, a total of 26,811 negative blood slides randomly drawn for recheck every month were rechecked, accounting for 26.17% of the total blood slides, and no positive results were found; the coincidence rate of the rechecking was 100% (Table 6.16).

During the assessment of the pilot project of the NMEP in 2012, the blood slides of all malaria cases were rechecked once again, and 30 negative slides were randomly drawn for recheck for each PCD; the results showed that the coincidence rate of microscopic examination of blood slides in all pilot counties was 100%.

4.3.8 Supervision of the process of the NMEP
Through two-level supervision in Shanghai and Zhejiang, the implementation of all the measures for the pilot project of the NMEP and its progress

Table 6.16 Recheck of blood slides in pilot counties/districts (PCDs) in 2009–2012

Pilot	No. blood examination	No. rechecked	Percentage of recheck (%)	No. coincided	Coincidence rate (%)
Zhabe, Shanghai	25,832	7980	30.89	7980	100
Songjiang, Shanghai	33,769	10,691	31.66	10,691	100
Haiyan, Zhejiang	17,752	3212	18.09	3212	100
Anji, Zhejiang	25,098	4928	19.64	4928	100
Total	102,451	26,811	26.17	26,811	100

were inspected, and the problems found were rectified in a timely manner. In Shanghai, supervision of the NMEP work covering all the districts were carried out twice a year, and supervision of the pilot counties was also carried out in Zhejiang once a year. The progress of personnel training, blood examination of febrile cases, case reporting, case rechecking and focus disposal, health education, and filing of documents and data and other aspects were inspected respectively. The operations of malaria microscopic examination stations were guided and inspected.

4.4 Evaluation and certification

4.4.1 Assessment score

Through the scoring by an evaluation team, the total score of the four pilot counties was 95 points or more (Tables 6.17–6.19), met the requirements of the assessment programme (the score was more than 90 points).

4.4.2 Summary of assessment and evaluation

Through 2 days' listening to the report, data access, capacity verification, on-site surveys, discussion and evaluation and scoring, the evaluation team considered that under the unified leadership of local government, the pilot counties had further improved the regional public health network and surveillance system, and strongly ensured the NMEP work in policy, funding and personnel aspects, and completed the work required in the 'Programme of pilot project of NMEP in China' (trial version) (MOH, 2009) with integrated information, detailed data, clear and complete archives; and no local infected malaria case was found in consecutive 3 years. Based on the 'Criteria for malaria control and elimination (GB26345-2010)'

Table 6.17 Detail of assessment score of pilot counties/districts in 2012

Item (points)	Details (points)	Assessment score			
		Zhabei	Songjiang	Haiyan	Anji
Policy guarantee (10)	Policy (2)	1	2	2	2
	Setup of institution (3)	3	3	3	3
	Personnel (2.5)	2.5	2.5	2.5	2.5
	Fund (2.5)	2.5	2.5	2.5	2.5
Management of data (15)	Working plan and summary (6)	6	6	6	6
	Report form (6)	6	6	6	6
	Filing archives (3)	3	3	3	3
Training (15)	Training materials (12)	12	12	12	12
	Coverage of training (3)	3	3	3	3
Blood examinations (20)	Coverage of blood examinations (4)	4	4	4	4
	Number of blood examinations (10)	10	9.5	10	10
	Number of blood examinations in transmission season (3)	3	1.5	3	3
	Quality control of blood examinations (3)	3	3	3	3
Disposal of malaria cases (20)	Timely report (2)	2	2	2	2
	Laboratory examination rate (3)	3	3	3	3
	Case survey (4)	4	4	4	4
	Standard treatment (2)	2	2	2	2
	Record and report of radical cure of vivax malaria in intermittent period (2)	1.5	2	2	2
	Epidemic focus survey (4)	4	4	4	3
	Vector control (3)	3	3	3	3
Supervision (10)	Supervision (10)	10	10	9.5	9.5
Health education (10)	Planning and summary (3)	3	3	3	3
	Health education (7)	7	7	6.8	6.8
Total		98.5	98.0	99.3	98.3

Note: 1. Sociological data, such as population, will be subject to the local bureau of statistics data released in 2010.
2. Indicators, such as case report and investigation and disposal of epidemic focus; if no cases in the county assessed, the score would be 90%.
3. If all the scores for an item have been deducted, no negative score would be given.

Table 6.18 Score of on-site assessment of pilot counties/districts in 2012

Details (points)	Assessment score			
	Zhabei	Songjiang	Haiyan	Anji
Microscopic examination skill (30)	30	30	30	27.5
Knowledge of malaria control (20)	20	20	20	20
Recheck of blood film (30)	30	30	30	30
Investigation of mosquito-proof measures (10)	10	10	10	10
Investigation of awareness rate of residents' knowledge about malaria control (10)	9.8	8	10	10
Total	99.8	98.0	100	97.5

Table 6.19 Assessment score summary of pilot counties/districts in 2012

Items	Weights	Score summary			
		Zhabei	Songjiang	Haiyan	Anji
Data assessment	0.5	98.5	98.0	99.3	98.3
On-site assessment	0.5	99.8	98.0	100	97.5
Total		99.15	98.00	99.65	97.30

(National Criteria, 2010), it has been certified that Zhabei District and Songjiang District of Shanghai and Haiyan County and Anji County of Zhejiang Province have reached the criteria for the NMEP.

After receiving the evaluation report, Zhejiang Provincial Department of Health and Shanghai Municipal Bureau of Health replied in a timely manner, and issued documents to Haiyan County, Anji County, Zhabei District and Songjiang District on 16 November 2012 and 18 December 2012, respectively, confirming on behalf of the municipal and provincial government that the above counties/districts had reached the criteria for NMEP.

Thus, Zhabei District and Songjiang District of Shanghai, and Haiyan County and Anji County of Zhejiang Province, have first reached the goal of eliminating malaria in China.

4.5 Experiences

4.5.1 Disease control network

Both Shanghai and Zhejiang have a disease control network covering the villages; independent institutions of disease control and prevention in provinces, cities, prefectures and counties; community health service centers in townships; and village doctors in administrative villages. Both places not only have complete organizational setups but also have complete personnel security. The township has a big community health service center with a large number of disease control personnel; for instance, Songjiang District has a population of 349,400, 15 township institutions, 575 disease control personnel at the township level, and 358 village doctors in 251 neighborhood or village committees. The salary of all these disease control personnel were paid by the local governments and their duty is to support the control and prevention against the disease related to the people in the administrative area. These grassroots disease control organizations and personnel ensure the good implementation of the NMEP and realization of the predetermined goal.

4.5.2 Sector cooperation

Malaria control or elimination might be considered just the work of the health sector, but for better implementation of disease control from a longer perspective, namely to get a maximum output for the NMEP by using minimum input (fund) and not have recurrence after elimination (new local infected malaria cases occurring after elimination), multisectoral cooperation must be the key. For instance, the education sector involved in transferring the malaria control knowledge, to teach students an understanding of how to control malaria, and bring the disease control knowledge to their families. Publicity departments can not only tell a broad swath of citizens how to control malaria, but also inform them about the progress of the NMEP, and mobilize them to take part in the NMEP work. Planning and finance departments could work in accordance with the requirements of the NMEP, giving timely financial support. It was through such multisectoral cooperation that the four pilot counties finally reached the goal of the NMEP on schedule.

4.5.3 Finding malaria cases

If every patient with any kind disease could be found, diagnosed and treated in early stage, it would be possible to control and eventually eliminate this disease. During the implementation of pilot project of malaria elimination, the strategy adopted was to find malaria case as early as possible. The main measures about the strategy were to find out different kinds of malaria cases by all means, to treat the target patient, to investigate the population around the target patient including persons from abroad, and to carry out the mosquito density control in standardized approach. Its purpose was to find out all possible conditions for malaria transmission within the scope of epidemic focus, so as to adopt different effective measures to interrupt malaria transmission.

In order to discover every malaria case in time, in addition to improving the microscopists' skill, clinicians' understanding about the malaria distribution in the world and patients' initiative to inform the doctor are also indispensable factors. Since the PCD has already been a hypo-endemic area of malaria, many doctors working today do not see malaria patients in routine outpatient work; their knowledge about malaria is only from textbooks and their memories, and they only know the typical clinical manifestation of malaria. Therefore, every year in combination with the 'National Malaria Day', hospitals arranged the activities to train the doctors of major sections on the malaria situation, inform them of the current global and domestic epidemic situation of malaria, and tell them what populations should be examined for malaria, as well as the untypical clinical manifestations of malaria, in order to find as may patients as possible in a timely manner.

4.5.4 Funding mechanism
In order to effectively implement the NMEP's pilot project, supported by the central, provincial/municipal and local governments, the working funds for the pilot project had been ensured. The funds were not only from the governments at different levels, but also from NGOs, such as, GFATM, etc. Due to our funding guarantees, the work of the pilot project was implemented smoothly, and eventually reached the goal of the NMEP.

4.6 Lessons
4.6.1 Number of blood examinations
One of the main means for finding malaria cases is blood examination of malaria parasites among the febrile cases in the hospitals, with the purpose of finding almost all malaria cases. In the past, special attention was paid to the blood examination of febrile patients. The rate of blood examination in whole population was taken as indicator of implementation in malaria control. The '1983–1985 National Malaria Control Programme' provided that the number of blood examinations in different endemic counties must be more than 5% of the total population of the county (MOH, 1983). The programme provided that in different endemic areas, by township/town as a unit, the annual blood examination rate of the clinically diagnosed malaria cases, suspected malaria cases and febrile patients with unexplained reasons (referred to as 'three fever patient', hereinafter) must be 50% at the end of

2007, and 60% and 70% respectively at the end of 2010 and 2015 (MOH, 2006). In the 'National malaria surveillance programme' (MOH, 2005) and 'Malaria control technical scheme (trial version)' (MOH, 2007), it was provided that in areas with relatively high incidence (annual incidence over than 100 per 100,000), the annual number of blood examinations of febrile cases must not be less than 5% of the total population; in areas with unstable incidence (annual incidence 10 per 100,000–100 per 100,000), the annual number of blood examinations must not be less than 2% of the total population; and in areas with relatively low incidence (annual incidence <10 per 100,000), the annual number of blood examinations must not be less than 1% of the total population (MOH, 2007). In the 'Action plan of the NMEP in China (2010–2020)', based on the history of malaria epidemics and the case reporting in 2006–2008, the 2858 counties in P.R. China are divided into four types based on the incidence rate of malaria during three years, and different indicators of blood examination are provided for the four types. In type 1 and 2 counties, by township/town as a unit, the annual number of blood examinations for malaria in 'three fever patients' must be not less than 2% and 1% of the total population in the administrative areas, respectively, and not less than 2‰ in type 3 counties; and no corresponding required number of blood examinations for type 4 counties (MOH, 2010). But in the 'Technical specification of NMEP (trial version)' published by China CDC in 2011, in type 1 and 2 counties, by township as a unit, the number of blood examinations of febrile patients with unexplained reason not less than 1‰; and in type 3 counties, by county as a unit, the number of blood examinations of febrile cases with unexplained reason not less than 0.5‰ (China CDC, 2011). However, in the WHO malaria surveillance, there is no exact quantitative index of blood examination for any country or region, but object indicators, namely active and passive case detection (WHO, 2012).

In the four pilot counties, during the 4 years, the accumulated case number was only 23. For finding these cases, 102,541 person-times blood examination were performed, one malaria case could be found in 4458.3 person-times blood examination on average. Especially in 2011 and 2012, a total of 77,571 person-times were examined, accounting for 75.65% of the total number of 4 years, but only one case could be found in 6664.25 person-times blood examination in average. However, although the number of blood examinations had reduced greatly in 2012, and the number of blood examinations in the four PCDs was 19,081 person-times, accounting for 0.89% of the total population, four malaria case were still found, being one malaria case found among 4770.25 febrile cases on average (Table 6.13).

Thereby, the number of blood examinations does not mean the discovery of malaria cases.

A large amount of blood examination do not only find malaria cases and show the progress of implementation of malaria control, but also give persons practically training on microscopic skill. Without a large amount of blood examinations, the microscopic examination skill might not be improved greatly. Malaria surveillance of WHO gave the corresponding objects the blood examination (WHO, 2012). Although the corresponding objects can be clearly defined, where was the corresponding objects, no person can be answered. Especially when malaria cases were rare and in sporadic distribution, since we did not know when a malaria case would appear and clinicians' concept of malaria was very vague, not one malaria case has been found for several years in some hospitals. This situation leads to a vicious cycle, namely, doctors will not pay attention to it, laboratory technicians do not know it, patients will not be found, and again doctors will not pay attention to it, thus delayed diagnosis and even death of malaria cases will occur.

During the implementation of a pilot project, some workers considered that the number of persons required for blood examination was too large, which would not only increase the workload of hospitals, but also waste the limited medical sources, enhance the specific staff members' aversion to blood examination and extend the time for medical treatment in some hospitals. Therefore, in 2012 China CDC made a large adjustment in the number of blood examinations in different counties, namely, in type 1 and 2 counties, the number of blood examination would be about 1‰, and in type 3 counties about 0.5‰ (China CDC, 2011). At the same time, in view of the convenience degree of traffic and the accessibility to medical treatment in P.R. China, the author suggested to include the function of malaria diagnosis and treatment into the county hospitals or some other hospitals at the same level, not at town/countryside level, to enhance the utilization rate of resources and keep and maintain malaria detection ability (Zhou et al., 2011; Zofou et al., 2014).

4.6.2 Selection of blood examination methods

For malaria diagnosis, the classic method is microscopic examination of blood film to find the malaria parasite. This method needs a relatively high technical skill and the personnel need to be trained for a long time, and its operation is time consuming and needs endurance (Berry et al., 2008; Zheng et al., 2013). At present, RDT technique is a relatively easier method. It detects the antigen in the sample by inclusive antibody in the test kit. This technique is easy to operate, does not need additional equipment and operation time is short. A study was done to compare RDT with microscopic examination (Cai et al., 2012, and the result showed that the time for completion of examination was 14.7 and 29.0 min, respectively, likeness degree was 60% and 30%, respectively. The cost of maintaining was lower, without use of equipment in RDT. In the two pilot counties of Shanghai, one used RDT as the main blood examination method (Songjiang), and the other used the routine microscopic examination (Zhabei) as the main blood examination method; by comparing the completion rate of the two methods, and the number of persons receiving blood examination and the rate of blood examination in the four years, the latter was lower than the former. It is indicated that RDT was more welcomed by the actual operation. The author considers that RTD could save test time and perform more tests in grade two and three hospitals, and make up for the inadequacy of microscopic examination ability in grade one hospitals. Therefore, in the NMEP in Shanghai, RDT has been adopted as the method of first choice, namely after collecting a blood sample, first to make an RDT test, and make a blood smear at the same time. If the RDT test was positive, the blood smear would be stained and microscopically examined immediately, and sent to county CDC for recheck; if the RDT test was negative, the blood films would be collected, and concentrated staining would be made regularly, and kept (Health Bureau of Shanghai, 2010). This process has been widely implemented in Shanghai area, and has achieved the expected effect.

4.6.3 People's awareness of malaria control knowledge

People's behaviour comes after awareness. During the process of health promotion, people experienced a process of knowing, understanding and implementing. With the process of malaria control, more and more action was taken, for example the active participation of a wide range of people, the density control of malaria mosquito, actively seeking the medical help for the febrile patients, and protecting from mosquito biting. Based on the implementation of activity mention above, the malaria epidemic situation was significantly controlled.

Along with the control of the malaria epidemic situation, the percentage of local infected malaria cases is declining, imported malaria cases from abroad are increasing, malaria cases are sporadically distributed, and the possibility of local malaria infection tends to be zero. Under current situation, if we still carry the previous method, providing people with huge amount of malaria knowledge, spending a lot of energy to carry out health education activities, and giving it more credit, the assessment score might be raised, but it might not help the NMEP to a great extent. Therefore, based on the assessment of the pilot project, we have suggested to delete the item of knowledge of malaria control in the 'Evaluation programme of pilot project of the NMEP' that has been endorsed by the Ministry of Health. Therefore, in the 2013 edition and 2014 edition of the 'Evaluation programme of pilot project of the NMEP', this item is not presented there (MOH, 2013, 2014).

At present, malaria cases are mainly those who get infected in the endemic areas abroad and have onset after returning home. Those who have been in the endemic areas abroad are mainly domestic people and foreign people, the former includes organized laborers, business travelers, those visiting relatives, students, etc. and the latter includes foreign students, business delegations, exchange of personnel between government service, foreign trainees, etc. All these objects are specific groups of people, not the general population, therefore they are not included in the routine health education. Thus, the general sense health education is not conducted to the target object. The authors suggest that health education should be more targeted, such as the specific sites with foreign labor service enterprises, schools with foreign students, so as to raise the efficiency of health education.

4.6.4 Malaria vector surveillance

Vector surveillance is an important part of malaria surveillance; its purpose is to know the species of *An. sinensis*, its bionomics and role in malaria transmission, to provide a basis for working out malaria control measures, and at the same time to make a correct evaluation of the effect of vector control measures. In the four pilot counties of the NMEP, besides a vector species survey conducted in the baseline survey, no other surveillance was made, such as seasonal variation of mosquito density, its role in malaria transmission and sensitivity of *An. sinensis* to insecticides (Tambo et al., 2014).

Although at present China has not been listed as early phase elimination, it is still at control stage (WHO, 2014), and most areas of China have entered the early phase elimination or consolidation and maintenance phase. At the same time malaria-transmitting vectors still exist in most areas of China (Tang et al., 2012). If the imported malaria cases are not effectively diagnosed and treated and the vectors in the epidemic focus are not effectively controlled, the possibility of local infection still exists. Therefore, the authors suggest that vector surveillance is still necessary after elimination of malaria, and the surveillance work can be carried out in some sentinel points and implemented by provincial institutions.

5. CONCLUSIONS

Through a half century's efforts, especially the recent 3–4 years' intensive implementation of the NMEP malaria P.R. has been eliminated in the four PCDs which reaching the objectives of the NMEP. This achievement is an inevitable result of the local government's consistent attention and active control and prevention, and the long-term battle against malaria, not afraid of hard and selfless dedication spirit of several generations' control and research personnel. The realization of the NMEP's objectives in the pilot counties has increased confidence and hope for the rest of the country, and greatly promoted the process of the NMEP in P.R. China (Tambo et al., 2014).

The experience in the process of achieving the goal of the NMEP suggests that under the current economic and technical conditions, it is possible to eliminate malaria if we can grasp some key measures, and also suggests that as long as we join hands and work together, the goal to eliminate malaria will certainly be implementable and realizable (Zheng et al., 2013).

ACKNOWLEDGEMENTS

The project was supported by the National S & T Major Programme (grant no. 2012ZX10004220), by the National S & T Supporting Project (grant no. 2007BAC03A02) and by China UK Global Health Support Programme (grant no. GHSP-CS-OP1).

REFERENCES

Berry, A., Benoit-Vical, F., Fabre, R., Cassaing, S., Magnaval, J.F., 2008. PCR-based methods to the diagnosis of imported malaria. Parasite 15, 484–488.

Bi, Y., Tong, S., 2014. Poverty and malaria in the Yunnan province, China. Infect.Dis. Poverty 3, 32.

Cai, L., Ma, X.B., Zhu, M., Jin, Y.J., Jiang, X.J., 2012. Control and research of parasitic diseases in Shanghai. In: Tang, L.H., Xu, L.Q., Chen, Y.D. (Eds.), Control and Research of Parasitic Diseases in China. Beijing Science and Technology Press, pp. 1044–1060.

China Center for Disease Control and Prevention (China CDC), 2011. Technical Guideline for ME in China, 2011ed. http://www.chinacdc.cn/jkzt/crb/nj/jszl_2223/201109/t20110906_52144.htm.

Gavernment of Anji. Overview of Anji. http://www.anji.gov.cn/default.php?mod=article&do=detail&tid=23542.

Government of Haiyan County, 2009. Statistical Bulletin on National Economic and Social Development of Haiyan County in 2009. http://www.haiyan.gov.cn/art/2010/3/30/art_8_9434.html.

Government of Songjiang District. About Songjiang. http://www.songjiang.gov.cn/website/pages/intro_85.htm?cid=85.

Government of Zhabei District. About Zhabei. http://en.shzb.gov.cn/about1.asp.

Investigation Term of Zhejiang, 2010. Zhejiang Provincial Statistics Yearbook—2010. (Beijing).

Liu, W.D., 1992. Forty years of study on mosquitoes in Shanghai. Shanghai Preventive Med. 4 (7), 4–6.

Liu, D.Q., 2014. Surveillance of antimalarial drug resistance in China in the 1980s-1990s. Infect. Dis. Poverty 3, 8.
Liu, Y., Hsiang, M.S., Zhou, H., Wang, W., Cao, Y., Gosling, R.D., et al., 2014. Malaria in overseas labourers returning to China: an analysis of imported malaria in Jiangsu Province, 2001–2011. Malaria Journal 13, 29.
Ministry of Health of the People's Republic of China (MOH), 1983. National Malaria Control Programme in 1983–1985. Beijing (internal document, in Chinese).
Ministry of Health of the People's Republic of China (MOH), 2005. National Malaria Surveillance Plan in China. http://www.chinacdc.cn/ztxm/ggwsjc/jcfa/200507/t20050727_41345.htm.
Ministry of Health of the People's Republic of China (MOH), 2006. National Malaria Control Programme in 2006–2015. Beijing (internal document, in Chinese).
Ministry of Health of the People's Republic of China (MOH), 2007. Technical Guidelines for Malaria Control. Beijing (internal document, in Chinese).
Ministry of Health of the People's Republic of China (MOH), 2009. Work-plan for malaria elimination pilot project in China. http://wsb.moh.gov.cn/mohjbyfkzj/s3594/200911/44698.shtml. Beijing (internal document, in Chinese).
Ministry of Health of the People's Republic of China (MOH), 2010a. Action plan of malaria elimination in China (2010–2020). http://wsb.moh.gov.cn/mohjbyfkzj/s3593/201005/47529.shtml. Beijing (internal document, in Chinese).
Ministry of Health of the People's Republic of China (MOH), 2010b. Chinese Statistics Yearbook about Health. Press of China Union Medical University, Beijing, pp. 1–424.
Ministry of Health of the People's Republic of China (MOH), 2012. Detailed rules for the malaria elimination certification in pilot counties. Beijing (internal document, in Chinese).
Ministry of Health of the People's Republic of China (MOH), 2013. Certification plan for the malaria elimination in China, 2013ed. Beijing (internal document, in Chinese).
Ministry of Health of the People's Republic of China (MOH), 2014. Certification plan for the malaria elimination in China, 2014ed. Beijing (internal document, in Chinese).
National Criteria of People's Republic of China, 2010. Criteria for Control and Elimination of Malaria. GB 26345–2010. Beijing, Chin. Standard. Press.
Ning, X., Qin, L., Jinchuan, Y., Jiuping, Y., Xintian, L., 1999. Surveillance of risk factors from imported cases of falciparum malaria in Sichuan, China. Southeast Asian J. Trop. Med. Pub. Health 30, 235–239.
Shang, L.Y., 2008. Distribution. In: Tang, L.H. (Ed.), *Anopholes anthropophagus* in China: Biology and Control. Shanghai Science and Technology Press, pp. 83–89.
Shang, L.Y., Li, G.Q., Zheng, B., Gu, Z.C., Tang, L.H., Gao, Q., Xu, L.S., Zhang, S.Y., Chen, G.W., Wang, S.Q., Xia, Z.G., Yang, H.L., Pan, B., Wu, K.S., 2012. Malaria research. In: Tang, L.H., Xu, L.Q., Chen, Y.D. (Eds.), Control and Research of Parasitic Diseases in China. Beijing Science and Technology Press, pp. 73–160.
Shanghai Municipal Health Bureau, 2010. Shanghai Municipal Action Plan for ME (2010–2020). http://wsj.sh.gov.cn/wsj/n429/n432/n1485/n1496/u1ai80306.html. Shanghai (internal document, in Chinese).
Shanghai Statistics Bureau, 2010. Shanghai Municipal Statistics Yearbook—2010. China Statistics Press, Beijing. 1
Tambo, E., Ai, L., Zhou, X., Chen, J.H., Hu, W., Bergquist, R., et al., 2014. Surveillance-response systems: the key to elimination of tropical diseases. Infect. Dis. Poverty 3, 17.
Tang, L.H., Gao, Q., Gu, Z.C., Zhou, S.S., Yang, Q.G., Zhu, C.L., Xu, L.S., Wang, S.Q., Zhang, Z.X., Shang, L.Y., 2012. Malaria prevalence. In: Tang, L.H., Xu, L.Q., Chen, Y.D. (Eds.), Control and Research of Parasitic Diseases in China. Beijing Science and Technology Press, pp. 73–102.
WHO, RBM, 2012. Disease Surveillance for Malaria Elimination: An Operational Manual. Switzerland, Geneva.

WHO, 2014. World malaria report. Switzerland, Geneva. 2013.

Yao, L.N., Yu, K.G., 2012. Control and research of parasitic diseases in Zhejiang province. In: Tang, L.H., Xu, L.Q., Chen, Y.D. (Eds.), Control and Research of Parasitic Diseases in China. Beijing Science and Technology Press, pp. 1086–1114.

Yao, L.N., Zhang, L.L., Ruan, W., Chen, H.L., Lu, Q.Y., Yang, T.T., 2013. Species identification in 5 imported cases previously diagnosed as vivax malaria by parasitological and nested PCR techniques. Chin. J. Parasitol. Parasit. Dis 31 (221–223), 234.

Zhang, X.H., Ni, Q.X., Chen, Y., Yao, L.N., Lu, Q.Y., Yu, X.H., et al., 2013. Diagnosis and analysis of the first imported ovale malaria case in Wenzhou City. Chin. J. Parasitol. Parasit. Dis 31, 380–381.

Zheng, Q., Vanderslott, S., Jiang, B., Xu, L.L., Liu, C.S., Huo, L.L., et al., 2013. Research gaps for three main tropical diseases in the People's Republic of China. Infect.Dis. Poverty 2, 15.

Zheng, S.N., Chen, Y.F., Ma, Q.Q., Yuan, B.C., Zhou, Z.Q., Xu, F.L., 1989. Surveillance of imported falciparum malaria in Yongkang County, Zhejiang Province. Chin. J. Parasitol. Parasit. Dis 7, 122–124.

Zhou, S.M., Wang, C.X., Wu, K., Mao, C.X., Yang, Y., 2011. Application of nested PCR in diagnosis of imported malaria. Chin. J. Parasitol. Parasit. Dis 29, 43–45.

Zofou, D., Nyasa, R.B., Nsagha, D.S., Ntie-Kang, F., Meriki, H.D., Assob, J.C., et al., 2014. Control of malaria and other vector-borne protozoan diseases in the tropics: enduring challenges despite considerable progress and achievements. Infect.Dis. Poverty 3, 1.

CHAPTER SEVEN

Surveillance and Response Strategy in the Malaria Post-elimination Stage: Case Study of Fujian Province

Fa-Zhu Yang[1], Peiling Yap[2], Shan-Ying Zhang[1,*], Han-Guo Xie[1], Rong Ouyang[1], Yao-Ying Lin[1], Zhu-Yun Chen[1]

[1]Fujian Center for Disease Control and Prevention, Fujian, People's Republic of China
[2]Department of Epidemiology and Public Health, Swiss Tropical and Public Health Institute, Basel, Switzerland; University of Basel, Basel, Switzerland
*Corresponding author: E-mail: zsy@fjcdc.com.cn

Contents

1. Introduction	184
2. Background	185
2.1 The epidemic situation of malaria in Fujian	185
2.2 Stages of malaria control in Fujian	186
2.2.1 Stage of pilot projects on malaria control (1950–1955)	186
2.2.2 Stage of an integrated malaria control programme (1956–1976)	186
2.2.3 Stage of management of individual malaria cases (1977–1985)	189
2.3 Stage of malaria pre-elimination in Fujian (1986–2000)	189
2.4 Stage of elimination and post-elimination (2000–present)	192
2.5 Geographic patterns of malaria epidemic areas	194
3. Surveillance after Malaria Elimination	195
3.1 Population surveillance	196
3.2 Mosquito vector surveillance and control	196
3.3 Genetic monitoring	197
4. Response in post-elimination	197
4.1 Local cases and epidemic status	198
4.2 The distribution pattern of malaria cases	199
4.3 Establishment of malaria monitoring sites	200
4.3.1 Monitoring sites in areas originally endemic with malaria	200
4.3.2 Monitoring imported malaria in the monitoring sites	200
5. Experiences and Lessons	201
6. Conclusion	202
Acknowledgements	202
References	202

Abstract

Malaria used to be a serious health problem in Fujian province in the past, but no local malaria transmission has been found since 2000. In order to eliminate the potential residual cases and prevent re-introduction of malaria so as to achieve the final goal of malaria elimination in Fujian province, various strategy and intervention approaches were tailored to the local settings. For instance, the monitoring of febrile patients by blood smear examinations and vector surveillance and control were strengthened in addition to the routine intervention in the mountainous area of Fujian province, where malaria was highly endemic and the mosquito *Anopheles anthropophagus* distributed with a high vectorial capacity. There were two local cases who got infected due to imported cases found in the building site of an expressway in 2004 and 2005, respectively. All other imported malaria cases were detected during post-elimination stage through surveillance system. Based on results from post-transmission surveillance, malaria transmission has been interrupted in Fujian province for 13 years. Therefore, post-transmission surveillance and response is an important intervention to maintain the malaria elimination achievements in Fujian province.

1. INTRODUCTION

Malaria has a long history in Fujian province, which is known as "a place of miasma", located in Southern China (Su and Shen 1956). The earliest record of malaria in Fujian province was traced back to the Tang Dynasty, 2000 years ago (Feng, 1932). For a long time, only a few scholars had completed research on the epidemiology of malaria cases, and they failed to take effective prevention and control measures against the disease. This led to malaria being one of the most serious infectious

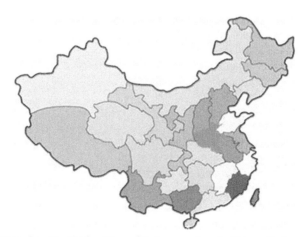

Figure 7.1 *Map showing the location of Fujian province in China.* Red area is the location of Fujian Province.

diseases in Fujian over a long period of time (Hemenway, 1930). After the founding of the People's Republic of China (P.R. China) in 1949, a malaria control institute was established in Fujian province (Zhang, 1984). Pilot projects on the epidemiology and control of malaria were actively carried out (Xia, 1958). These interventions had allowed Fujian to effectively control the malaria epidemic and eliminate malaria within the province (Figure 7.1).

2. BACKGROUND
2.1 The epidemic situation of malaria in Fujian

The modern literature of malaria in Fujian began in 1871, when the British scholar, Dr. Patrick Manson, practiced medicine in Xiamen, Fujian Province (Chen, 1986). His survey found a malaria incidence of 16.2% in 1871. Subsequently, many Chinese and foreign scholars have reported malaria in Fujian Province afterwards (Dang, 1935; Helena, 1937; Lee 1932).

Based on epidemiological data in history, areas endemic with malaria in Fujian are confined to the hilly region, or northwest regions of the province. Once the environmental or social factors change, incidence rises sharply in a short time to more than the annual level of cases and thus produces an outbreak of malaria (Zheng, 1985). The range of such an epidemic is often beyond the original endemic area. It might further spread throughout the province, causing a pandemic. For example, malaria broke out in Fujian from 1937 to 1938 (Chen, 1940). The annual incidence was 3076.9 per 10,000. Later, two pandemics occurred throughout the province, with peaks in cases appearing in 1954 and 1974, where the incidence was 266.43 per 10,000 and 238.25 per 10,000, respectively (Wu et al., 1998).

Three species of *Plasmodium* protozoan parasites had been found in Fujian, namely *Plasmodium vivax*, *P. falciparum* and *P. malariae*. *P. vivax* is the most widely distributed, followed by *P. falciparum* (Gu, 1940; Lahann, 1921). The distribution of *P. malariae* is scattered throughout the province. More than 18 species *Anopheles* had been found in Fujian, but the main vectors for malaria transmission are *Anopheles anthropophagus*, *An. sinensis* and *An. minimus* (Hu, 1037). *An. anthropophagus* has an affinity for human smells and blood. Their medium energy is 19.2 times higher than that of *An. sinensis*, which are most widely distributed throughout Fujian, especially in the low endemic areas (Xu, 2009). *An. sinensis* suck the blood of both human and animals, but they prefer the blood of cattle. They can breed in rice paddies and other types of water bodies, thus resulting in their large population number. The density fluctuation of *An. sinensis* coincides with malaria epidemic curve (Xu and Wu, 1981).

2.2 Stages of malaria control in Fujian

After the founding of the P.R. China in 1949, Fujian underwent three stages to effectively control malaria until 1986, e.g. (1) establishing pilot projects on malaria control, (2) launching a comprehensive malaria prevention and control programme and (3) performing management of individual malaria cases. Having achieved these three milestones, the malaria incidence of Fujian dropped to below 1 per 10,000 in 1986 (Wu and Huang, 1986), which could be considered an initial victory for malaria control in China. Then two stages, including the pre-elimination stage and post-elimination stage, were implemented with an emphasis on the surveillance from 2000 to present. With the efforts during the stages, local malaria transmission has been interrupted eventually.

2.2.1 Stage of pilot projects on malaria control (1950–1955)

In the early 1950s, the national economy was faced with a period of convalescence. Work on the prevention and control of malaria was faced with challenges, for instance, a large number of patients and a lack of experience and expertise in the whole province. Therefore, the main focus for malaria control during this stage was (1) establishing pilot projects on malaria control, (2) investigating the epidemic situation, (3) launching work in key areas and (4) summarizing experiences. These were the provisions, which eventually led to a comprehensive control strategy against malaria in the province.

During this period, workstations were established to monitor the epidemiological status of malaria in Jianning, Yongan and Ninghua, respectively. Observation of the effect of different methods in those workstations provided the basis for the provincial malaria control programme. Although antimalarial work was successful in various pilot projects, in-depth work in the entire province had not yet been carried out. Thus, the epidemic was still increasing year by year, and reached its peak in 1954 and 1955, where the incidence rate was 2.67% and 2.13%, respectively (Figure 7.2).

2.2.2 Stage of an integrated malaria control programme (1956–1976)

In this period from 1956 to 1976, a large number of control principles were guided by the experience of the pilot projects, where comprehensive

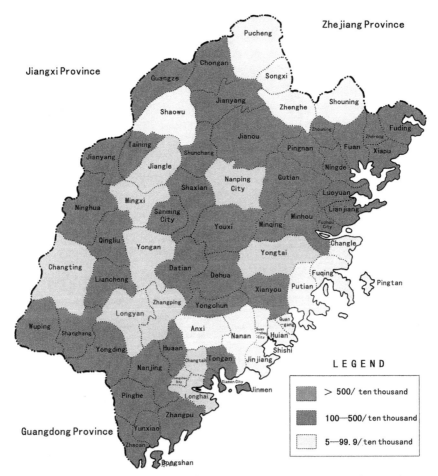

Figure 7.2 *Geographic distribution of malaria in Fujian in 1954.* Dark lines represent regional boundaries in Fujian; green areas represent the incidence rate of malaria was over 500 per 10,000; red represents areas where the incidence rate was between 100 and 500 per 10,000; yellow areas show where the rate was between five and 99.9 per 10,000. (For interpretation of the references to color in this figure legend, the reader is referred to the online version of this book.)

measures that mainly eliminated the source of infection were carried out through case management. In addition, intensive treatment and monitoring of patients and preventive medicine in transmission season were also performed. A public health campaign was carried out throughout the province for reducing mosquito density. In the key endemic areas, pesticides were used in indoor residual spraying against the mosquitoes to control the

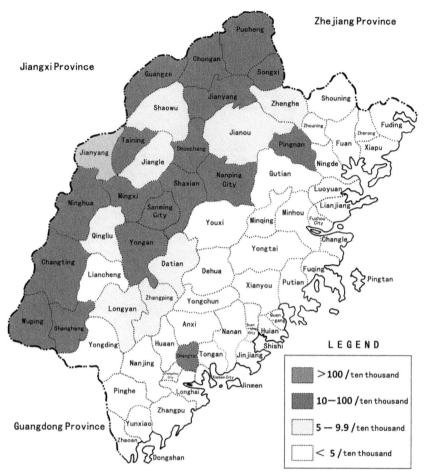

Figure 7.3 *Geographic distribution of malaria in Fujian in 1964.* Green areas represent the incidence rate of malaria was over 100 per 10,000; red areas represent the incidence rate was between 10 and 100 per 10,000; yellow areas show the malaria incidence rate was between five and 9.9 per 10,000; white areas present the incidence rate at less than five per 10,000. (For interpretation of the references to color in this figure legend, the reader is referred to the online version of this book.)

epidemic. In 1964, the annual number of malaria cases decreased to 11,489, which was 89% less than in 1956 (Figure 7.3).

In 1966, malaria broke out in some villages in Shunchang and Shaowu located in Northern Fujian. However, due to political reasons, malaria control activities were in a state of paralysis. The malaria epidemic grew rapidly, and the epidemic areas expanded each year. In 1972, the number of malaria cases were increased up to 512,550 people. The annual incidence rate was 238.25 per 10,000, which was close to the level in 1954. In 1973, health professionals who had taken lessons on the prevention and control of malaria discussed and formulated the countermeasures in the next period of time. Thus, a three-year working plan (1973–1975) for prevention and control of malaria was made and aimed to reduce the morbidity rate of malaria by 40% compared to the previous year. According to statistical data from 1973 to 1976, more than 3,530,000 people were treated, over 9,320,000 took antimalarial prophylaxis, 1,660,000 living rooms were sprayed with pesticides and 1,030,000 mosquito-breeding sites were cleaned up. These measures were able to control the malaria epidemic effectively and the annual incidence declined substantially to 11.2 per 10,000 in 1976 (Figure 7.4).

2.2.3 Stage of management of individual malaria cases (1977–1985)

In this stage from 1977 to 1985, the malaria incidence rate was decreased significantly in most parts of the province, so prevention efforts were essential to the places where a high incidence of malaria still occurred, in order to decrease the incidence rate further year by year. In areas where malaria has been successfully controlled, the monitoring and management schemes need to be maintained in the late stage of malaria control, including the investigation of infectious sources, management of positive patients and the mobile populations and foci responses, etc. (Huang, 1986). Since 1980, monitoring work, such as blood smear examination of febrile patients (which was the main measure to identify the infection source), treatment of patients, case investigation and vector surveillance, was done to maintain the previous achievements of malaria control. In 1985, the incidence rate of malaria dropped to 1.02 per 10,000 in Fujian (Figure 7.5) (Wu et al., 1991).

2.3 Stage of malaria pre-elimination in Fujian (1986–2000)

After 1986, malaria endemic areas in Fujian were gradually shrinking, and the incidence rate was maintained at less than 1%. But because malaria infection has the characteristics of a short transmission cycle with a high reinfection rate, being fast spreading where was a wide distribution of *An. anthropophagus*, it was difficult to control the transmission and local continues to exist for a long term. To achieve elimination, "antimalarial defense cooperation networks" were set up in 22 counties in two cities, namely

Figure 7.4 *Geographic distribution of malaria in Fujian in 1972.* Green areas represent the incidence rate of malaria was over 500 per 10,000; red represents areas where the incidence rate was between 100 and 500 per 10,000; yellow areas show the rate was between five and 99.9 per 10,000; white areas present the rate at less than five per 10,000. (For interpretation of the references to color in this figure legend, the reader is referred to the online version of this book.)

Nanping and Sanming of Fujian. Two kinds of measures to prevent and control malaria in these areas were adapted to local conditions. First, in areas where *An. anthropophagus* was widely distributed, comprehensive measures against the vector were taken to interrupt transmission effectively and control outbreaks rapidly. Second, surveillance and management in a postcontrol setting were also actively carried out. Distinct parameters of blood

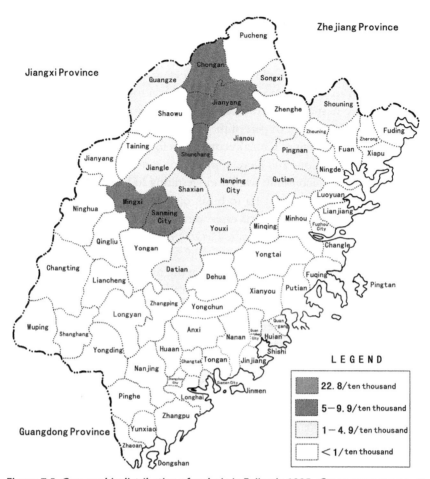

Figure 7.5 *Geographic distribution of malaria in Fujian in 1985.* Green areas represent the incidence rate of malaria was 22.8 per 10,000; red areas represent the incidence rate was between five and 9.9 per 10,000; yellow areas show the rate was between 15 and 4.9 per 10,000; white areas present the rate at less than one per 10,000. (For interpretation of the references to color in this figure legend, the reader is referred to the online version of this book.)

smear examinations were made in different endemic areas, which not only saved health resources but also allowed identification of infectious sources. Due to such measures, the incidence decreased every year, dropping from 1593 cases in the beginning of 1988 to 59 cases in 1995, achieving a reduction of 96.3%. The total proportion of cases dropped from 75.5% in 1988 to 6% in 1995. *An. anthropophagus* had also been effectively controlled and only eight mosquitoes of *An. anthropophagus* were found in two villages in 1995. Incidence rate of the 22 counties (cities, districts) within the "antimalarial defense cooperation networks" all fell to below 1%, which reached the criteria of malaria pre-elimination (Xu et al., 1995).

Although malaria in most of Fujian was controlled effectively by 1986 with a feature of sporadic cases distributed in the province, changes in infection trends in the southern part of the province were subsequently detected. With the development of China's economic reform and the formation of the Southern Fujian Economic Development Zone in the mid 1980s, there was a change in social economic conditions, resulting in a large influx of mobile population and causing malaria to break out locally on a regular basis after 1985. For example, 1156 malaria cases were found from 1984 to 1994 in the south of Fujian, most of which were *P. vivax* malaria cases. Blood smear examinations of febrile patients at the point of outbreak found that the positive rate was as high as 49.7%. In 1995, most of the malaria cases were distributed in 10 particular villages, towns and farms of Southern Fujian, accounting for 83.6% of 983 cases total in the province. After a massive malaria prevention and control campaign, the epidemic was effectively controlled and the number of cases decreased gradually again. With the above-mentioned control efforts, the last outbreak was detected in 1997. In 1999, only 28 patients were found in the whole province and the incidence rate dropped to the lowest in history (Zhang et al., 1999). After 1998, no more new outbreaks occurred in Fujian, and only the import cases were sporadically detected.

2.4 Stage of elimination and post-elimination (2000–present)

In order to strengthen the prevention, control and surveillance of malaria, "antimalarial defense cooperation networks" were expanded to nine counties of three cities, namely Xiamen, Zhangzhou and Quanzhou, in Southern Fujian in 1996, where the malaria epidemic was still serious in mobile population after social economic development. The regional antimalarial and monitoring network was established to provide timely information about the epidemic status of malaria. It allowed health officials to grasp malaria dynamics and manage control efforts among the mobile population.

At this stage, three activities were strengthened, such as the blood examination among febrile patients, prompt treatment to the confirmed cases and active searches for the source of infection. Once malaria outbreak occurred, the following response measures were undertaken immediately. First, mosquitoes in households of each village where *An. anthropophagus/An. minimus* were distributed were treated by providing insecticide-treated bed nets, residual indoor spraying and breeding transformation techniques. Second, management of the mobile population was strengthened with registration of patients, investigation of infection history, blood smear examinations and treatment of confirmed cases. Third, the mobile population was surveyed by understanding the time of migration into the local place and their living condition. Fourth, confirmed cases were then timely treated and followed up. Thanks to the intensive surveillance, only two local cases, who received the infection through the imported cases, were found in 2004 and 2005, respectively. The imported malaria cases started to increase gradually during this transition period of time (Zhang et al., 1999).

In summary, great achievements on the malaria control and elimination programme have been gained in Fujian since 1950, causing a gradual decline in the epidemic with the following milestones. (1) *Plasmodium malariae* was eliminated throughout the province in 1965, (2) locally infected cases with *P. falciparum* were not detected after 1984, (3) the epidemic area of *P. vivax* was significantly reduced with the decline of infection rate, and no local malaria transmission has been found since 2000 and (4) the mosquito population of *A. anthropophagus* and *A. minimus* was also effectively controlled and the malaria annual incidence was less than one per 10,000 after 1986 (Figure 7.6).

Figure 7.6 *Logarithmic value of annual incidence (malaria case per 10,000 population) from 1995 to 2000.*

2.5 Geographic patterns of malaria epidemic areas

According to the epidemiological characteristics of malaria and the 60-year prevention and control efforts in Fujian province, three types of epidemic regions can be divided among 85 counties (cities/districts) in the province (Table 7.1).

The first type of the region consists of areas severely endemic with malaria and is correlated to the density of *An. anthropophagus*, which included 22 counties in two cities of Sanming and Nanping in Northern Fujian. In the early 1970s, the incidence rate reached as high as 124%, and *An. anthropophagus* was the main vector in the malaria transmission. After years of efforts, the malaria incidence rate was reduced remarkably. This region has been considered as the malaria elimination area after evaluation in 1998.

The second type of the region consists of areas with unstable malaria transmission and is correlated to the density of *An. minimus*. This region covered 28 counties in three cities of Xiamen, Zhangzhou and Quanzhou in Southern Fujian. At the beginning of the 1970s, the malaria epidemic only affected a part of these areas, and malaria transmission had been controlled by the early 1980s. However, with the social economic development and population mobilization, the malaria epidemic had presented locally since the beginning of the 1990s, or malaria resurgence occurred in the region.

The third type of the region consists of areas with sporadic cases of malaria covering 35 counties in four cities of Ningde, Fuzhou, Putian and Longyan. In the early 1970s, malaria only affected a few counties in the region and was controlled in the late 1970s. No outbreak or infected case of malaria has been found since the end of the 1980s. Only a few imported cases from other counties were frequently detected in recent years (Figure 7.7).

Table 7.1 The three types of malaria areas in Fujian province

Types	Geography	Vector (s)	Risk of transmission
Severe area	North of Fujian	*An. anthropophagus* and *An. sinensis*	High
Unstable area	South of Fujian	*An. minimus* and *An. sinensis*	Moderate
Sporadic area	Others	*An. sinensis*	Low

Figure 7.7 *The malaria distribution in three types of region in Fujian.* Red represents a severe area, which is in the north of Fujian and is correlated to the density of *Anopheles anthropophagus*; yellow represents an area of unstable transmission, which is in the south of Fujian and is correlated to the density of *Anopheles minimus*; blue represents a sporadic area, in other places of Fujian). (For interpretation of the references to color in this figure legend, the reader is referred to the online version of this book.)

3. SURVEILLANCE AFTER MALARIA ELIMINATION

In order to consolidate the achievements of the malaria prevention and control programme, activities in the monitoring and blood smear examinations of three types of febrile patients also strengthened in Fujian from the year 2000 onwards. Patients were followed and treated with the standard protocol in a

Table 7.2 Blood examinations of three types of febrile patients in Fujian 2000–2012

Year	No. blood examination	No. positive	Local case	No. patients in Fujian	Ratio of local patients to the total cases (%)	Death cases
2000	35,721	35	3	7	20.00	4
2001	30,025	41	3	9	21.95	0
2002	33,105	85	1	7	8.24	1
2003	29,347	113	0	48	42.48	0
2004	29,107	61	1	12	19.67	0
2005	27,862	44	2	11	25.00	0
2006	30,809	49	0	9	18.37	1
2007	30,405	44	0	7	15.91	0
2008	22,374	79	0	9	11.39	0
2009	15,497	63	0	22	34.92	0
2010	27,436	68	0	31	45.59	1
2011	39,482	68	0	30	44.12	0
2012	63,068	54	0	31	57.41	1
Total	414,238	804	10	233	28.98	8

timely manner. A system to register and report patients was built to understand the epidemic and its dynamics. Three to five villages in each county, where *An. anthropophagus* and *An. minimus* were originally distributed, were surveyed/reviewed every year. The risk of malaria was also regularly assessed to determine corresponding measures for imported malaria cases (Ouyang et al., 2013).

3.1 Population surveillance

The population surveillance was implemented by blood examinations of three types of febrile patients, including those patients who were diagnosed as malaria cases or suspicious malaria cases, or febrile patients without any explanation. A total of 414,238 febrile patients were tested for blood examination from 2000 to 2012. A total of 804 confirmed malaria cases were found among febrile patients, including 10 local cases, 233 (accounting for 29%) imported cases of *Falciparum* malaria and eight deaths (Table 7.2).

3.2 Mosquito vector surveillance and control

In the malaria monitoring system, vector surveillance had been carried out in Fujian. Most vectors were *An. sinensis*, except for two mosquitoes of (Table 7.2) *anthropophagus* found in 2001 and 14 mosquitoes of (Table 7.2) *minimus* found in 2001 and 2002, respectively (Ouyang et al., 2004, 2006). In areas with high mosquito density, insecticide-treated bed nets were

Table 7.3 Malaria vector surveillance and control in Fujian 2000–2012

Year	An. anthropophagus	An. sinensis	An. minimus
2000	0	3026	0
2001	2	3210	2
2002	0	766	12
2003	0	6233	0
2004	0	4546	0
2005	0	2915	0
2006	0	3928	0
2007	0	3824	0
2008	0	1585	0
2009	0	2741	0
2010	0	2444	0
2011	0	1093	0
2012	0	1497	0
Total	**2**	**37,808**	**14**

distributed and indoor residual spraying was still carried out to protect the population and prevent malaria (Table 7.3).

3.3 Genetic monitoring

In addition to the monitoring of malaria patients, the molecular approaches have also been applied in the malaria monitoring. For example, the genotyping and molecular epidemiological survey of imported malaria cases were investigated by using the merozoite surface protein 1 (PvMSP-1) of *P. vivax* (Zhang et al., 2004). By different approaches, such as genotyping, sequencing, sequence analysis and alignment, the relationship of the different genetic sources was shown and it was also speculated that the imported cases were from Myanmar (Zhang et al., 2001a,b).

4. RESPONSE IN POST-ELIMINATION

After the elimination of malaria transmission, where no local case was found in Fujian since 2000, malaria surveillance as an intervention was continued to perform by the two major approaches. One was monitoring management in large construction sites, especially in places where more mobile populations were located (Liu, et al., 2012; Marks, et al., 2014). The other one was response to malaria resurgence due to the imported cases, according to different local conditions so the introduction of new malaria cases were effectively controlled (Tambo, et al. 2014).

4.1 Local cases and epidemic status

When carefully investigating the malaria epidemic status in Fujian after 2000, it was found that the major risk of malaria epidemic was introduced by the imported cases who came from other provinces or countries, either directly or indirectly. Due to quick and correct responses, no local transmission occurred, which was demonstrated in the following three events.

In the first event, from 2000 to 2002, seven patients were detected from Southern Fujian where (Table 7.2) *minimus* were distributed, including five cases in Nanan (three cases found in 2000 and two cases in 2001), one case in Tongan and one case in Longhai (Zhang et al., 2005). The survey found that these seven local cases were all migrant workers or had close contact with migrant workers. With prompt treatment and surveillance of the local communities, no other new malaria case has been found thus far.

In the second event, two malaria patients were diagnosed in Xudun hospital in September, 2004. These patients were originally resident in Jianou of Northern Fujian, including one migrant worker from Honghu hydropower station and one villager from Hongtang of the town of Xudun (Zhang et al., 2007). These cases were initially identified to be caused by imported infections. The epidemic prevention station of Jianou immediately reported the cases to Nanping Center for Disease Control and Prevention and Jianou Health Bureau. At the same time, an epidemiological investigation and response were also carried out. Thirty-two neighborhood people underwent blood examination, but all turned out to be negative. In addition, 19 of *An. sinensis* were captured, but all of them were negative of infection. After taking a series of effective prevention and control measures, no other malaria case occurred.

In the third event, three malaria patients, one migrant and two local workers, were detected in June, 2005. All of them were from a highway construction site in Taining county. After the cases were reported, field epidemiological surveys were carried out in Sanming and Taining. Sixty-four close contacts were examined for *Plasmodium* infections, but no new case was discovered. The confirmed cases received chemotherapy promptly, while the contacts were given prophylaxis treatment. After much publicity among the local communities and other comprehensive prevention and control efforts, no new local infection was detected (Gui and Chen, 2009).

Based on the above-mentioned local epidemics caused by migrant workers, the Northern Fujian continues to face a high risk for the resurgence of malaria and thus case management, especially on imported cases, should be strengthened in the future (Ouyang et al., 2006).

4.2 The distribution pattern of malaria cases

When tracing back to the 1980s, the distribution of malaria cases in Northern Fujian, which was the focal epidemic area, appeared as three patterns. The first pattern was that the annual incidence from 1986 to 2000 was sustainably reduced to a low level, less than 1 per 10,000. For instance, in 1986, the annual incidence of malaria in Fujian was below 1per10,000. Malaria cases were mainly distributed in severely endemic areas in Northern Fujian (Zheng et al., 1995). There were 1007 cases found in 1987 in Northern Fujian, accounting for 64.9% of the total number of malaria cases in the whole province. In addition, there were seven counties with an incidence rate more than 10 per 1000. In 1995, there were only 59 confirmed cases from Northern Fujian, accounting for only 6% of the total number of cases in the province. But in the south there were 822 cases, accounting for 83.62%. The key monitoring efforts of malaria cases in Fujian therefore switched from north to the south of the province.

The second pattern occurred after 2000. Malaria in Northern Fujian largely appeared as a sporadic pattern at a very low level of incidence (less than 1 per 100 thousands), accounting for more than 50% of the total number of cases in the province (Lin et al., 2009) (Figure 7.8).

The third pattern occurred after 2005. No local cases were found but the imported cases found were increasing year by year in the province. For instance, *Falciparum* malaria made up a rising share of the imported cases (Figure 7.9). Within 54 imported malaria cases in 2012, 30 (55.6%) of them were due to infection with *P. falciparum*. The majority of the 68 imported cases were came from Africa (46 cases; 67.7%), followed by Southeast Asia (18 cases; 26.5%).

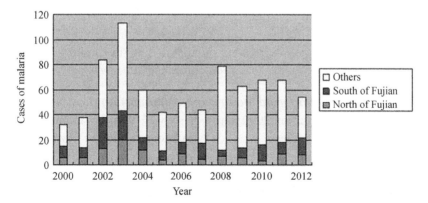

Figure 7.8 *Malaria cases distribution in each of three type regions in Fujian after 2000.*

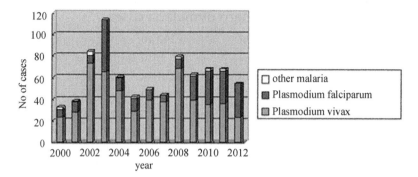

Figure 7.9 *Variations on imported cases of malaria in Fujian from 2001 to 2012.*

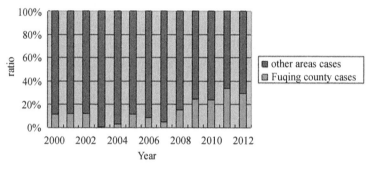

Figure 7.10 *Ratio of Fuqing county cases to the total cases of the province after 2000.*

4.3 Establishment of malaria monitoring sites

4.3.1 Monitoring sites in areas originally endemic with malaria

For a better understanding of the malaria epidemic situation, Fujian has established monitoring sites in Jimei in the south, and in Jianyang and Jiangle in Northern Fujian, where originally endemic with malaria. The major monitoring works in the sites were monitoring the occurrence of malaria cases and vector population dynamics in the local communities. The results show that no new case was found in the monitoring sites, indicating malaria transmission in the province has been controlled effectively.

4.3.2 Monitoring imported malaria in the monitoring sites

Malaria monitoring from 2000 onwards showed that the number of the imported cases were increasing, especially in Fuqing county of Fuzhou, where more international trades happened in an increasing pattern. So that the annual number of local cases has been about 23% of the total number of cases in the province, in 2009 (Figure 7.10), due to a lot of the imported cases from personnel coming back from abroad. A monitoring site was set up in Fuqing in 2012 for the surveillance of patients and the screening of returnees. Survey efforts of vectors from 2013 have also been performed to determine the species and density of malaria vectors there.

5. EXPERIENCES AND LESSONS

After the control and surveillance of malaria in Fujian province for more than 60 years, the following results have been achived in Fujian. (1) Interruption of malaria transmission: Fujian used to be one of the most serious malaria endemic areas in P.R. China. Through the great efforts for many years, the incidence has decreased year by year. There was basically no local infection in 2000. After 2005, only imported cases were detected. In other words, the goal of eliminating malaria has achieved in Fujian. (2) Effective control of the vector had great impact to the reduction of malaria transmission. After years of efforts, including clearing breeding sites of *An. anthropophagus* and *An. minimus*, of which both have a high capacity for malaria transmission, the vector transmission capacity has been effectively controlled. It has never been found any of these two mosquitoes in Fujian since 2002. Malaria transmission has thus been interrupted effectively since 2005. At the same time, there's also an increased awareness of reducing the mosquito population through public campaigns. (3) The malaria monitoring system has been established in Fujian through professional's training. After years of practice, a large number of malaria control personnel has been trained, which constructed malaria prevention and monitoring teams, thereby the malaria monitoring system was improved significantly. The system can effectively discover a patient, treat and report the case in a timely manner, provide an early warning signal and perform the necessary antimalaria response (Yadav, et al., 2014; Zofou, et al., 2014). This will help to ensure that malaria does not resurgence.

Two important experiences from the elimination of malaria in Fujian are highlighted herewith. The first one is that in order to control malaria effectively, it is essential to find out the status and characteristics of the epidemic patterns. The response approaches are tailored to local circumstances based on specific guidance for different localities. The second one is that malaria elimination and surveillance efforts should receive strong support from the government and the local people. If the government does not increase funds or the communities do not support or understand the efforts, a lot of work cannot be implemented, carried out or completed.

6. CONCLUSION

Although the transmission of malaria has been interrupted in Fujian province, *An. sinensis* still exists in the province. Imported cases of malaria have also increased with an influx of mobile population. Therefore, the risk of malaria resurgence still exists. As the number of imported cases surged, population infected with *P. falciparum* also increased, which caused more serious challenges to the prevention of malaria reintroduction in the province. These infections can often lead to death due to lack of timely treatments. Hence, the surveillance and management of malaria can not slack up our further efforts. To maintain the malaria elimination status, investment in health resources need to be increased through the strengthening of local health personnel, specialized in antimalarial efforts (Zheng, et al., 2013).

ACKNOWLEDGEMENTS

This work was supported in part by the Foundation of National Science and Technology Major Programme (Grant No. 2012ZX10004-220) and by China UK Global Health Support Programme (grant no. GHSP-CS-OP1).

REFERENCES

Chen, G.Z., 1940. Malaria in Fujian Province, China. Chin. Med. J. 26 (1), 1011 (in Chinese).
Chen, G.Z., 1986. The history of parasitic disease in Fujian. Chin. Med. Doctor J. 16 (3), 162–166 (in Chinese).
Dang, M.G., 1935. Report of nine hundred and sixty cases of malaria in Mintsing, Fukien. Chin. Med. J. 49 (1), 1235–1240.
Feng, L.Z., 1932. The studies of malaria and transmission in Fujian. Chin. Med. Doctor J. 18 (3), 370–395.
Gu, R.Y., 1940. The mosquitoes of Shaxian. Fujian Acta Physiol. College 1 (5), 1–2 (in Chinese).
Gui, S.G., Chen, S.H., 2009. Analysis on survey of malaria in Sanming Region between 2001-2008. Cross-strait J. Prevent. Med. 15 (6), 46–47 (in Chinese).
Helena, W., 1937. Some notes on malaria in Foochow. Chin. Med. J. 51, 385–390.
Hemenway, R.V., 1930. One hundred malaria cases (Mintsing Fukien). Chin. Med. J. 44 (11), 1118–1123.
Hu, S. M. K. (1937). A brief mosquitoes survey of Foochow region, South China. Ling. Sci. J. 16(4): 579–584.
Huang, B.F., 1986. The measure of monitor and manage after control of malaria. Fujian Pub. Health 14, 70–71.
Lahann, W.A., 1921. The mosquitoes of some parts of China and Japan. Bull. Ent Res. 12, 401–409.
lee, C.U., 1932. A survery of mosquitoes and their breeding habits in the Amoy region. Mar. Biol. Ass'n China 75–77.
Lin, Y.Y., Zhang, S.Y., et al., 2009. Analysis of imported malaria cases in Fujian in 2008. Chin. J. Pathogen Bio. 4 (8), 639 (in Chinese).
Liu, J., Yang, B., Cheung, W.K., Yang, G., 2012. Malaria transmission modelling: a network perspective. Infect Dis Poverty 1, 11.
Marks, M., Armstrong, M., Walker, D., Doherty, T., 2014. Imported falciparum malaria among adults requiring intensive care: analysis of the literature. Malar J 13, 79.

Ouyang, R., Xie, H.G., et al., 2013. Analysis on malaria in Fujian province between 2006-2012. Cross-strait J. Prevent. Med. 19 (4), 73–74 (in Chinese).

Ouyang, R., Xie, H.G., et al., 2006. Analysis on malaria in Fujian province in 2005. Cross-strait J. Prevent. Med. 12 (6), 29 (in Chinese).

Ouyang, R., Zhang, S.Y., et al., 2004. Analysis on malaria in Fujian province in 2003. Dis. Surveill. 19 (8), 301–302 (in Chinese).

Su, S., Shen, G., 1956. The midicine of Su and Shen. People's Health Press, Beijing, pp. 1–100.

Tambo, E., Ai, L., Zhou, X., Chen, J.H., Hu, W., Bergquist, R., et al., 2014. Surveillance-response systems: the key to elimination of tropical diseases. Infect Dis Poverty 3, 17.

Wu, J.J., Huang, B.F., 1986. Observation on the malaria control in Fujian Province. Fujian Pub. Health 15, 74–77.

Wu, J.J., Huang, B.F., et al., 1991. Effect of control malaria for 40 years in Fujian Province. Chin. J. Parasit. Dis. Control 4 (3), 165–166 (in Chinese).

Wu, J.J., Xu, L.S., et al., 1998. Diagnoses on effect and contermeasure of crotrol of malaria in Fujian Province. Med. Animal Control 14 (6), 46–48 (in Chinese).

Xia, Z.X., 1958. The report of malaria in Fujian Province. Fujian Med. J. 1 (1), 43–49 (in Chinese).

Xu, L.S., 2009. Developmment report on medical parasitology in Fujian Province. Cross-Strait Sci. 25 (1), 81–90 (in Chinese).

Xu, L.S., Wu, J.J., 1981. The relation between *Anopheles sinensis* and malaria. Fujian Pub. Health 5 (1), 37–43 (in Chinese).

Xu, L.S., Wu, J.J., et al., 1995. Observation on the efficacy of deltamethrin against mosquitos in a large scale for malaria control. Cross-strait J. Preventive Med. 1 (1), 9–11 (in Chinese).

Yadav, K., Dhiman, S., Rabha, B., Saikia, P., Veer, V., 2014. Socio-economic determinants for malaria transmission risk in an endemic primary health centre in Assam, India. Infect Dis Poverty 3, 19.

Zhang, S.Y., Lu, H.M., et al., 2001a. Applied study on detection of plasmodium in blood samples by multiplex polymerase chain reaction. Cross-strait J. Preventive Med. 7 (1), 7–10 (in Chinese).

Zhang, S.Y., Lu, H.M., et al., 2001b. Evaluation of the effect of multiplex polymerase chain reaction for malaria surveillance in southern Fujian. Chin. J. Zoonoses 17 (2), 72–74 (in Chinese).

Zhang, S.Y., Xu, L.S., et al., 1999. The malaria epidemic posture and its control in recent years in Fujian Province. Chin. J. Parasit. Parasit. Dis. 12 (3), 167–169 (in Chinese).

Zhang, S.Y., Xu, L. s., et al., 2005. Study on the epidemic characteristics of malaria evaluation on the intervention effects in Southen Fujian. Cross-strait J. Preventive Med. 11 (5), 1–4 (in Chinese).

Zhang, S.Y., Xu, L.S., et al., 2004. Analysis on introduced malaria cases in Fujian by *Plasmodium vivax* merozoite surface protein 1(PvMSP-1) gene sequence. Cross-strait J. Preventive Med. 10 (3), 4–8 (in Chinese).

Zhang, Z.F., 1984. The investigation of parasitic dieases in Fujian Province. Fujian J. Med. 5 (5), 68.

Zhang, Z.P., Zhou, M.Y., et al., 2007. Report of one case of malaria. Chin. J. Zoonoses 21 (11), 1166–1167 (in Chinese).

Zheng, Q., Vanderslott, S., Jiang, B., Xu, L.L., Liu, C.S., Huo, L.L., et al., 2013. Research gaps for three main tropical diseases in the People's Republic of China. Infect Dis Poverty 2, 15.

Zheng, Z.J., 1985. Discussion on outbreak of malaria in Fujian Province. Fujian Pub. Health 12, 87–89.

Zheng, Z.J., Xu, L.S., et al., 1995. Study on malaria in Southen Fujian. Chin. J. Pathog. Bio. 8 (2), 145–149 (in Chinese).

Zofou, D., Nyasa, R.B., Nsagha, D.S., Ntie-Kang, F., Meriki, H.D., Assob, J.C., et al., 2014. Control of malaria and other vector-borne protozoan diseases in the tropics: enduring challenges despite considerable progress and achievements. Infect Dis Poverty 3, 1.

CHAPTER EIGHT

Preparation of Malaria Resurgence in China: Case Study of Vivax Malaria Re-emergence and Outbreak in Huang-Huai Plain in 2006

Hong-Wei Zhang[1], Ying Liu[1], Shao-Sen Zhang[3,4], Bian-Li Xu[1], Wei-Dong Li[2], Ji-Hai Tang[3,4], Shui-Sen Zhou[3,4], Fang Huang[3,4,*]

[1]Henan Center for Disease Control and Prevention, Zhengzhou, People's Republic of China
[2]Anhui Center for Disease Control and Prevention, Hefei, People's Republic of China
[3]National Institute of Parasitic Diseases, Chinese Center for Disease Control and Prevention, Shanghai, People's Republic of China
[4]Key Laboratory of Parasite and Vector Biology, Ministry of Health; WHO Collaborating Centre for Malaria, Schistosomiasis and Filariasis, Shanghai, People's Republic of China
*Corresponding author: E-mail: ipdhuangfang@163.com

Contents

1. Introduction	206
2. Background	206
3. Vivax Malaria Re-emergence and Outbreak in Huang-Huai Plain	209
3.1 Distribution by district of malaria outbreaks in the Huang-Huai Plain in 2006	210
3.2 Distributions by sex, age and occupation of malaria cases in outbreak spots in 2006	210
3.3 Distribution by month of malaria cases in outbreak spots in 2006	211
4. Characteristics of Malaria Re-emergence and Outbreak in Huang-Huai Plain	213
4.1 Aggregation of malaria cases	213
4.2 Sole vector of *An. sinensis*	213
4.3 Main cases located around water bodies	214
5. Factors of Malaria Re-emergence and Outbreak in Huang-Huai Plain	215
5.1 Land use has nothing to do with re-emergance and outbreak in the Huang-Huai Plain	215
5.2 Climate warming and increasing rainfall extending malaria transmission	215
5.3 Increasing vectorial capacity of *An. sinensis* strengthens malaria transmission	216
5.4 Low capacity of diagnosis leading to accumulation of infectious sources	217
6. Response to Re-emergence and Outbreak in Huang-Huai Plain	218
6.1 Mass drug administration	218
6.1.1 MDA-'spring treatment'	*219*
6.1.2 MDA-'chemoprophylaxis on transmission season'	*221*

6.2 Case management	221
6.3 Vector control	222
6.4 Training supported by global fund and government	223
7. Effects on Response to Re-emergence and Outbreak	223
8. Challenge of Malaria Elimination in Huang-Huai Plain	224
8.1 Imported malaria	224
8.2 Long incubation time of *P. vivax*	224
8.3 Insecticide resistance	225
9. Conclusions	225
Acknowledgements	226
References	226

Abstract

This chapter reviews the patterns of malaria re-emergence and outbreak that occurred in the Huang-Huai Plain of China in 2006, and the way of quick response to curtail the outbreak by mass drug administration and case management. The contribution of the each intervention in quick response is discussed. Particularly due to the special ecological characteristics in the Huang-Huai Plain, the intervention of vector control is not implemented. Finally, the challenges in the elimination of malaria in this region are highlighted.

1. INTRODUCTION

Vivax malaria was historically epidemic in the Huang-Huai Plain of China and *Anopheles sinensis* was the main vector. There were two major epidemics in which these malaria cases accounted for 93.1% and 91.2% of total cases of the entire country in 1960 and 1970, respectively. Vivax malaria cases were reduced significantly in the Huang-Huai Plain by the end of the 1980s, and malaria incidence was below 1/10,000 in most areas. However, the incidence of vivax malaria had a resurgence pattern in the early twenty-first century, and malaria outbreaks were found in 2006. This chapter reviews a case study of vivax malaria outbreak in the Huang-Huai Plain in 2006, and the way of quick response to cut down the outbreak by mass drug administration (MDA) and case management.

2. BACKGROUND

Malaria is an important cause of death and illness in children and adults in tropical countries. According to World Malaria Report 2012 (WHO, 2012), there were an estimated 219 million cases of malaria (range 154–289 million) and 660,000 deaths (range 610,000–971,000) in 2010. Of the four human *Plasmodium* species, *Plasmodium vivax* has the widest global distribution and accounts for most malaria infections outside Africa (Mendis et al., 2001).

The greatest proportion of the worldwide vivax malaria burden almost certainly lies in south and southeast Asia (probably more than 80% of infections), with perhaps underappreciated numbers of infection in Africa (Mueller et al., 2009). Despite significant reductions in the overall burden of malaria in the twentieth century, the disease still represents a significant public health problem in China. Vivax malaria was historically epidemic in the Huang-Huai Plain, and there were two major epidemics in which the number of malaria cases accounted for 93.1% and 91.2% of total cases of the entire country in 1960 and 1970, respectively.

Huang-Huai Plain (also called Huang-Huai River region) is located 32°–36° N and 112°–122° E in central China, between south of Yellow River and north of Huai River. The average altitude of this area is 50 m above sea level. The climate of the region is temperate. The area receives about 700 mm of rainfall per annum with 50–80% occurring during the summer from June to September. The average annual temperature is 14–15 °C, ranging from −1 °C to −4 °C in January and from 25 °C to 27 °C in July. The frost-free period is 200–240 days. The main crops are wheat, rice, corn, potato, soybean, cotton and rapeseed. The population of these areas is above 50 million, which covers 116 counties in four provinces of Henan, Anhui, Shandong and Jiangsu (Figure 8.1).

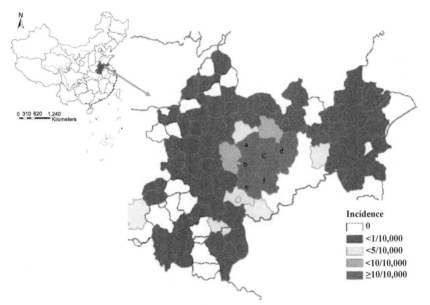

Figure 8.1 Geographic distribution of malaria incidence in the Huang-Huai Plain in 2006 (Different color represents the various malaria incidences at county level. a. Yongcheng; b. Woyang; c. Suixi; d. Yongqiao; e. Lixin; f. Mengcheng).

Malaria in the Huang-Huai Plain can be classified as seasonally unstable and epidemic which means malaria transmission in this region is still existed with a higher incidence. Poor farming communities bear the greatest burden of disease. A major epidemic of *P. vivax* occurred in the 1960s and at the beginning of the 1970s. The total malaria cases in these areas were 9.5 million, accounting for 93.1% of the total reported cases in the country in 1960 (Zhou, 1991). There may be three main reasons that caused the malaria epidemic in 1960. The first and the main reason was a considerable amount of the wheat fields (dry land) changing into rice fields (wet land) in the late 1950s. New irrigating projects were set up; a lot of irrigation channels and a reservoir were built. As a result, the ground water area expanded considerably, stagnant water areas remained constant. The population of *An. sinensis* increased dramatically. An investigation in Suyang County of Jiangsu Province showed that density of *An. sinensis* was dozens to a hundred times higher than that both in house and cattle shelf before changing the dry land to wet land. At that time, local houses were generally simple and crude. It was hot and muggy indoors in summer and autumn, and local residents had the habit of sleeping outside during the summer. This increased the chance of mosquitoes biting residents and strengthened malaria transmission. The second reason may have been a decreasing animal barrier. Animals such as cattle and buffalo play a zooprophylactic role for malaria transmission due to attracting larger numbers of mosquitoes and thus decreasing human vector contact (Manh et al., 2010). Most households have dogs and chickens in this region. Cattle and buffalo free range during the day, when not used for work, but at night are penned or tethered near the owner's house. When People's Community was organized after 1957, the model of livestock feed was changed to feeding together from separate feed, and no cattle were fed near the residents' houses. The chance of local people being exposed to mosquitoes increased significantly due to this decreasing animal barrier. The third reason was that China did not have the ability to produce a large amount of chloroquine and primaquine in 1960, so most MDA of pyrimethamine and proguanil were used to treat patients and give chemoprophylaxis at that time, in which malaria epidemics could be controlled to a low prevalent level, but plasmodium parasites could not be clearly eliminated.

Another major vivax malaria epidemic occurred in 1970. The main reason was the epidemic occurred in the special event of the Cultural Revolution (beginning 1965–1968), during which time civil disturbances resulted in the abandonment of antimalarial activities. Malaria prevalence was controlled by MDA of using chloroquine and primaquine after 1970.

With active implementation of malaria control measures (integrated vector control measures and appropriate treatment of malaria cases) for more than

30 years, considerable success had been achieved and human cases infected with *P. vivax* have been reduced significantly in the Huang-Huai Plain of China in the new millennium. Malaria incidence in the areas where *An. sinensis* was the only vector had been reduced significantly to a low level, for example, a total of 1321 counties reached the standard of the basic malaria elimination (the incidence is less than 1/10,000) in 1999 (Xu et al., 1994, 2006; Tang and Gao, 2013). According to the national data on malaria transmission, the average annual incidences of vivax malaria in 1998 in Anhui and Henan provinces were below 1 per 100,000 and 0.03 per 100,000, respectively (Qian and Shang, 1999).

However, the incidence of vivax malaria had a resurgence in the Huang-Huai Plain in the early part of this century, and malaria outbreaks were found in Anhui and Henan provinces in 2006. About 66.4% of malaria cases in the country were located in the Huang-Huai Plain, especially in Anhui and Henan Provinces (Zhou et al., 2007b).

3. VIVAX MALARIA RE-EMERGENCE AND OUTBREAK IN HUANG-HUAI PLAIN

From 2000 to 2006, there was a substantial increase in malaria cases due to re-emerging vivax malaria in the Huang-Huai Plain. The total number of annual malaria cases was 18,762 and 40,991 in four provinces of Anhui, Henan, Jiangsu and Shandong in 2005 and 2006, respectively, with 118.48% of an increase rate between two years. In 2006, the number of malaria cases in Anhui and Henan provinces accounted for 62.45% of the total cases in the country (Zhou et al., 2007b), and malaria incidence increased by 123.10% and 120.92% compared to that in 2005 in Anhui and Henan provinces, respectively (Table 8.1).

Dramatic resurgence appeared in Yongcheng and Xiayi counties in the east of Henan Province, where malaria incidences in 2006 were increased by

Table 8.1 Malaria situation in the Huang-Huai Plain in 2006 and 2005

	No. of malaria cases			Malaria incidence (1/10,000)		
Province	Year 2006	Year 2005	Increase/decrease (±%)	Year 2006	Year 2005	Increase/decrease (±%)
Anhui	34,984	15,681	+123.10	6.40	2.45	+161.22
Henan	5090	2304	+120.92	0.52	0.24	+116.67
Jiangsu	767	651	+17.82	0.11	0.10	+10.00
Shandong	150	126	+19.05	0.02	0.02	+0
Huang-Huai plain	40,991	18,762	+118.48	1.38	0.65	+111.74

307.04% and 360.94% compared to that in 2005, respectively. In Yongcheng County, malaria incidence was 1.23 per 10,000 in 2004 and 5.19 per 10,000 in 2005, so that the malaria incidence of the county in 2006 were 13 and 3.22 times higher than that of the previous year, respectively (Zhang et al., 2007). In Jiangsu Province, malaria cases in 2006 were mainly in the three cities of Xuzhou, Suqian and Huai'an, where a total of 422 malaria cases reported, which accounted for 55.02% of all cases in Jiangsu Province (Wang et al., 2007). There were 150 malaria cases reported in Shandong Province in 2006, a 19% increase over 2005 (Zhou et al., 2007b).

3.1 Distribution by district of malaria outbreaks in the Huang-Huai Plain in 2006

According to the 'Malaria emergency preparedness plan of action in China' published by the Ministry of Heath, a malaria emergency event is defined as described herein: in township level where a malaria case was reported in past three years, 10 or more local malaria cases are found in the same administrative village within one month (MOH, 2006). In this chapter, the definition of the malaria outbreak is the same as the malaria emergency event.

In 2006, a total of 302 outbreaks occurred in 63 townships of 11 counties in the Huang-Huai Plain. All outbreaks were at the administrative village level, of which 275 outbreaks were found in 10 counties of Anhui Province and 36 were reported in 1 county of Henan Province (Zhou et al., 2007b; Su et al., 2008). Totally 280 outbreaks were found in four counties of Guoyang, Lixin, Yongcheng and Mengcheng counties, accounting for 92.72% of total outbreaks (Figure 8.1; Table 8.2).

In Yongcheng County, 36 malaria outbreaks and 1825 cases were found in four townships, accounting for 63.1% of total malaria cases in the county in 2006. Average malaria incidence of four townships was 106.73 per 10,000 and 4.8 times higher than that of Yongcheng County in 2005. In 36 outbreak spots located at administrative villages, 1265 cases were found and the average malaria incidence was 1.8%. The highest malaria incidence was up to 4.0% in a village with 43 malaria cases (Zhou et al., 2008).

3.2 Distributions by sex, age and occupation of malaria cases in outbreak spots in 2006

There were 9678 malaria cases reported in 302 outbreak spots, accounting for 31.27% of total cases (30,951 cases) in 11 counties with outbreaks and 23.85% of total cases (40,574 cases) in the Huang-Huai Plain. It was found that there were 582 patients less than 5 years old, accounting for 4.82% of 9678 malaria cases reported in 302 outbreak spots; 17.67% were

Table 8.2 Malaria situation in 11 counties with outbreaks in the Huang-Huai Plain in 2006

Name of county	Number of cases	Malaria incidence (1/10,000)	Number of townships with outbreak	Number of outbreaks
Lixin	5422	38.86	17	69
Lieshan	728	47.16	1	1
Lingbi	802	7.36	1	2
Mengcheng	3539	30.54	8	30
Jiaocheng	790	5.60	3	3
Suixi	5456	46.05	5	12
Guoyang	7950	58.38	20	145
Xiangshan	163	5.19	1	1
Xiaoxian	996	8.01	2	2
Yongqiao	2214	13.27	1	1
Yongcheng	2891	22.29	4	36

Table 8.3 Distribution of malaria cases by age, sex and occupation in 302 outbreak spots in the Huang-Huai Plain in 2006

Age				Sex		Occupation				
<5	5–15	15–65	≥65	Male	Female	Farmer	Student	Children	Labor	Others
466	1710	6047	1455	5524	4124	6654	2073	725	172	54

between 5 and 15 years old; 62.48% were between 15 and 65 years old; and 15.34% were over 65 years old (Table 8.3). In contrast to holo-endemic and hyperendemic regions in Africa where greater morbidity and mortality occur in children and infants, only a limited proportion of the cases presented in children <5 years of age. But 22.49% of cases occurred in children <15 years old, which may indicate transmission in the household and school neighborhood. The ratio of male to female population was 1.33:1. Farmers were the main portion of patients, accounting for 68.75%, followed by students at 21.42%.

3.3 Distribution by month of malaria cases in outbreak spots in 2006

There were significant seasonal characteristics of malaria transmission in the Huang-Huai Plain. Normally, malaria transmission was from the mid-May to late October. No malaria transmission occurred from early November to next late April, although here were a few reported malaria cases during January to March. Malaria cases increased in April and there was a small pick-up from May to June, malaria cases increased significantly after July, and a high peak can be found from August to September (Figure 8.2).

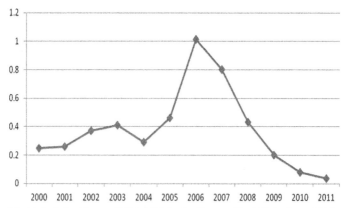

Figure 8.2 Malaria incidence (1/10,000) in the Huang-Huai Plain from 2000 to 2011.

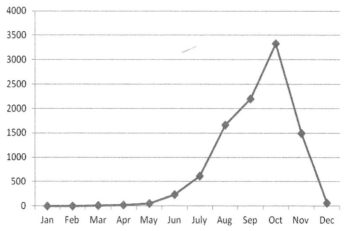

Figure 8.3 Distribution of malaria cases by month in 302 outbreak spots in the Huang-Huai Plain in 2006.

Malaria cases decreased quickly after October. More than 85% of malaria cases were reported from June to October (Su et al., 1995).

In 2006, malaria cases were mainly reported during August and November; 8681 cases were reported, accounting for 89.7% of total cases found in outbreak spots (Figure 8.3). The peak of the epidemic occurred in October with 3334 cases reported. Transmission season was longer than that of before. There is a continuously increasing trend of malaria cases from June to October, forming a straight rising line (Table 8.4). This shows that malaria transmission was strengthened in the Huang-Huai Plain. Monthly *P. vivax* malaria incidence showed a seasonal pattern, whose peak period

Table 4 Distribution of malaria cases by month in 302 outbreaks in the Huang-Huai Plain in 2006

Month	No. of cases
January	3
February	2
March	5
April	22
May	52
June	235
July	618
August	1665
September	2192
October	3334
November	1490
December	60

was from July to November, when nearly 86.58% of total malaria cases were reported in Yongcheng County.

4. CHARACTERISTICS OF MALARIA RE-EMERGENCE AND OUTBREAK IN HUANG-HUAI PLAIN

4.1 Aggregation of malaria cases

From Figure 8.1, we can find that there was an aggregation of malaria cases distributed at the boundary of north Anhui Province and Henan Province, including Guoyang, Mengcheng, Suixi, Yongqiao, Lixin and Yongcheng counties. The center of malaria aggregation was Guoyang County with the highest malaria incidence of 58.38 per 10,000. Zhou et al., (2007a) studied the spatial distribution of malaria in the Huang-Huai Plain based on the kriging method, showed that the distribution of malaria was autocorrelated in space, and the range of the malaria cluster area was 98,928 m. It means that re-emerging malaria will appear in the areas within nearly 100 km around high incidence of malaria areas.

4.2 Sole vector of *An. sinensis*

As a vector-borne disease, malaria transmission has a close relationship with the biological and natural factors due to the anopheles vector distribution features. There are four major vectors of malaria transmission in China, *An. minimus* and *An. dirus* are mostly distributed in Yunnan, Hainan Guangxi and Guizhou provinces of southern China. *An. sinensis* and *An. anthropophagus* are relatively more widely distributed in China.

An. sinensis is the most widely distributed, found in over 29 provinces and regions in China. This species has a large population in most regions, and is the predominant vector (Tang and Gao, 2013). *An. sinensis* is campestral and its blood feeding habit is on cattle and pigs, and its larval habitats are the small water bodies such as ponds, paddy fields, gullies, and so on (Hu et al., 1988; Gou et al., 1998). Since the development of the economy and popularity of agricultural mechanization, the farmers no longer keep pigs in their houses, and the number of farm cattle has also been greatly reduced. Currently, *An. sinensis* is the sole or major vector in the Huang-Huai Plain (Zhou et al., 2010; Liu et al., 2011). One hundred fifty-four *An. sinensis* were captured and no *An. anthropophagus* were found in vector surveillance in Yongcheng County in 2005 (Zhang et al., 2007). The vector species that transmitted malaria in Yongcheng County was morphologically identified as *An. sinensis*, with PCR results of a random sample of 30 mosquitoes confirmed as *An. sinensis* (Zhang et al., 2012b).

4.3 Main cases located around water bodies

The term *water bodies* refers to the paddy fields, gullies and little streams that are suitable breeding sites for *An. sinensis*. Distribution of water bodies may be an important factor influencing the occurrence and distribution of malaria cases in the *An. sinensis* distributed areas. Three hundred fifty-seven malaria cases and their georeferecing data recorded by geographic position system (GPS) as well as surrounding water bodies were collected and analyzed in Anhui Province. The distances from households of cases to the nearest water bodies had a positive-skew distribution, the median of the distrance was 60.9 m, 74.28% (182/246) of cases were distributed at a distance ≤60 m, 16.34% (40/246) of them were located at a distance of 60–120 m, and 9.38% (182/246) were scattered over 120 m. The risk rate of people living within the extent of 60-m proximity to the water bodies and presenting with malaria was significantly higher than others. The results imply that the scope and population within 60 m around water bodies are at risk and could be a targeted population for case management of malaria (Zhou et al., 2012). Other research about relationship between mosquitoes and malaria distributions had similar results, with 80% and 90% of the marked *An. sinensis* recaptured within a radius of 100 m from the release point in study sites I and II, respectively, with a maximum dispersal range of 400 m within the period of the study. The results indicate that local *An. sinensis* may have limited dispersal ranges. Therefore, control efforts should target breeding and resting sites in proximity to the villages (Liu et al., 2012a).

5. FACTORS OF MALARIA RE-EMERGENCE AND OUTBREAK IN HUANG-HUAI PLAIN

5.1 Land use has nothing to do with re-emergance and outbreak in the Huang-Huai Plain

According to historical records, there was malaria outbreak in the Huang-Huai Plain in the early 1960s. One factor related to the outbreak was due to the increased temperatures and rainfall. Another factor was the change of land use status. In the late 1960s, large areas of wheat fields were changed to paddy fields, which changed the local ecological environment, and then induced a malaria outbreak in the region (Zhou, 1991). But the data on changes of land use and land cover of the Huaiyuan and Yongcheng counties in 1996–2000 showed no significant relationship with malaria re-emergence and outbreak (Zhang et al., 2012a).

5.2 Climate warming and increasing rainfall extending malaria transmission

Climate warming can affect the geographic distribution and intensity of the transmission of vector-borne diseases such as malaria. The transmitted parasites usually benefit from increased temperatures as both their reproduction and development are accelerated. It has been investigated that temperature and rainfall played the determinant role of environmental factors in the transmission of malaria. It has been demonstrated that temperature increase would improve the survival chances of anopheles mosquitos and thus contribute to malaria transmission (Bai et al., 2013). Rainfall often leads to small puddles that serve as mosquito breeding sites and it increases humidity, which enhances mosquito survival (Bi et al., 2003). Temperature and rainfall may not influence the transmission of malaria in a linear and direct way (Gou et al., 1998; Qu et al., 2000). Typically, temperatures lower than 16 °C or higher than 30 °C have a negative impact on the development and activities of mosquitoes. Excessive rainfall often leads to small puddles serving as mosquito breeding sites, therefore increasing malaria transmission. But heavy rain may destroy existing breeding places and flush the eggs or larvae out, leading to reduced transmission (Bi et al., 2003; Hui et al., 2009). A negative effect of rainfall on malaria spread was detected in Anhui Province; the results showed that every 1 mm of annual rainfall increase corresponded to a 27% decrease of malaria cases (Wang et al., 2009).

A study in Anhui Province showed that the average monthly temperature and rainfall may be the key factors of malaria transmission; 75.3% changes of monthly malaria incidence contributed to the average monthly temperature, the average temperature of last two months and the average rainfall of current month (Zhou et al., 2010). Another study in Yongcheng County showed that monthly *An. sinensis* density, temperature, humidity, and rainfall had significant positive correlations with malaria incidence with delays of 0 to 3 months, while wind velocity showed negative correlations with delays of 0 and 1 month. There was not a significant association observed between duration of sunshine and malaria incidence (Zhang et al., 2012). A study conducted in Jinan, which is a temperate city in northern China, showed that a 1 °C rise in maximum temperature may be related to a 7.7–12.7% increase in the number of malaria cases, while a 1 °C rise in minimum temperature may result in approximately 11.8–12.7% increase in number of malaria cases (Zhang et al., 2010b).

Moreover, the distribution of mosquitoes is also dependent on relative humidity, and then determines the extent of malaria transmission. Relative humidity exerted an influence on the survival of mosquito eggs and adults and the moderate increase in malaria risk associated with average humidity. No malaria transmission will occur where the monthly average relative humidity is lower than 60%. Conversely, in regions with relative humidity >60%, temperature is the major driver of malaria transmission intensity (Yang et al., 2010).

5.3 Increasing vectorial capacity of *An. sinensis* strengthens malaria transmission

While the level of *P. vivax* transmission is significantly related to two main factors, one is vector capacity and the other one is the number of infectious sources or human cases (Cohuet et al., 2010). It is believed that transmission intensity of human malaria is highly dependent on the vector capacity and competence of local mosquitoes. Vector capacity is a good indicator to assess the transmission level of vivax malaria (Garrett-Jones, 1964). The vector capacity was estimated from the calculation of three indicators: mosquito biting rate (MBR), human blood index (HBIs), and the parous rate (M) (Ree et al., 2001).

Pan et al. (2012) investigated vectorial capacities of *An. sinensis* in 2007; the vectorial capacities of *An. sinensis* in Huaiyuan and Yongcheng counties were 0.7740 and 0.5502, respectively. The results showed that the vector capacity was about 2.3 and 1.7 times higher than 0.331 in the 1990s (Qian et al., 1996), and were 4.6 and 3.3 times higher than 0.1686 in Henan in 1996–1998 (Qu et al., 2000), respectively. Another study got the same results. The vectorial capacities of *An. sinensis* of Huaiyuan and Yongcheng counties were 0.6969 and 0.4689, which were 4.12 and 2.78 times higher compared to that of the 1990s (Zhou et al., 2010). The results demonstrated that the ability of *An. sinensis* to transmit *P. vivax* had been obviously enhanced.

HBIs in Huaiyuan and Yongcheng counties in 2007 were 0.6667 and 0.6429, respectively and the HBI values were more than 12 times higher than that in the historical records. The increased probability of *An. sinensis* feeding on human blood indicates that mosquitoes change their behavior caused by changes in number of livestock (Habtewold et al., 2004). The significant changes in man-biting habit reflected that the number of livestock as sources of infection changed, and this change in pattern is closely related to the reduction of livestock numbers. Artificial infection of *P. vivax* to *An. sinensis* and *An. anthropophagus* had been carried out. The results showed that the susceptibility of *An. sinensis* to *P. vivax*-infected blood is similar to *An. anthropophagus* (Zhu et al., 2013c).

Mosquito nets and mosquito-repellent incense were generally used in the local villages, and most families had screened doors, and/or screened windows. An investigation showed that 93.0% of 80 households in Huaiyuan and 89.3% of 192 households in Yongcheng County had anti-mosquito facilities. It was interesting that the most patients in Wuhe County lived in brick or cement buildings without mosquito-proof doors or windows (Wang et al., 2013c). But the villagers had a habit of enjoying cool outdoor fresh air before midnight, and they preferred sleeping outside during summer. These behaviors increased the chance of mosquitos biting residents, which led to being infected by the mosquitoes.

5.4 Low capacity of diagnosis leading to accumulation of infectious sources

Diagnosis is usually based on parasite identification on a thin–thick smear. There were deficiencies in providing timely diagnosis and treatment due to insufficient microscopists in rural areas. Among 9678 cases reported in 302 outbreak spots in 2006, 57.13% (5529 cases) were confirmed by blood smear and 42.87% (4149 cases) were diagnosed by clinical symptoms; 11.2% (1084 cases) were treated within 1 day, 29.62% (2867 cases) were treated within 3 days, and 59.18% (5727 cases) were treated over 3 days, with average treatment within 6 days. Of 767 malaria cases in Jiangsu Province in 2006, only 6% (46) were confirmed with 1 day, 72.10% (553) were confirmed within 10 days, and the longest one needed 113 days to be confirmed (Wang et al., 2007). Of 710 malaria cases in Yongcheng County in 2005, patients needed (6.27 ± 5.58) days to be confirmed (Zhang et al., 2007).

Of the human malaria parasites, only *P. vivax* and *P. ovale* have the ability to delay the development of a fraction of the infectious load of sporozoites in the liver. The activation of dormant hypnozoites results in the relapse of the disease after the primary infection is cleared from the bloodstream. Low patient compliance for the full primaquine treatment poses a serious challenge to eliminating the disease, with relapses known to occur for up to 5 years (Roy et al., 2013). An investigation of 703 malaria patients in Anhui Province in 2001 showed a relapse rate of 10.67% (Xu et al., 2003). Of 767 malaria cases in Jiangsu Province in 2006, the relapse rate was 9.00% (Wang et al., 2007).

Parasitaemia surveys were carried among 22,500 and 4518 residents in Henan and Anhui provinces, and the parasitaemia rate was 0.47% and 0.58%, respectively, in 2006 (Liu et al., 2008); 87.8% of the antibody positive cases were asymptomatic, indicating that there were potential infection sources with symptomatic parasitaemia in the Huang-Huai Plain (Zheng et al., 2008). The underreporting rate was 5% in outbreak spots, and 417 suspect cases were found in Yongcheng County in 2006 (Zhou et al., 2008).

6. RESPONSE TO RE-EMERGENCE AND OUTBREAK IN HUANG-HUAI PLAIN

6.1 Mass drug administration

MDA has been a key strategy for controlling or eliminating highly prevalent neglected tropical diseases (NTDs) such as lymphatic filariasis (Mwakitalu et al., 2013), soil-transmitted helminthes (Parikh et al., 2013), onchocerciasis (Tekle et al., 2012), and schistosomiasis (Omedo et al., 2012; Wu and Huang, 2013). The simultaneous administration of essential medicines to target high-prevalence NTDs has two main functions: to treat prevalent infection and subsequently to reduce further transmission within the population.

The administration of antimalarial drugs to whole populations, termed MDA, was one of the measures for malaria elimination programmes in the 1950s, and was once again attracting interest as a malaria elimination tool. Mass antimalarial drug administration, defined as the empiric administration of a therapeutic course of an antimalarial regimen to an entire population at the same time without screening or diagnostic testing prior to administration, has been used for malaria control since the early 1930s; it was advocated by the World Health Organization (WHO) in the 1950s as a tool in situations where other more conventional control measures had failed (Poirot et al., 2013). MDA of chloroquine could reduce parasitaemia prevalence and potentially reduce malaria transmission by inhibiting asexual intraerythrocytic stages of the parasite, thereby reducing the number of parasites that can progress to form gametocytes. In addition, MDA of primaquine could have a direct effect on gametocytes and the liver stages of the parasite to inhibit the sporogonic cycle in the mosquito. If every member of a given population is treated by antimalarial MDA, then one would expect an immediate reduction in asexual parasite prevalence in the population, and possibly a sustained reduction in the population parasitaemia prevalence if there was a concomitant reduction in transmission.

China had extensive experience with MDA for malaria control in the 1970s and 1980s. Large-scale MDA was implemented and associated with declines in high *P. vivax* malaria transmission in Jiangsu province (Hsiang et al., 2013); About 148.17 million residents were given chloroquine and primaquine before or at the start of the malaria transmission season that typically begins in April, and 202.75 million were given chloroquine in transmission season in the Huang-Huai Plain in 1974–1986. The number of malaria cases was 231,000 in 1986 with a decrease of 98.2% compared with that in 1973 (Zhou, 1991). The main antimalaria measures included health education in at-risk villages, MDAs in at-risk populations staying overnight in the mountains, following up malaria cases for implementing radical cure, but without using traditional residual spraying or impregnating bed-nets with insecticides in the mountainous areas of Hainan Province. The annual parasite incidence (API) of malaria declined from 3.5% in 1994 to 1.1% in 1996 and 0.8% in 1997, and the API of falciparum malaria declined from 1.0% to 0.3% and 0.3%, respectively (Chen et al., 1999).

6.1.1 MDA-'spring treatment'

P. vivax places dormant forms in the liver (hypnozoites) that cause repeated attacks called relapses. Primaquine, an 8-aminoquinoline, has been in widespread clinical use for over 60 years and remains the only treatment option for eradication of liver hypnozoites, or the prevention of relapses. High doses of primaquine are associated with lower rates of relapse of *P. vivax* infection (Townell et al., 2012). An effective relapse treatment level can have a bigger impact when transmission in the previous year is low (e.g. cumulative cases below 1000 between August and November), with the prevention of 10% of relapses reducing cases in the following transmission season by as much as 70% (Roy et al., 2013).

The goal of seasonal MDA with primaquine was to treat potential reservoirs of *P. vivax* blood and liver stage infection before the start of the malaria transmission season. MDA was administered before or at the start of the malaria transmission season that typically begins in April, so that it is called as 'spring treatment'. From 2006 to 2010, 3.9 million residents were given the 'spring treatment' of chloroquine plus primaquine in the Huang-Huai Plain. In Yongcheng County, all the residents in four townships with the highest malaria incidence and outbreaks were targeted; 80,438 residents received MDA in 2006 and 59,457 in 2007; 1.71 million of residents in 15,459 natural villages received MDA for the spring treatment in Anhui Province in 2008. In other areas, people with a history of infection during the past year and their families and neighbors were given pretransmission season treatment with chloroquine and primaquine for the radical cure, or 'spring treatment' of *P. vivax* malaria (Hsiang et al., 2013).

Community health workers and/or local public health doctors administered medications daily by directly observed therapy and documented this by collecting patient signatures. Informed consent was conducted verbally and through a handout. Villagers were instructed to have food before taking the medication. The administration of drugs was by directly observed therapy and documented by collecting patient signatures. Exclusion criteria for MDA targets included: age <1 year, pregnancy, serious heart, liver or kidney disease, fever, and history of cyanosis, systemic bleeding, or dark-colored urine. Medication was stopped if serious adverse effects such as dark urine, cyanosis or haemolysis occurred.

A major challenge of primaquine use for MDA is the risk of haemolysis in patients with underlying glucose-6-phosphate dehydrogenase (G6PD) deficiency. The most comparable MDA was conducted in 300,000 U.S. soldiers returning from Korea on troopships in 1952–1954 who received 210 mg of primaquine over 2 weeks. The severe haemolytic reaction rate was estimated at 4 per 100,000 (Alving, 1954). Fok et al., (1985) screened G6PD deficiency by fluorescent spot test on cord blood samples of 1228 Chinese neonates, revealing an incidence of 4.4% in males and 0.35% in females. G6PD deficiency is common in Guangdong, Taiwan, Guangxi and other parts of South China. Yan et al., (2006) investigated 4704 individuals in Guangxi, the mutation frequency of male G6PD-deficient individuals was observed to be 7.43% in population. For ethnic Han Chinese, the detection rate was 0.7%, which was lower than the majority of ethnic minorities. In Henan Province, 500 adults were investigated in 1977, and no G6PD deficiency was found (Su et al., 1995). The reported prevalence of G6PD deficiency among Han, the predominant ethic group in Jiangsu, is low, <5% among males (Jiang et al., 2006). Five reports documenting G6PD deficiency-related severe adverse events from primaquine treatment were identified in Jiangsu Province when they carried out MDA of spring treatment in 1973–1983 (Hsiang et al., 2013). No irreversible adverse events were reported in the Huang-Huai Plain in 2006–2010.

6.1.2 MDA-'chemoprophylaxis on transmission season'

The aim of chemoprophylaxis on transmission season was to interrupt transmission, where antimalarial drugs are administered throughout transmission season. Chemoprophylaxis has been found to be highly effective at reducing mortality and morbidity from malaria in highly endemic areas (Greenwood, 2004). Residents who lived near water bodies and patients were targeted for chemoprophylaxis on transmission season in Anhui Province (Zhu et al., 2013a). From August to October in 2007, piperquine (600 mg) was given once a month to residents who lived near a water body in 24 counties of Anhui Province; 4.43 million population in 15,740 natural villages received the administration, covering 98.69% of the population. Compared to that in the last month, malaria incidence of these areas decreased by 2.0%, 46.3%, and 60.2% from August to October in 2007, respectively. Malaria incidence in 2007 decreased by 21.82% compared to that in 2006.

6.2 Case management

Malaria case management is focused on early detection, i.e. diagnosis of malaria cases, prompt and effective treatment of symptomatic patients. Doctors at all levels make malaria diagnosis according to epidemiological history and clinical manifestations, such as fever, chill, rigor, headache and body ache etc. Malaria cases are confirmed by blood smear examination carried out on three kinds of fever cases, including suspected cases, clinically diagnosed cases and patients with fever of unknown origin. A total of 7,756,119 febrile individuals recieved blood smear examination in 2006–2010 in the Huang-Huai Plain. 91,900 local doctors and microscopists were trained for malaria diagnosis in the Huang-Huai Plain from 2006 to 2010. The number and ratio of laboratory-confirmed malaria cases obviously increased as a result of this work, and all malaria cases were laboratory-confirmed in 2012.

Chloroquine and primaquine is still the first-line treatment for vivax malaria. Clinical chloroquine-resistant *P. vivax* was firstly reported from Papua New Guinea in 1989 (Peters, 1990; Schuurkamp et al., 1992;

Murphy et al., 1993), followed by multiple reports from Indonesia, Myanmar, India, Guyana, Brazil, Colombia, and more recently in Ethiopia and South Korea (Baird et al., 1991; Collignon, 1991; Phillips et al., 1996; Singh, 2000; Soto et al., 2001; de Santana Filho et al., 2007; Guthmann et al., 2008; Teka et al., 2008; Ketema et al., 2009; Lee et al., 2009). Although some hypotheses suggest that the chloroquine–primaquine combination, which has been the first-line of treatment for *P. vivax* for almost 50 years, may not be equally effective nowadays, there is no evidence of this decrease in effectiveness of malaria treatment in China.

In 2005–2006, drug susceptibility in vitro was measured for 42 clinical *P. vivax* isolates by using a schizont maturation inhibition technique. Geometric means of 50% inhibitory concentrations (IC50s) and 95% confidence intervals (CIs) were 10.87 (4.50–26.26) ng/mL for chloroquine. The IC50 for chloroquine was lower than those obtained from isolates from Thailand and South Korea, suggesting that chloroquine remained effective against *P. vivax* malaria in central China (Lu et al., 2011).

Thirty-nine monoinfection vivax malaria patients were enrolled to evaluate efficiency of chloroquine treatment of vivax from 2008 to 2009 in Suixi County, Jiangsu province. No recrudescence or danger signs were observed within the 28-day follow-up, which showed vivax was still susceptible to chloroquine in China (Zhu et al., 2013b).

6.3 Vector control

An. sinensis is slightly exophagic, which means *An. sinensis* tends toward outdoor resting after indoor blood feeding. This behavior has made vector control of this species more difficult. Insecticide-treated nets (ITNs) and indoor residual spraying (IRS) were not useful for vector control of *An. sinensis*. Biologic control measures against mosquito larvae were carried out in 31 administration villages of five townships in Yongcheng County with malaria outbreaks in 2006. 8 mL/m^2 suspending agent of *Bacillus sphaericus* were nebulized on surfaces of ponds in/or around the villages at intervals of 15 days. Density of *An. sinensis* larvae and adults decreased by 75.6–100%, 50–100%, respectively, two days after nebulizing. No outbreak was found in Yongcheng County in 2007. Three hundred twenty-three malaria cases were reported and malaria incidence was 0.48% in villages where carrying biologic control measures in 2007, dropped by 51.3% of malaria incidence compared to 0.98% of previous year. *B. sphaericus* had the effect of decreasing density of *An. sinensis* larvae and adult. Biologic control measures against mosquito larvae can effectively decrease malaria incidence (Zhou et al., 2009).

6.4 Training supported by global fund and government

Social and economic status significantly changed since the 1990s in central parts of China, and malaria control interventions also transferred from vectorial control, such as IRS, ITNs combined with case management, to enhancing case detection with health education particularly on risks and vulnerable populations. A total of 91,900 local doctors and microscopists were trained on malaria diagnosis in the Huang-Huai Plain from 2006 to 2010. Residents in epidemic areas received knowledge of malaria through a range of media, such as radio and TV broadcasts, newspaper articles, and posters, especially for the annual Malaria Day Campaign all over the epidemic areas on 26 April every year.

Malaria control was supported by 5 rounds of Global Fund to Fight AIDS, Tuberculosis and Malaria (GFATM) programmes since 2003. This indicates the great efforts of the country to decrease malaria transmission, and during the last couple of years more timely diagnosis and treatment as well as prevention and control activities are being deployed jointly supported by the government and the GFATM programmes. This approach appears to be producing a beneficial impact. In the context of the GFATM programmes, the malaria control strategy is successfully leading to community participation in prevention and control activities that therefore become more efficient and cost-effective.

Currently, greater efforts are being invested to maintain the decreasing trend of malaria transmission in the Huang-Huai Plain, China, although long-term monitoring to avoid epidemic outbreaks is still required, i.e. by prioritizing critical control activities such as effective vector management and educational activities as well as greater efforts to overcome administrative hurdles that limit a more robust diagnosis and treatment system.

7. EFFECTS ON RESPONSE TO RE-EMERGENCE AND OUTBREAK

The re-emergence of malaria was effectively controlled by an extensive control strategy on patient management supported by the central government and GFATM programmes. Malaria incidence dropped dramatically in the Huang-Huai Plain after 2007. A total of 769 vivax malaria cases from the Huang-Huai Plain were reported in 2011. Only 30 local vivax malaria cases were reported in Anhui Province, and no local transmission vivax malaria was reported in Henan, Jiangsu, and Shandong provinces in 2012.

8. CHALLENGE OF MALARIA ELIMINATION IN HUANG-HUAI PLAIN

8.1 Imported malaria

Imported malaria has been found in more provinces in China, which requires urgent information to be dispensed in a setting of increasing travel activity and migration. In 2011, 4479 malaria cases were reported in China, of which 66.4% were imported cases that were mainly distributed in Yunnan (36.5%), Jiangsu (12.0%), Henan (6.2%), Sichuan (5.8%) and Hunan (4.8%) provinces. The number of malaria deaths due to falciparum malaria increased to 33 from 19 in 2010 (Xia et al., 2012). A total of 84 imported falciparum malaria cases with one death were reported in Henan Province in 2005–2009. The ratio of males to females was 20:1(80/4). The average age was (34 ± 11) years old. Seventy-five patients returned from Africa, accounting for 89.3% of total number of imported cases. The number of annual imported cases reported in 2008 (24 cases) and 2009 (32 cases) was higher than that in 2005–2007 (9–10 cases). The top three high-risk populations were farmers (34 cases), workers (17 cases) and the cadre (13 cases), which accounted for 76.2% of all cases (Zhang et al., 2010a). A total of 233 overseas imported malaria cases were reported in Jiangsu Province, and 226 cases (97.0%) were back from African countries. A total of 208 cases (89.3%) were falciparum malaria, and 224 cases (96.1%) were laboratory confirmed. The imported malaria cases were young adults who were mainly migrant farmers and skilled male workers (Liu et al., 2013). It was necessary to further strengthen the professional training and multisectoral cooperation, establish the collaborative investigation mechanism for high-risk groups, and take effective prevention and control measures, in order to reduce the risk of malaria resurgence in the Huang-Huai Plain. due to overseas-imported malaria.

8.2 Long incubation time of *P. vivax*

P. vivax was endemic in temperate areas in historic times up to the middle of the last century. *P. vivax* has a long incubation time of up to 8–10 months, which partly explains how it can be endemic in temperate areas in the cold winter season. Lately *P. vivax* has been seen along the demilitarized zone in South Korea, replicating a high endemicity in North Korea. The potential for transmission of *P. vivax* still exists in temperate zones, but reintroduction of a larger scale of *P. vivax* to areas without present transmission requires large population movements of *P. vivax*-infected people. The highest threat at present is from refugees from *P. vivax* endemic North Korea entering China and South Korea in large numbers (Petersen et al., 2013).

8.3 Insecticide resistance

Insecticide resistance in malaria vectors is a growing concern in many countries and requires immediate attention because of the limited chemical arsenal available for vector control. Bioassays were performed on F1 progeny of *An. sinensis* reared from wild-caught females using the standard WHO susceptibility test with diagnostic concentrations of 0.25% deltamethrin and 4% DDT. The results indicated that *An. sinensis* was completely resistant to both deltamethrin and DDT, and resistance to pyrethroid has risen strikingly compared to that recorded during 1990s. The results highlight the importance of longitudinal insecticide resistance monitoring and the urgent need for a better understanding of the status of insecticide resistance in this region (Wang et al., 2013). The same results were found in Henan, Jiangsu and Anhui provinces (Li et al., 2011; Wu et al., 2011; Liu et al., 2012b; Zhang et al., 2012; Chang et al., 2013).

9. CONCLUSIONS

Unlike *P. falciparum*, *P. vivax* rarely causes severe disease in healthy travelers or in temperate endemic regions, and has been regarded as readily treatable with chloroquine in the Huang-Huai Plain. However, in tropical areas, recent reports have highlighted severe and fatal disease associated with *P. vivax* infection. Studies from Indonesia, Papua New Guinea, Thailand and India have shown that 21–27% of patients with severe malaria have *P. vivax* mono-infection. The clinical spectrum of these cases is broad with an overall mortality of 0.8–1.6%. Major manifestations include severe anaemia and respiratory distress, with infants being particularly vulnerable. Severe, fatal and multidrug-resistant vivax malaria challenges our perception of *P. vivax* as a benign disease. Strategies to understand and address these phenomena are urgently needed if the global elimination of malaria is to succeed (Price et al., 2009).

In order to avoid death from the imported malaria cases and reduce the risk of secondary transmission, malaria screening and health education for those returned from malaria-endemic countries should be strengthened, and the diagnosis and treatment capabilities must be improved at county medical units (Chen et al., 2012).

ACKNOWLEDGEMENTS

This work was supported by Project of Medical Science and Technique of Henan, China (No. 092102310007), and the Special Funding of the Henan Health Science and Technology Innovation Talent Project. The project was also supported by the National S & T Major Programme (grant no. 2012ZX10004220), by the National S & T Supporting Project (grant no. 2007BAC03A02) and by China UK Global Health Support Programme (grant no. GHSP-CS-OP1).

REFERENCES

Alving, A., 1954. Clinical treatment of malaria. U.S. Army Med. Sci. 22, 210–218.
Bai, L., Morton, L.C., Liu, Q., 2013. Climate change and mosquito-borne diseases in China: a review. Global Health 9, 10.
Baird, J.K., Basri, H., Purnomo, Bangs, M.J., Subianto, B., Patchen, L.C., Hoffman, S.L., 1991. Resistance to chloroquine by *Plasmodium vivax* in Irian Jaya, Indonesia. Am. J. Trop. Med. Hyg. 44 (5), 547–552.
Bi, P., Tong, S., Donald, K., Parton, K.A., Ni, J., 2003. Climatic variables and transmission of malaria: a 12-year data analysis in Shuchen County, China. Public Health Rep. 118 (1), 65–71.
Chang, X.L., Xue, Y.Q., Zhang, A.D., Zhu, G.D., Fang, Q., 2013. Deltamethrin resistance, metabolic detoxification enzyme and kdr mutation in *Anopheles sinensis* in region along Huaihe River in Anhui Province. Chin. J. Epi. 25 (3), 263–267 (in Chinese).
Chen, W.Q., Su, Y.P., Deng, Y., Zhang, H.W., 1999. A pilot study on malaria control by using a new strategy of combining strengthening infection source treatment and health education in mountainous areas of Hainan province. Chin. J. Parasitol. Parasit. Dis. 17 (1), 1–4 (in Chinese).
Chen, W., Wu, K., Lin, M., Tang, L., Gu, Z., Wang, S., Lan, C., Lan, X., Li, H., Huang, M., Chen, X., Sheng, H., 2012. Epidemiological analysis of imported malaria in Henan Province in 2011. Chin. J. Parasitol. Parasit. Dis. 30 (5), 387–390 (in Chinese).
Cohuet, A., Harris, C., Robert, V., Fontenille, D., 2010. Evolutionary forces on *Anopheles*: what makes a malaria vector? Trends Parasitol. 26 (3), 130–136.
Collignon, P., 1991. Chloroquine resistance in *Plasmodium vivax*. J. Infect. Dis. 164 (1), 222–223.
Fok, T.F., Lau, S.P., Fung, K.P., 1985. Cord blood G-6-PD activity by quantitative enzyme assay and fluorescent spot test in Chinese neonates. Aust. Paediatr. J. 21 (1), 23–25.
Garrett-Jones, C., 1964. Prognosis for interruption of malaria transmission through assessment of the mosquito's vectorial capacity. Nature 204, 1173–1175.
Gou, G.X., Li, D.F., Shang, L.Y., Guo, X.S., 1998. The study on ecological habits of *Anopheles sinensis* in Guantang, Luyi county from 1971 to 1996. Chin. J. Vector Bio. Con. 9 (2), 133–134 (in Chinese).
Greenwood, B., 2004. The use of anti-malarial drugs to prevent malaria in the population of malaria-endemic areas. Am. J. Trop. Med. Hyg. 70 (1), 1–7.
Guthmann, J.P., Pittet, A., Lesage,, A., Imwong, M., Lindegardh, N., Min Lwin, M., Zaw, T., Annerberg, A., de Radigues, X., Nosten, F., 2008. *Plasmodium vivax* resistance to chloroquine in Dawei, southern Myanmar. Trop. Med. Int. Health 13 (1), 91–98.
Habtewold, T., Prior, A., Torr, S.J., Gibson, G., 2004. Could insecticide-treated cattle reduce Afrotropical malaria transmission? Effects of deltamethrin-treated Zebu on *Anopheles arabiensis* behaviour and survival in Ethiopia. Med. Vet. Entomol. 18 (4), 408–417.
Hsiang, M.S., Hwang, J., Tao, A.R., Liu, Y., Bennett, A., Shanks, G.D., Cao, J., Kachur, S.P., Feachem, R.G., Gosling, R.D., Gao, Q., 2013. Mass drug administration for the control and elimination of *Plasmodium vivax* malaria: an ecological study from Jiangsu province, China. Malar. J. 12 (1), 383.

Hu, Y.X., Miao, Y.G., Fan, T.B, 1988. The further study on ecological habits of *Anopheles sinensis* in the area along Yellow River and Huai River. Chin. J. Parasitol. Parasit. Dis. S1, 135 (in Chinese).

Hui, F.M., Xu, B., Chen, Z.W., Cheng, X., Liang, L., Huang, H.B., Fang, L.Q., Yang, H., Zhou, H.N., Yang, H.L., Zhou, X.N., Cao, W.C., Gong, P., 2009. Spatio-temporal distribution of malaria in Yunnan Province, China. Am. J. Trop. Med. Hyg. 81 (3), 503–509.

Jiang, W., Yu, G., Liu, P., Geng, Q., 2006. Structure and function of glucose-6-phosphate dehydrogenasedeficient variants in Chinese population. Hum. Genet. 119, 463–478.

Ketema, T., Bacha, K., Birhanu, T., Petros, B., 2009. Chloroquine-resistant *Plasmodium vivax* malaria in Serbo town, Jimma zone, south-west Ethiopia. Malar. J. 8, 177.

Lee, K.S., Kim, T.H., Kim, E.S., Lim, H.S., Yeom, J.S., Jun, G., Park, J.W., 2009. Short report: chloroquine-resistant *Plasmodium vivax* in the Republic of Korea. Am. J. Trop. Med. Hyg. 80 (2), 215–217.

Li, J.L., Zhou, H.Y., Cao, J., Zhu, G.D., Wang, W.M., Gu, Y., Liu, Y., Cao, Y., Zhang, C., Gao, Q., 2011. Sensitivity of *Anopheles sinensis* to insecticides in Jiangsu Province. Chin. J. Schisto. Control, 23 (3), 296–300 (in Chinese).

Liu, Q., Liu, X., Zhou, G., Jiang, J., Guo, Y., Ren, D., Zheng, C., Wu, H., Yang, S., Liu, J., Li, H., Li, Q., Yang, W., Chu, C., 2012a. Dispersal range of *Anopheles sinensis* in Yongcheng City, China by mark-release-recapture methods. PLoS One 7 (11), e51209.

Liu, X.B., Liu, Q.Y., Guo, Y.H., Jiang, J.Y., Ren, D.S., Zhou, G.C., Zheng, C.J., Zhang, Y., Liu, J.L., Li, Z.F., Chen, Y., Li, H.S., Morton, L.C., Li, H.Z., Li, Q., Gu, W.D., 2011. The abundance and host-seeking behavior of culicine species (Diptera: Culicidae) and *Anopheles sinensis* in Yongcheng city, People's Republic of China. Parasit. Vectors 4, 221.

Liu, Y., Chen, J.S., Zhou, R.M., Qian, D., Chen, Q.W., Xu, B.L., Zhang, H.W., 2012b. Investigation on the sensitivity of *Anopheles sinensis* to insecticide. Chin. J. Parasitol. Parasit. Dis. 30 (4), 309–311 (in Chinese).

Liu, Y., Su, Y.-P., Zhang, H.-W., 2008. Analysis of malaria situation in five-provinces of Jiangsu, Shandong, Henan, Anhui and Hubei in 2006. Chin. J. Trop. Med. 8 (2), 190–192 (in Chinese).

Liu, Y.B., Cao, J., Zhou, H.Y., Wang, W.M., Cao, Y.Y., Gao, Q., 2013. Analysis of overseas imported malaria situation and implication for control in Jiangsu Province, PR China. Chin. J. Schisto. Control. 25 (1), 44–47 (in Chinese).

Lu, F., Gao, Q., Chotivanich, K., Xia, H., Cao, J., Udomsangpetch, R., Cui, L., Sattabongkot, J., 2011. In vitro anti-malarial drug susceptibility of temperate *Plasmodium vivax* from central China. Am. J. Trop. Med. Hyg. 85 (2), 197–201.

Manh, C.D., Beebe, N.W., Van, V.N., Quang, T.L., Lein, C.T., Nguyen, D.V., Xuan, T.N., Ngoc, A.L., Cooper, R.D., 2010. Vectors and malaria transmission in deforested, rural communities in north-central Vietnam. Malar. J. 9, 259.

Mendis, K., Sina, B.J., Marchesini, P., Carter, R., 2001. The neglected burden of *Plasmodium vivax* malaria. Am. J. Trop. Med. Hyg. 64 (Suppl. 1–2), 97–106.

MOH, 2006. Malaria Emergency Preparedness Plan of Action in China. Ministry of Heath.

Mueller, I., Galinski, M.R., Baird, J.K., Carlton, J.M., Kochar, D.K., Alonso, P.L., del Portillo, H.A., 2009. Key gaps in the knowledge of *Plasmodium vivax*, a neglected human malaria parasite. Lancet Infect. Dis. 9 (9), 555–566.

Murphy, G.S., Basri, H., Purnomo, Andersen, E.M., Bangs, M.J., Mount, D.L., Gorden, J., Lal, A.A., Purwokusumo, A.R., Harjosuwarno, S., 1993. Vivax malaria resistant to treatment and prophylaxis with chloroquine. Lancet 341 (8837), 96–100.

Mwakitalu, M.E., Malecela, M.N., Pedersen, E.M., Mosha, F.W., Simonsen, P.E., 2013. Urban lymphatic filariasis in the metropolis of Dar es Salaam, Tanzania. Parasit. Vectors 6 (1), 286.

Omedo, M.O., Matey, E.J., Awiti, A., Ogutu, M., Alaii, J., Karanja, D.M., Montgomery, S.P., Secor, W.E., Mwinzi, P.N., 2012. Community health workers' experiences and perspectives on mass drug administration for schistosomiasis control in western Kenya: the SCORE Project. Am. J. Trop. Med. Hyg. 87 (6), 1065–1072.

Pan, J.Y., Zhou, S.S., Zheng, X., Huang, F., Wang, D.Q., Shen, Y.Z., Su, Y.P., Zhou, G.C., Liu, F., Jiang, J.J., 2012. Vector capacity of *Anopheles sinensis* in malaria outbreak areas of central China. Parasit. Vectors 5, 136.

Parikh, D.S., Totanes, F.I., Tuliao, A.H., Ciro, R.N., Macatangay, B.J., Belizario, V.Y., 2013. Knowledge, attitudes and practices among parents and teachers about soil-transmitted helminthiasis control programs for school children in Guimaras, Philippines. Southeast Asian J. Trop. Med. Public Health 44 (5), 744–752.

Peters, W., 1990. Plasmodium: resistance to antimalarial drugs. Ann. Parasitol. Hum. Comp. 65 (Suppl. 1), 103–106.

Petersen, E., Severini, C., Picot, S., 2013. *Plasmodium vivax* malaria: a re-emerging threat for temperate climate zones? Travel Med. Infect. Dis. 11 (1), 51–59.

Phillips, E.J., Keystone, J.S., Kain, K.C., 1996. Failure of combined chloroquine and high-dose primaquine therapy for *Plasmodium vivax* malaria acquired in Guyana, South America. Clin. Infect. Dis. 23 (5), 1171–1173.

Poirot, E., Skarbinski, J., Sinclair, D., Kachur, S.P., Slutsker, L., Hwang, J., 2013. Mass drug administration for malaria. Cochrane Database Syst. Rev. 12, CD008846.

Price, R.N., Douglas, N.M., Anstey, N.M., 2009. New developments in *Plasmodium vivax* malaria: severe disease and the rise of chloroquine resistance. Curr. Opin. Infect. Dis. 22 (5), 430–435.

Qian, H., Shang, L., 1999. Malaria situation in the People's Republic of China in 1998. Chin. J. Parasitol. Parasit. Dis. 17 (4), 193–195 (in Chinese).

Qian, H.L., Tang, L.H., Tang, L.Y., 1996. Preliminary estimation on the critical value of man biting rate and vectorial capacity of *Anopheles sinensis*. Practical Preventire Med. 3, 1–2 (in Chinese).

Qu, C.Z., Su, T.Z., Wang, M.Y., 2000. Vectorial capacity of malaria transmission of *Anopheles sinensis* in Zhengzhou in nature. J. Henan Medieal Univ. 35, 394–396 (in Chinese).

Ree, H.I., Hwang, U.W., Lee, I.Y., Kim, T.E., 2001. Daily survival and human blood index of *Anopheles sinensis*, the vector species of malaria in Korea. J. Am. Mosq. Control Assoc. 17 (1), 67–72.

Roy, M., Bouma, M.J., Ionides, E.L., Dhiman, R.C., Pascual, M., 2013. The potential elimination of *Plasmodium vivax* malaria by relapse treatment: insights from a transmission model and surveillance data from NW India. PLoS Negl. Trop. Dis. 7 (1), e1979.

de Santana Filho, F.S., Arcanjo, A.R., Chehuan, Y.M., Costa, M.R., Martinez-Espinosa, F.E., Vieira, J.L., Barbosa, M.D., Alecrim, W.D., Alecrim, M.D., 2007. Chloroquine-resistant *Plasmodium vivax*, Brazilian Amazon. Emerg. Infect. Dis. 13 (7), 1125–1126.

Schuurkamp, G.J., Spicer, P.E., Kereu, R.K., Bulungol, P.K., Rieckmann, K.H., 1992. Chloroquine-resistant *Plasmodium vivax* in Papua New Guinea. Trans. R. Soc. Trop. Med. Hyg. 86 (2), 121–122.

Singh, R.K., 2000. Emergence of chloroquine-resistant vivax malaria in south Bihar (India). Trans. R. Soc. Trop. Med. Hyg. 94 (3), 327.

Soto, J., Toledo, J., Gutierrez, P., Luzz, M., Llinas, N., Cedeno, N., Dunne, M., Berman, J., 2001. *Plasmodium vivax* clinically resistant to chloroquine in Colombia. Am. J. Trop. Med. Hyg. 65 (2), 90–93.

Su, S.Z., Ma, Y.X., Wang, Z. (Eds.), 1995. Malaria Study and Control in Henan Province. Zhongyuan Nongmin Press, Zhengzhou (in Chinese).

Su, Y.P., Zhang, H.W., Liu, Y., 2008. Evaluation on malaria control and malaria situation in Henan province in 2006. J. Pathogen Bio. 3 (9), 670–672 (in Chinese).

Tang, L.H., Gao, Q., 2013. Malaria Control and Eliminate in China. Shanghai Scientific & Technical Publishers, Shanghai. 42 (in Chinese).

Teka, H., Petros, B., Yamuah, L., Tesfaye, G., Elhassan, I., Muchohi, S., Kokwaro, G., Aseffa, A., Engers, H., 2008. Chloroquine-resistant *Plasmodium vivax* malaria in Debre Zeit, Ethiopia. Malar. J. 7, 220.

Tekle, A.H., Elhassan, E., Isiyaku, S., Amazigo, U.V., Bush, S., Noma, M., Cousens, S., Abiose, A., Remme, J.H., 2012. Impact of long-term treatment of onchocerciasis with ivermectin in Kaduna State, Nigeria: first evidence of the potential for elimination in the operational area of the African Programme for Onchocerciasis Control. Parasit. Vectors 5, 28.

Townell, N., Looke, D., McDougall, D., McCarthy, J.S., 2012. Relapse of imported *Plasmodium vivax* malaria is related to primaquine dose: a retrospective study. Malar. J. 11, 214.

Wang, D.Q., Xia, Z.G., Zhou, S.S., Zhou, X.N., Wang, R.B., Zhang, Q.F., 2013a. A potential threat to malaria elimination: extensive deltamethrin and DDT resistance to *Anopheles sinensis* from the malaria-endemic areas in China. Malar. J. 12, 164.

Wang, L.P., Fang, L.Q., Xu, X., Wang, J.J., Ma, J.Q., Cao, W.C., Jin, S.G., 2009. Study on the determinants regarding malaria epidemics in Anhui province during 2004-26. Chin. J. Epi. 30 (1), 38–41 (in Chinese).

Wang, W.M., Gao, Q., Jin, X.L., Zhou, H.Y., 2007. Malaria epidemic situation in Jiangsu Province in 2006. Chin. J. Parasitol. Parasit. Dis. 25 (3), 258–259 (in Chinese).

Wang, X.M., Xia, L.H., Fang, Q., Tao, Z.Y., Jiao, Y.M., Xia, H., Sun, X., 2013b. Malaria epidemiologic characteristics in Wuhe County of Anhui Province from 2009 to 2011. Chin. J. Schisto. Control 25 (1), 70–72 (in Chinese).

WHO, 2012. World Malaria Report 2012. World Health Organization, Geneva.

Wu, S., Liu, Q., Yang, F.T., 2011. Investigation of *Anopheles sinensis* resistance to deltamethrin in the malaria-endemic northern part of Anhui Province, China. J. Pathog. Biol. 6 (12), 920–921 (in Chinese).

Wu, W., Huang, Y., 2013. Application of praziquantel in schistosomiasis japonica control strategies in China. Parasitol. Res. 112 (3), 909–915.

Xia, Z.G., Yang, M.N., Zhou, S.S., 2012. Malaria situation in the People's Republic of China in 2011. Chin. J. Parasitol. Parasit. Dis. 30 (6), 419–422 (in Chinese).

Xu, B., Li, H., Webber, R.H., 1994. Malaria in Hubei Province, China: approaching eradication. J. Trop. Med. Hyg. 97 (5), 277–281.

Xu, B.L., Su, Y.P., Shang, L.Y., Zhang, H.W., 2006. Malaria control in Henan Province, People's Republic of China. Am. J. Trop. Med. Hyg. 74 (4), 564–567.

Xu, F.N., Shen, Y.Z., Jia, S.Y., 2003. Feature analysis of local malaria re-rise in *Plasmodium vivax* endemic area in anhui province. Chin. J. Parasit. Dis. Control 16 (1), 20–21 (in Chinese).

Yan, T., Cai, R., Mo, O., Zhu, D., Ouyang, H., Huang, L., Zhao, M., Huang, F., Li, L., Liang, X., Xu, X., 2006. Incidence and complete molecular characterization of glucose-6-phosphate dehydrogenase deficiency in the Guangxi Zhuang autonomous region of southern China: description of four novel mutations. Haematologica 91 (10), 1321–1328.

Yang, G.J., Gao, Q., Zhou, S.S., Malone, J.B., McCarroll, J.C., Tanner, M., Vounatsou, P., Bergquist, R., Utzinger, J., Zhou, X.N., 2010. Mapping and predicting malaria transmission in the People's Republic of China, using integrated biology-driven and statistical models. Geospat. Health 5 (1), 11–22.

Zhang, H.W., Su, Y.P., Zhao, X.D., Yan, Q.Y., Liu, Y., Chen, J.S., 2007. Re-emerging malaria in yongcheng city of Henan Province. Chin. J. Vector Bio. Control 18 (1), 42–44 (in Chinese).

Zhang, H.W., Su, Y.P., Zhao, X.D., Yan, Q.Y., Liu, Y., Chen, J.S., 2010a. Imported falciparum malaria situation in Henan Province during 2005–2009. Chin. J. Parasitol. Parasit. Dis. 28 (6), 476–477 (in Chinese).

Zhang, S.S., Zhou, S.S., Tang, L.H., Huang, F., Zheng, X., 2012a. Study on the correlation between land use and cover change and malaria transmission in the areas along the Yellow River and Huai River. Chin. J. Parasitol. Parasit. Dis. 30 (2), 102–108 (in Chinese).

Zhang, Y., Bi, P., Hiller, J.E., 2010b. Meteorological variables and malaria in a Chinese temperate city: a twenty-year time-series data analysis. Environ. Int. 36 (5), 439–445.

Zhang, Y., Liu, Q.Y., Luan, R.S., Liu, X.B., Zhou, G.C., Jiang, J.Y., Li, H.S., Li, Z.F., 2012b. Spatial-temporal analysis of malaria and the effect of environmental factors on its incidence in Yongcheng, China, 2006–2010. BMC Public Health 12, 544.

Zhang, Y.Q., Liu, J.Q., Guo, X.S., Tang, Z.Q., Zhao, X.D., Zhou, R.M., Zhao, Q., 2012c. Resistance of *Anopheles sinensis* to insecticides in Huaibin County of Henan Province. Chin. J. Parasitol. Parasit. Dis. 30 (6), 493–495 (in Chinese).

Zheng, X., Fang, W., Huang, F., Shen, Y.Z., Su, Y.P., Huang, G.Q., Zhou, S.S., 2008. Evaluation on malaria situation in areas along Yellow River and Huaihe River by indirect fluorescent antibody test. Chin. J. Parasitol. Parasit. Dis. 26 (6), 417–421 (in Chinese).

Zhou, G.C., Zhang, H.W., Su, Y.P., 2008. Analysis of malaria outbreak in Yongcheng county of Henan province. J. Trop. Med. 8 (4), 381–383 (in Chinese).

Zhou, G.C., Zhang, H.W., Su, Y.P., 2009. Study on control of malaria outbreak and prevalence by biologic control measures against mosquito larvae in Yongcheng County. Chin. Trop. Med. 9 (2), 103–105 (in Chinese).

Zhou, S.S., Huang, F., Wang, J.J., Zhang, S.S., Su, Y.P., Tang, L.H., 2007a. Study on the spatial distribution of malaria in Yellow River and Huai River areas based on the "kriging" method. J. Pathog. Biol. 2 (3), 204–206.

Zhou, S.S., Huang, F., Wang, J.J., Zhang, S.S., Su, Y.P., Tang, L.H., 2010. Geographical, meteorological and vectorial factors related to malaria re-emergence in Huang-Huai River of central China. Malar. J. 9, 337.

Zhou, S.S., Wang, Y., Tang, L.H., 2007b. Malaria situation in the People's Republic of China in 2006. Chin. J. Parasitol. Parasit. Dis. 25 (6), 439–441 (in Chinese).

Zhou, S.S., Zhang, S.S., Wang, J.J., Zheng, X., Huang, F., Li, W.D., Xu, X., Zhang, H.W., 2012. Spatial correlation between malaria cases and water-bodies in *Anopheles sinensis* dominated areas of Huang-Huai Plain, China. Parasit. Vectors 5, 106.

Zhou, Z.-J., 1991. Malaria Control and Study in China. People Health Press, Beijing (in Chinese).

Zhu, D.S., Wang, J.J., Xu, X., Zhu, J., Li, H.Z., 2013a. Evaluating the effect of preventive medicine for residents living around mosquito breeding water during rest period of malaria. Chin. J. Preventive Med. 47 (1), 44–48 (in Chinese).

Zhu, G., Lu, F., Cao, J., Zhou, H., Liu, Y., Han, E.T., Gao, Q., 2013b. Blood stage of *Plasmodium vivax* in central China is still susceptible to chloroquine plus primaquine combination therapy. Am. J. Trop. Med. Hyg. 89 (1), 184–187.

Zhu, G., Xia, H., Zhou, H., Li, J., Lu, F., Liu, Y., Cao, J., Gao, Q., Sattabongkot, J., 2013c. Susceptibility of *Anopheles sinensis* to *Plasmodium vivax* in malarial outbreak areas of central China. Parasit. Vectors 6 (1), 176.

CHAPTER NINE

Preparedness for Malaria Resurgence in China: Case Study on Imported Cases in 2000–2012

Jun Feng[1], Zhi-Gui Xia[1], Sirenda Vong[2], Wei-Zhong Yang[3], Shui-Sen Zhou[1], Ning Xiao[1,*]

[1]National Institute of Parasitic Diseases, Chinese Center for Disease Control and Prevention; Key Laboratory of Parasite and Vector Biology, MOH; WHO Collaborating Centre for Malaria, Schistosomiasis and Filariasis; Shanghai, People's Republic of China
[2]World Health Organization, China Representative Office, Beijing, People's Republic of China
[3]Chinese Preventive Medicine Association, Beijing, People's Republic of China; Chinese Center for Disease Control and Prevention, Beijing, People's Republic of China
*Corresponding author: E-mail: ningxiao116@126.com

Contents

1. Introduction	232
2. Background	232
3. Epidemiological Situation	234
3.1 General status	234
3.2 Species classification	235
3.3 Morbidity and mortality	236
3.4 Source of the imported cases	237
3.5 Geographical distribution of the imported cases in P.R. China	238
3.6 Gender and age characteristics	239
3.7 Monthly and seasonal distribution	240
4. High Risk Areas and Risk Factors	241
4.1 Border regions	241
4.1.1 China–Myanmar border	242
4.1.2 China–North Korea border	243
4.2 Eastern coastal regions	244
4.2.1 Zhejiang Province	244
4.2.2 Jiangsu Province	244
4.3 Central China	245
4.3.1 Henan Province	245
4.3.2 Anhui Province	246
5. Strategies on Prevention of Malaria Reintroduction	247
5.1 Imported malaria determination in China	248
5.2 Case management and surveillance	248
5.2.1 Management of imported malaria cases	249

Advances in Parasitology, Volume 86
ISSN 0065-308X
http://dx.doi.org/10.1016/B978-0-12-800869-0.00009-3
Copyright © 2014 Elsevier Ltd.
All rights reserved.

5.2.2 Diagnosis and treatment	252
5.2.3 Surveillance	253
5.3 Preparation for malaria resurgence through the imported cases	254
5.3.1 Training	254
5.3.2 Lessons learned	255
6. Case Studies	257
6.1 Treatment of an imported severe *P. falciparum* case	257
6.2 Treatment of imported mixed *P. falciparum* and *P. vivax* case	257
7. Challenges	258
8. The Way Forward	260
Acknowledgements	261
References	261

Abstract

Malaria is the most important parasitic protozoan infection that has caused serious threats to human health globally. China has had success in reducing the morbidity and mortality of malaria to the lowest level through sustained and large-scale interventions. Although the total number of malaria cases declined gradually, the burden of the imported malaria cases mainly from Southeast Asian and African countries has increased substantially since 2000, posing a severe threat to public health in China. This review explores and analyses the epidemiological characteristics of the imported malaria based on data from 2000 to 2012, in order to provide theoretical bases and insights into effective prevention, avoid the resurgence of malaria in malaria-susceptible areas and develop appropriate strategies to protect people's health in China. This review also intends to offer the useful information of innovative approaches and tools that are required for malaria elimination in various settings.

1. INTRODUCTION

With the dramatic decrease of autochthonous malaria cases, imported infection has accounted for most of the reported malaria cases in the People's Republic of China (P.R. China). In 2012, for instance, the imported cases account for 91.02% of the total malaria cases reported in China. The significantly increasing imported malaria may bring out the reemerging risk to the previous endemic areas, which put heavy pressure on malaria diagnosis and treatment. This chapter reviews the patterns of imported malaria for 13 years based on data from 2000 to 2012, summarizes our understanding of the prevalence trend and seek to obtain the overview study of the epidemiological characteristics of the imported malaria in P.R. China.

2. BACKGROUND

Malaria is a severe parasitic disease caused by different species of *Plasmodium* spp., including *Plasmodium falciparum*, *P. vivax*, *P. ovale*, *P. malariae* and

P. knowlesi, which was mainly via bites from infected *Anopheles* spp. mosquitoes (Butcher 2004). Severe malaria is mainly caused by *P. falciparum*. About 26 million cases were reported from 99 countries and territories in 2011, while the estimation number of malaria cases based on a true incidence may be of 219 million cases in 2010. Similarly, while about 100,000 deaths were reported globally in 2011, the World Health Organization (WHO) estimated that the true number of deaths worldwide was 600,000 in 2010. It is estimated that 3.3 billion people were at risk of malaria globally in 2011, with populations living in sub-Saharan Africa having the highest risk of malaria infection (WHO, 2012a). Risk also spreads with increasing global travel to and from endemic malaria countries. Approximately 80–90 million people from non-endemic countries annually travel to high risk areas of malaria transmission. Despite slowly decreasing global malaria burden, the imported malaria cases have increased even in the United States and United Kingdom, exhibiting more severe forms of the disease (Schlagenhauf and Petersen, 2008). For early diagnosis, every febrile patient with a travel history from these endemic areas should be suspected of malaria (Trampuz et al., 2003). The fatality rates among returning travelers with *P. falciparum* malaria are high and severe malaria can reach nearly 20%, even though they have been managed in intensive care (Bartoloni and Zammarchi, 2012; Nadjm and Behrens, 2012).

Since the 1980s, Chinese people have increasingly travelled around the world for tourism, business or visiting their relatives (Yin et al., 1995). In the provinces around Huaihe River Basin, covering Anhui, Henan, Hubei and Jiangsu provinces, the reemergence of local transmission was observed from 2000 to 2006 (Gao et al., 2002; Huang et al., 2007). After substantial efforts were undertaken to strengthen malaria control at all levels, the epidemics have been well controlled. The activities included the implementation of the 2006–2015 National Malaria Control Programme (NMCP) (MOH, 2006) and China Malaria Elimination Action Plan (2010–2020) (MOH, 2010). In addition, China also profited from the malaria control project supported by the Global Fund Fighting against AIDS, tuberculosis and malaria (GFATM) which contributed to decrease of malaria incidence significantly in China from 2002 to 2012 (Ding, 2012).

Imported malaria has been defined as an infection that was acquired in an endemic area by an individual (either a tourist or autochthonous native) but diagnosed in a non-endemic country after development of the clinical disease (Muentener et al., 1999). Therefore, the increasing number of domestic and international travellers to malaria-endemic countries should have a significant impact on malaria cases in the non-endemic countries

(Pavli et al., 2011). The imported malaria is now a great burden not only for people in but also for those in non areas (Schlagenhauf and Petersen, 2013).

However, in recent years there has been an increase of the imported malaria cases in P.R. China, which contrasts with a substantial decline of local malaria transmission (Zheng et al., 2013). Imported malaria has the risk of reintroducing the disease into the areas where malaria is no longer endemic but the environmental conditions suitable for the maintaining the life cycle of the parasite and its mosquito carrier. In the early 2000s there have been a handful of examples that the imported cases finally caused the secondary generation of malaria cases leading to local transmission of either *P. falciparum* or *P. vivax* in France, Greece, Italy, Spain, the United States (Florida), Jamaica, the Bahamas, Singapore and Oman (WHO, 2011).

As China's labour export and international trade continue to increase, the management of the imported malaria has and will become a prominent public health challenge (Liu et al., 2013). Especially in the non-endemic areas in P.R. China, surveillance was relatively weak, the medical and health institutions have little awareness of diagnosis and treatment of malaria, which may lead to misdiagnosis, thereby increasing the chances of secondary transmission. Herein, we have characterized the epidemiological features of the imported malaria cases in P.R. China from 2000 to 2012, and analized the current trends of the imported cases in order to improve or sustain our strategies to eliminate malaria in P.R. China.

The surveillance data were obtained from the national information reporting system and annual reporting system established by the Chinese Center for Disease Control and Prevention (China CDC) and analysed by ArcGIS 10.0 and SAS 9.2.

3. EPIDEMIOLOGICAL SITUATION

3.1 General status

China has made great progress on malaria control. The reported autochthonous cases and death cases were significantly decreased from 2000 to 2012, with a peak occurred in 2006. After 2006, then the total number of autochthonous cases showed decrease pattern successively. But the imported malaria cases were showed in various patterns.

During 2000–2012, a total of 46,772 imported malaria cases were reported from 31 provinces/municipalities/autonomous regions (P/M/A), accounting for 12.29% of all reported malaria cases in P.R. China (Gu and Zheng, 2001; Sheng et al., 2003; Zhou et al., 2005, 2007, 2008; Zhou et al., 2006a, 2006b; Zhou et al., 2009a; Zhou et al., 2011a, 2011b; Xia et al., 2012, 2013). During

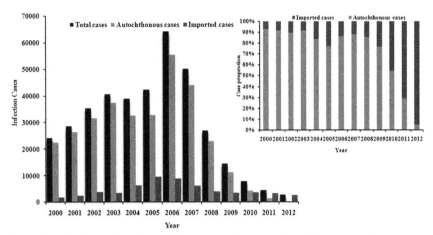

Figure 9.1 *Total number of autochthonous and imported malaria cases and proportion of imported cases per year, reported by national surveillance, 2000–2012, P.R. China.*

this period, malaria increased first with a peak in 2005–2006, particularly occured in central China. The number of imported cases also peaked in 2005 with 9593 cases, which accounted for 22.67% of malaria cases that year. In 2012, this proportion was up to 91.02% (n = 2474) of all reported malaria cases (Figure 9.1) (Xia et al., 2013).

From 2011 to 2012, 353 clinically-diagnosed cases (6.48%) and 5095 laboratory-confirmed cases (93.52%), confirmed by microscopy, PCR detection and rapid diagnostic tests (RDTs) respectively, were reported. Additionally, for *Plasmodium* species detected in laboratory, *P. falciparum* and *P. vivax* with laboratory-confirmed cases accounted for 55.32% and 40.91%, respectively (Xia et al., 2012, 2013).

3.2 Species classification

Four malaria species including *P. falciparum*, *P. vivax*, *P. malariae* and *P. ovale* were identified from 2000 to 2012; however, over 80% of all reports corresponded to *P. vivax* or *P. falciparum* infections. In detail, *P. falciparum* infection accounted for 9.28% of all malaria cases, and 31.85% of the total imported cases were *P. falciparum* infection.

The proportion patterns of *P. falciparum* in total malaria cases can be divided into three stages (Figure 9.2):
1. In 2000–2003, the proportion of imported *P. falciparum* in total imported malaria cases was steadily maintained at the same level in this stage;
2. In 2004–2008, the proportion was decreased with the observed lowest in 2007, even though the proportion of the imported *P. falciparum*

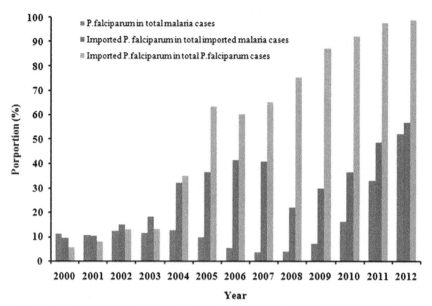

Figure 9.2 *The proportion of* Plasmodium falciparum *and imported* P. falciparum *in total malaria cases, total imported malaria cases and total* P. falciparum *cases, respectively.*

cases in total imported cases was increased. This may due to the largely increased autochthonous *P. vivax* cases;
3. In 2009–2012, the proportion was sharply increased, and this trend was also observed as the proportion of the imported *P. falciparum* in total imported malaria cases, which was due to the increasing number of Chinese labourers returned from African countries.

From 2000 to 2010, the annual proportion of *P. vivax* malaria cases was increased by nearly 30%. However, since 2011, the proportion of *P. vivax* was dropped and the proportion of *P. falciparum* patients increased dramatically. The available data on Chinese patients showed that 51.71% of them were suffering from *P. falciparum* malaria in 2011 and 2022.

In total, 5448 imported cases were reported in 2011 and 2012, including 353 clinically diagnosed cases (6.48%), 2084 *P. vivax* cases (38.25%), 2817 *P. falciparum* cases (51.71%), 118 *P. malariae* and *P. ovale* cases (2.16%) and 76 mixed cases (1.40%) (Table 9.1).

3.3 Morbidity and mortality

The highest incidence rate for the imported malaria was observed in 2006 with 0.46 per 10,000, subsequently sharply declining by an average annual rate of 7.3% from 2007 to 2012, with an incidence of 0.02 per 10,000 in 2012. However, the *P. falciparum* malaria morbidity was higher than *P. vivax*

Table 9.1 Imported malaria cases reported from 2011 to 2012 by *Plasmodium* species

Year	Total cases	Clinical diagnosis	*Plasmodium vivax*	*Plasmodium falciparum*	*Plasmodium malariae* and *Plasmodium ovale*	Mixed infection
2011	2974	278	1183	1414	62	37
2012	2474	75	901	1403	56	39
Total	5448	353	2084	2817	118	76

Source: China's Information System for Disease Control and Prevention V2.0 (Xia et al., 2012, 2013)

malaria morbidity after 2011, which could be explained by the increasing imported *P. falciparum* malaria cases returning from Africa.

According to the information from the reporting system, nearly 407 deaths were reported from 2000 to 2012; the highest number of deaths was observed in 2003 with 52 cases. The number of fatal malaria cases reported in Yunnan province was 243 from 2000 to 2012, the highest number in China, which accounted for 59.71% of total malaria deaths.

3.4 Source of the imported cases

Individual case reports with sources of the imported cases have been characterized since 2011, as reported in the China Information System for Disease Control and Prevention V2.0 established by China CDC. Based on this information system, a total of 5448 malaria cases considered as imported from 2011 to 2012 were collected and carefully analyzed. The results showed that from 2011 to 2012 the largest number of the migrant workers returned from Africa (50.24%), followed by those from Southeast Asia (42.42%). Due to the increasing number of laborers to Africa, 40 countries were reported as the imported source countries, including Nigeria (8.08%, n = 440), Angola (7.78%, n = 424), Equatorial Guinea (7.71%, n = 420), Ghana (5.95%, n = 324) and Congo (including the Republic of the Congo and Democratic Republic of the Congo, 2.18%, n = 119) (Figure 9.3). These patients were predominantly infected with *P. falciparum*, which has widely found in many parts of China (Lin et al., 2009). The patients from Southeast Asian countries were considered as another major infection source, including Myanmar (35.74%, n = 1947), Cambodia (2.88%, n = 157), Indonesia (1.32%, n = 72), Laos (1.17%, n = 64) and Pakistan (0.88%, n = 48) (Figure 9.3), mainly distributed in Yunnan province, especially on the China–Myanmar border (Hui et al., 2009).

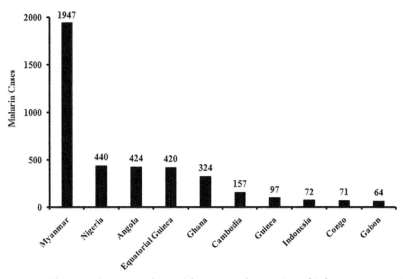

Figure 9.3 *Chinese migrant workers with presumed countries of infection among the imported malaria cases in P.R. China between 2011 and 2012.* Data were obtained from the China Information System for Disease Control and Prevention V2.0. The data were selected by dates of onset, reporting area and final review. Only the case acquisition as clearly described in the remarks column were obtained and analyzed for the imported cases. The number of total imported cases was selected from the top 10 countries as presumed countries of infection.

The percentage of migrant workers from Africa in 2012 accounted for nearly 60.05% of the imported cases (1458/2428), 81.41% of which were *P. falciparum* malaria cases (1187/1458). Among *P. falciparum* cases from Africa, Ghana (n = 207), Equatorial Guinea (n = 187) and Nigeria (n = 172) represent 17.44%, 15.75% and 14.49%, respectively. Nine hundred forty cases were reported from Southeast Asia, of which 81.28% originated from Myanmar (n = 764), Cambodia (5.21%, n = 49), Laos (3.83%, n = 36), Indonesia (3.72%, n = 35) and Pakistan (3.09%, n = 29).

3.5 Geographical distribution of the imported cases in P.R. China

From 2000 to 2012, imported cases were identified in 31 P/M/A, mainly distributed in the provinces of Yunnan (56.02%), Zhejiang (6.09%), Jiangsu (5.68%), Sichuan (3.87%) and Guangxi (3.70%). Imported *P. falciparum* malaria was mainly observed in Yunnan (39.64%), Jiangsu (12.60%), Zhejiang (10.93%), Sichuan (8.53%), and Guangxi (6.26%), respectively. While Imported *P. vivax* infection was mostly found in Yunnan province (73.63%), especially around the China–Myanmar border, following by Guizhou (4.65%), Zhejiang (4.61%),

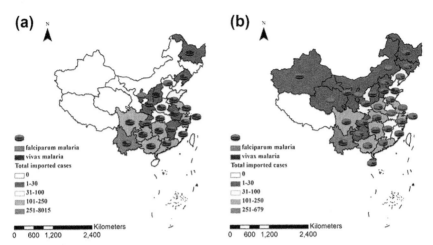

Figure 9.4 Variation trends of imported *Plasmodium falciparum* and *Plasmodium vivax* infection cases distribution in 2005 (a) and 2012 (b). Data were from the *National surveillance system (China Information System for Disease Control and Prevention V2.0)*.

Guangxi (3.04%) and Jiangsu (2.91%), respectively. Since the highest number of the imported malaria cases was observed in 2005 (n = 9593), we chose the malaria data of 2005 and 2012 for comparison. Results showed that in 2005, the imported malaria cases was clustered in south and central China, for example, Yunnan, Guizhou, Guangxi, Guangdong and Sichuan; these five provinces represent 90.81% (8711/9593) of all the cases reported in China (Figure 9.4(a)). In contrast, the data in 2012 indicated that the greatest number of infected patients, including autochthonous and imported cases (56.63%), were distributed in the provinces of Yunnan (27.45%), Guangxi (8.85%), Jiangsu (8.00%), Hunan (6.27%) and Henan (6.06%). As for the imported malaria, the imported cases have been found in almost all provinces, even in the western and northern China such as Xinjiang, Qinghai, Gansu, Inner Mongolia and Jilin, which did not report the imported cases in 2005 (Figure 9.4(b)).

Overall, imported malaria has been reported in 651 counties in 23 provinces in 2010, 760 counties in 26 provinces in 2011, and 598 counties in 29 provinces in 2012, respectively (Yin et al., 2013).

3.6 Gender and age characteristics

Among the imported malaria cases, the disease occurred more frequently in males (93.22%) than in females in 2011–2012 (Figure 9.5). This was possibly because men are the majority of the migrants and overseas workers (96.81%). The mean age of patients with imported malaria was 38.1 years of age, ranging

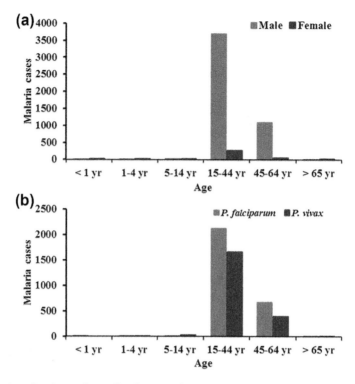

Figure 9.5 *Gender and age distribution of imported malaria from 2011 to 2012.* The data were collected from the China Information System for Disease Control and Prevention V2.0. The data were selected by dates of onset, reporting area and final review.

from 1 to 76 years of age; 76.37% of the cases were found in the 15–44 year age group, followed by the 45–64 year (22.07%) and 5–14 (0.81%) year age groups. The infectious species differed among ages (p value <0.0001), for example, *P. falciparum* was the predominant in the adult age groups of 15–44 and 45–64 age groups with 2125 cases (accounting for 40.82%) and 676 cases (accounting for 12.99%), respectively. While *P. vivax* predominately distributed in the age group of 15–44 (n = 1661), which accounted for 31.92% of the total imported cases, followed by age groups of 45–64 (7.65%, n = 398), 5–14 (0.58%, n = 30), 1–4 (0.13%, n = 7) and over 65 (0.12%, n = 6), and no case was observed in the age group below 1 year old.

3.7 Monthly and seasonal distribution

During 2011 and 2012, most cases (61.99%) were diagnosed between April and September, among whom imported malaria peaked in May (13.17%), June (12.56%) and July (9.68%). The trend of total cases was in accordance with

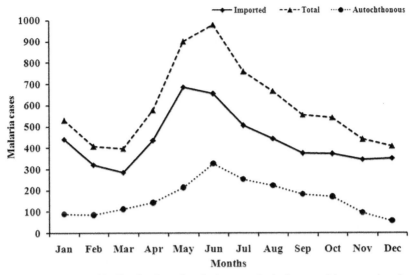

Figure 9.6 *Monthly distribution of malaria cases, including total, imported and autochthonous malaria, based on the data in 2011 and 2012 in P.R. China.*

imported cases, and the highest total number of cases was observed in June (Figure 9.6).

Regarding the seasonal distribution, the peak of malaria cases in almost all age groups was observed during the summer (May, June and July, 34.08%, 1774/5206) and the lowest number of malaria cases were observed during the spring (January, February and March, 20.09%, 1046/5206). In terms of parasite species found, the largest number of cases infected with *P. falciparum* was reported in summer (31.25%. 886/2835). Interestingly, no difference was observed between autumn (July, August and September, 24.34%), and winter (October, November and December, 23.88%); this was because most migrants labourers and overseas workers typically go back to China to have Spring Festival, a Chinese new year holiday. *P. vivax* was slightly different, the maximum number of cases was also observed in summer (38.26%, 805/2104), but the minimum was observed during winter (19.53%, 411/2104), which correlated to the variation in density of *Anopheles* mosquitoes.

4. HIGH RISK AREAS AND RISK FACTORS

4.1 Border regions

Several researches on risk assessment for the future malaria transmission potential in P.R. China have been reported. Three major risk areas in

P.R. China have been identified, such as border regions including China-Myanmar and China-North Korea borders, east coastal region and central region (Yang et al., 2012; Zhou et al., 2012 Zhou et al., 2013; Tambo et al., 2014).

4.1.1 China–Myanmar border

Malaria has been one of the most important public health problems in Yunnan Province (Yang, 2001). In 1953, the malaria morbidity incidence and mortality rates in Yunnan Province were 238.02 per 10,000 and 0.03 per 10,000, respectively. With sustained control efforts (Sheng et al., 2004), the malaria incidence in Yunnan Province dropped to 2.23 per 10,000 in 2000 (Li et al., 2003). But according to the national epidemic monitoring data, the total number of malaria cases and deaths in the most recent 10 years in Yunnan Province still ranked in the top first or second position among all of provinces in P.R. China, which suggests that the malaria situation in Yunnan Province is still a public health concern, especially for the counties on the China–Myanmar border (Li et al., 2006, 2008). Furthermore, frequent cross-border population movements have deteriorated the malaria situation, such as reported in the border counties of Tengchong and Longchuan (Shi and Qu, 2004; Li and Fan, 2005). Between 2002 and 2012, 23,568 imported malaria cases and 194 fatal cases were reported in Yunnan Province. *P. falciparum* malaria and *P. vivax* malaria infections represented over 80% of the reported cases, of which more than 56% were *P. vivax* malaria infection.

Therefore, the malaria control particularly the imported malaria on China-Myanmar border is the top issue for the NMEP in Yunnan Province. The highest number of *P. falciparum* and *P. vivax* infected cases in last decade in Yunnan Province were observed in 2005, and then the number of the imported cases declined but its percentage has increased gradually since 2008. However, in 2012, 690 cases were determined as imported malaria by the annual reporting system, including 506 *P. vivax* cases, 180 *P. falciparum* cases, 2 *P. malariae* cases and 2 mixed infections with co-infection of *P. falciparum* and *P. vivax*. Unlike in other malaria affected areas, the autochthonous cases still existed in Yunnan Province and imposed serious impacts on the prevention and treatment by the local authority. In 2012, 172 autochthonous cases were reported with 154 *P. vivax* cases and 18 *P. falciparum* cases, respectively. In addition, the autochthonous cases were mainly distributed in the counties on China–Myanmar border containing Gengma (13.95%), Yingjiang (10.47%), Mangshi (9.30%), Cangyuan (9.30%), Tengchong (6.98%) and Ruili (5.23%) in the age group of 25–60 of malaria cases males accounted for 63.23% and 68.61% of all imported malaria cases.

The challenge for the malaria control on China-Myanmar border remains in identifying and containing imported infections, particularly in the border areas of 25 counties. The World Health Organization (WHO) recommends border post screening for malaria not only to identify and treat infections but also to set up the response measures for preventing transmission throughout affected regions (WHO, 2010a). Moreover, since the borders are porous and migrant populations pass daily across the borders, additional strengthened measures may be needed at the border screening centers, which target longer-term migrant workers (Yangzom et al., 2012; Zhou et al., 2014). In order to tackle these problems, efforts including malaria surveillance, case management and vector control are urgently to implemented through a multi-country platform in border regions, as similar measures were done in Bhutan (Yangzom et al., 2012). It is worth noting that multidrug-resistant *P. falciparum* should also be continuously monitored at the Chinese borders, in addition to the population movement monitoring and effective vector control strategies (Chen et al., 2007; Liu, 2014).

4.1.2 China–North Korea border

Northeastern China was considered a low risk area for the malaria transmission where *P. vivax* was the only species, although the imported *P. falciparum* malaria occasionally caused epidemics in the histroy. For example, Liaoning Province, is one of three provinces in Northeastern China and the Yalu River marks its border with North Korea, was historically a malaria endemic area (Teng et al., 2013). Owing to great control efforts for many decades, Liaoning made great strides in malaria control up until 1988. Since then, imported malaria has become a growing threat to this border region owing to more labourers going abroad in recent years (Teng et al., 2013).

From 2002 to 2012, a total of 247 imported cases were laboratory-confirmed in Liaoning Province, of which 31.98% were *P. falciparum* malaria. The migrant workers from Southeast Asia (60.32%) are the majority of the imported cases, especially from Myanmar (70.21%), followed by Cambodia (8.24%). The imported cases, particularly of *P. falciparum* malaria, have gradually increased since 2008. In 2012, about 29 *P. falciparum* malaria cases were reported, which account for 76.32% of the imported cases. Most (>70%) of these cases were distributed in 11 counties of Dalian and Shenyang. Moreover, the imported cases in Liaoning Province were detected regardless of seasonal variations (McMichael, 1993; Rogers and Randolph, 2000).

4.2 Eastern coastal regions

Eastern coastal provinces have similar economic conditions with faster development for the modern socioeconomic socity than other regions of China, particularly in Jiangsu and Zhejiang, the two provinces where a large number of imported malaria cases reported recently. As mentioned before, imported malaria has also emerged as a public health threat to these two provinces (Ruan et al., 2007; Ye et al., 2004; Yao et al., 2009; Yuan et al., 2004; Chu et al., 2007; Cao and Wang, 2009).

4.2.1 Zhejiang Province

The annual malaria incidence has been below 1 per 10,000 in Zhejiang Province since 1993. However, imported malaria has increased steadily since 2003 (Jin et al., 2006).

A total of 2762 imported malaria cases were reported from 2002 to 2012 in Zhejiang Province, of which nearly 15.42% of the cases were considered as *P. falciparum* infection. Additionally, the number of the imported malaria peaked in 2007 with 610 imported cases, which accounts for 22.09% of all of the imported cases (610/2762). The number of the imported malaria cases from 2007 to 2011 were gradually reduced by 15.41% on average. In 2012, 135 imported cases were determined by laboratory diagnosis, including 89 *P. falciparum* malaria, 43 *P. vivax* malaria and 3 mixed malaria cases coinfected with both *P. falciparum* and *P. vivax*. In a word, the accumulation of the imported malaria patients may give rise to the possibility of the secondary malaria transmission leading to the reintroduction of malaria in local settings. The reintroduction of malaria need to be avoided based on strategy of the NMEP in whole China.

Unlike in other areas, mobile control programme, for example, the number of the imported malaria cases migrated from Anhui Province has accounted for a big proportion of the total number of malaria cases in Zhejing Province (Jiang et al., 2009). In 2005–2008, 87.63% of the immigrant labourers who came from Anhui Province were mainly found in 4 cities of Zhejiang Provinces, namely Ningbo, Taizhou, Jinhua and Hangzhou, and those imported cases played a serious threat to malaria control programme of Zhejing Province (Chen et al., 2011; Jiang et al., 2009).

4.2.2 Jiangsu Province

Historically, Jiangsu Province was an endemic area, prone to large outbreaks and classified as a malaria unstable region (Liu et al., 2014). *An. sinensis* and *An. anthropophagus* were the main vectors and *P. vivax* malaria continues to be present across the province (Zhou et al., 2009). In the 1980s, *P. falciparum* malaria transmission were reported in Jiangsu Province, however, no local *P. falciparum* infected cases have been present in Jiangsu Province since 1988 (Zhang et al., 1996). Given the decline of local *P. vivax* malaria transmission since 2002, *P. falciparum* malaria via imported cases has surpassed *P. vivax* malaria since 2005 (Zhou, et al. 2009b; Zhou et al., 2011c, 2012; Jin et al., 2006).

Of the 2853 imported malaria cases reported in Jiangsu Province between 2000 and 2012, *P. falciparum* and *P. vivax* infected cases accounted for 34.01% and 58.32%, respectively. The number of the imported malaria cases peaked in 2007 (n = 388) while the proportion of imported *P. falciparum* peaked in 2011 up to 96.52% (n = 361). In 2012, for instance, the total of 197 malaria cases were imported, including 169 were infected with *P. falciparum*, 4 with *P. vivax*, 2 with *P. malariae*, 20 with *P. ovale* and 2 with mixed malaria. One of the characteristics in Jiangsu Province was represented as the increasing *P. ovale* cases. This pattern of increasing *P. ovale* cases reflects the distribution of Chinese labours in Africa with an increasing migration to the west and central Africa, such as Angola, Nigeria and Congo.

4.3 Central China

Based on malaria prevalence and characteristics of classification by the national monitoring sites in the malaria surveillance system in China, the central China around Huanghuai basin are classified as 'unstable endemic and epidemic prone malaria' with a high malaria incidence (Diouf et al., 2011). These areas are warm-temperate and subtropical in relation to a unique geographical location, the Yangtze River and Huai River from west to east across a large territory. In addition, on average a rainy season lasts from 6 to 9 months per year. These conditions are well suited for the habitats of *An. anthropophagus* and *An. sinensi*, which pose a serious threat for the reintroduction of local malaria transmission (Gao et al., 2004; Gu and Zheng, 2001). There is evidence of a reemergence of malaria in central China covering the provinces of Anhui, Henan and Hubei (Jiao et al., 2013). In this study, Henan and Anhui provinces were chosen to identify common patterns and investigate the actual changes of the imported malaria brought to these areas.

4.3.1 Henan Province

In Henan Province, the malaria incidence was <1 per 10,000 until 1991 (Su et al., 1994). In 2005, it was reported to have sharply increased in the

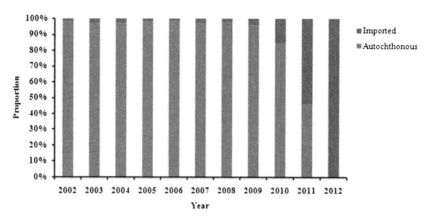

Figure 9.7 *Proportion of imported and autochthonous cases in Henan, 2002–2012.*

number of autochthonous malaria cases in Yongcheng, located in eastern Henan, bordering with Anhui Province, which was an endemic area during 50 years from 1950 to 2000 (Su et al., 2006; Zhang et al., 2007). The number of infected patients significantly increased since 2010 due to the increase of the domestic migrant population mainly coming back from Yunnan Province. From 2002 to 2012, there were 986 imported malaria cases reported in Henan Province, with 3.90% (986/25,253) of all malaria infections in P.R. China.

The total number of imported *P. falciparum* and *P. vivax* malaria cases were more than 780 cases in Henan Province from 2002 to 2012, accounting for 79.10% of all imported cases. The imported malaria cases had maintained a gradually increasing trend of 16.99% and reached a peak in 2007 with 105 cases (10.65%). Moreover, the number and proportion of the imported cases increased and while the number of autochthonous cases reduced sharply since 2009 (Figure 9.7). For instance, the highest number of the imported malaria cases was observed in 2011 with 192 cases, and in 2012, 148 imported malaria cases were determined by laboratory tests, composed of 21 *P. vivax*, 116 *P. falciparum*, 5 *P. malariae* and 6 *P. ovale* infections, respectively.

The number of infected people aged from 25 to 60 years old accounted for 68.33% of all malaria cases, and 73.14% of the infected patients were male. Of all of the imported malaria cases, the patients coming from Southeast Asia were 58.31%, following by Africa (38.13%).

4.3.2 Anhui Province

Four hundred eighty-five imported cases were identified in Anhui province from 2002 to 2012, which accounted for 0.39% of the reported malaria cases (n = 122,876) in the same period. In addition, the imported cases were mainly distributed in Hefei (52.03%), Bengbu (11.26%), Huainan (7.11%), Haozhou (7.11%) and Huaibei (4.13%). The majority of the imported cases were *P. falciparum* infections (47.01%) and *P. vivax* infections (49.7%). Between 2006 and 2012, though the incidence significantly declined from 6.40 per 10,000 to 0.02 per 10,000, the risk of transmission still existed.

In 2012, local malaria transmission still persisted in 17 counties with 32 autochthonous cases. A total of 64 cases were considered imported cases, of which 56 were *P. falciparum*, 4 *P. vivax*, 2 *P. malariae* and 2 *P. ovale* infections. The vast majority of those infected were located in the age group of 25–60 years (83.33%) and also male (71.21%). In addition, 96.97% of the imported cases were determined by laboratory tests; 54.55% of the patients had come to China from Southeast Asia, while 33.33% of the infected patients had come from Africa.

5. STRATEGIES ON PREVENTION OF MALARIA REINTRODUCTION

China has succeeded in preventing and controlling malaria in the past 13 years. The number of total malaria cases has significantly declined, especially for autochthonous cases. In contrast, the number of the imported malaria cases increased remarkably which have become a major threat to the NMEF in P.R. China (Gao, 2011).

The proportion of total imported cases reached a peak with 17,577 cases from 2004 to 2006, representing 37.58% of all imported cases from 2000 to 2012 according to the annual reporting system. Additionally, the fraction of the imported cases out of total malaria cases increased from 5.22% in 1992 to 14.01% in 2006. While in 2012, the percentage had reached 91.02%. Though great efforts have been made in some high incidence areas including the provinces of Yunnan, Hainan and Anhui, the risk of local and sustained transmission of malaria still persists, which could be translated into the occurrence of repeated localized epidemics probably triggered by the imported cases.

The threat posed by imported malaria to reintroduction and jeopardizing China's NMEP highlights the need to have good preparedness including the advanced preparation and planning in place at each risk area. This calls for active risk assessment to provide a clear estimation of the scale of the health threat such as high risk areas for occurrence of epidemics, while documenting the level of uncertainty. The approach is to use a structured identification of key information such as high susceptibility areas for sustained local transmission and systematic vulnerability (e.g. detection and diagnostic capacity, and early and effective treatment) (WHO, 2010b, Yadav et al., 2014).

5.1 Imported malaria determination in China

According to the National Criteria of Malaria Control and Elimination (MOH, 2011), the imported cases refer to non-local infection of malaria in county-level administrative regions, including foreign migrants and local residents infected in other places. But during the process of malaria elimination programmes considering China as a whole unit, the cases from the provinces, autonomous regions and municipalities within the scope of the cases of other counties (districts), shall not be regarded as the imported malaria case. After achieving the goal of national malaria elimination, only the malaria cases from overseas were considered as the imported cases.

Based on the Chinese technical guidline for the NMEP, malaria patient regarded as the imported case has to simultaneously meet the following criteria:
1. Come from malaria transmission area
2. Have a history of staying overnight during malaria transmission season
3. Presenting interval time from onset to entering China should be less than 25, 27, 35, and 16 days for *P. vivax*, *P. falciparum*, *P. malariae* and *P. ovale* in malaria transmission seasons, respectively. The imported cases diagnosed in the resting phase of *P. vivax* were not restricted by this time limitation.

Apart from this, there are additional two types of cases who should be classified as imported cases: (1) cases infected with imported *Anopheles* mosquitoes, often detected at airports, docks, frontier port stations; and (2) blood transfusion infections by overseas labourers, who has been detected as *Plasmodium* infection.

5.2 Case management and surveillance

The imported malaria may bring out a severe barrier to malaria control and elimination programmes. To tackle the issue, it is necessary to establish an early warning system to meet the requirement against the re-emergence of malaria, especially in the at-risk remote and resource-limitted areas. Effective control measures also rely on a prompt and proper diagnosis of the disease. This remains a challenge in poor areas that can be ill-equipped for diagnosis and where access to health care is low (Zheng et al., 2010).

In view of the fast spreading and high transmission capacity of malaria, the outbreak may occur in some endemic and non-endemic areas. Therefore, it was a key step to strengthen malaria surveillance in low transmission areas, and appropriate measures taking by travellers should be reinforced in a similar manner (Liu et al., 2012).

In non-endemic areas, malaria cases were found with a history of overnight residence and blood transfusion could be considered as the emergence of malaria, and need to further investigation. This has occurred under various circumstances including airports, ports, baggage, nosocomial transmission and transmission by local competent vectors which once reported in the United States (MMWR, 2004). The key recommendations for public health management in against malaria emergence in non-endemic areas are (Zoller et al., 2009): (i) rapid and regular assessment of new information by a team of clinicians, malaria experts, entomologists and public health experts; (ii) timeline of events with incubation times for different transmission scenarios to assess their probability, (iii) molecular genotyping of isolates from index and secondary cases; (iv) repeated interviews of index and secondary cases as well as all other personnel involved; (v) local entomological survey; (vi) selectively informing key health infrastructures and physicians; (vii) preparation of a press release in case of public media attention.

5.2.1 Management of imported malaria cases
5.2.1.1 Government duty
To maintain the malaria-free status in the country, the government should retain the services of experienced and qualified malaria personnel in a specialized service within the national epidemiological services, or a service for the prevention and control of vector-borne and parasitic diseases. As noted previously, the existing malaria control programme budget should not be reduced, so that all the requirements for malaria vigilance activities, procurement of drugs, reserve stocks of insecticides, spraying equipment, laboratory equipment and reagents, training and health education activities can be met. The government should ensure that screening for malaria and radical treatment remains free of charge for all suspected malaria patients, irrespective of their civil status, in all governmental health facilities.

As for China, the government needs to take responsibility to coordinate surveillance and control efforts in the high risk regions, such as border regions. Malaria control coordinating meeting on border region focussing on the joint actions against malaria transmission on the China–Myanmar border is suggested to be held once a year. In addtion, the sectorial cooperation is urgently to be strengthened aimed at the prevention of imported malaria in the country, particularly departments of education, agriculture, industry, forestry, immigration, tourism, defense and municipalities are necessary to be invited to involved in the NMEP with different resposibility (Table 9.2). Besides, it should also maintain close links with the private sectors (Thakor et al., 2010). It is equally important to obtain, by means of appropriate information, education and communication and social mobilization. Information and advice, through the media, should be given to persons entering or leaving malarious areas regarding the dangers to which they may be exposed.

China has approved '1-3-7 target elimination mode' treated with malaria cases in the malaria elimination phase. Once a malaria focus found with index cases who is either the imported or local cases, three kind of activities for prevention of further transmission will be taken promptly as follows: (i) potential malaria infections reported promptly including clinical and laboratory-confirmed cases within 24 hours; (ii) laboratory re-examination to confirm the species and infection source within 3 days; (iii) transmission risk assessment and adoption of effective treatment measures within 7 days (Cao et al., 2013; Zhou et al., 2014).

5.2.1.2 Management of entry and exit port

Frontier port is the important barrier against the imported malaria, the following two activities need to carry out in frontier port to against the imported malaria, one is case screening, and the other one is substantial control measures:

1. Case screening : all suspected cases of malaria with temperature over than 28°C need to take the screening, including (i) epidemiology study covering history of present illness, travel history, exposure history, past medical history, etc. (ii) taking rapid detection test (RDT) to confirm diagnosis; (iii) further investigation for those suspected malaria cases, covering close contacts registration, health education to encourage revisit doctor once suspected symptoms onset.
2. Subsequent control measures: once found the suspected malaria cases in the frontier port, some prevention and control activities need to take place, including disinfection of contact items by the suspected patients, mosquito investigation and control within the 400 m where suspected cases staying, and data tracking and storing.

Table 9.2 Main duties and responsibilities of departments in the NMEP in China

Departments	Main task
National Sectors	• Organize and make plans for imported malaria prevention and control • Provide technology solutions for imported malaria monitoring, analysis, risk and needs assessment; responsible for the cases of diagnosis, treatment and rescue of serious cases of illness
Ministry of Health	• Carry out imported malaria epidemiology investigation and treatment; strengthen the team construction, to provide the necessary equipment, the establishment of antimalarial drugs reserve system • Arranging the health education work • Training health care workers, to improve the diagnosis and treatment of malaria and management ability
Ministry of Public Security	• Provide relevant personnel entry and exit information, assist the health department in accordance with the implementation of malaria prevention and control measures.
Ministry of Commerce	• Improve the awareness and ability for people going abroad, accompanying physicians and antimalarial drugs, to provide timely diagnosis and treatment services.
Entry–Exit Inspection and Quarantine	• Organization for frontier port health and quarantine, epidemic disease monitoring, health supervision and sanitization job screening from high malaria patients in time, especially suspected cases should be transferred immediately to medical institutions
Propaganda Department	• Coordination of radio, film and television, press and publications and other timely reports of the public health emergency that can happen from imported malaria, strengthen the publicity, psychological crisis intervention and popularized knowledge about prevention
Ministry of Education	• Malaria prevention knowledge propaganda with focus on schoolchildren and teachers
Ministry of Transportation	• Malaria prevention knowledge propaganda in airports, stations, at wharfs and other vehicles
Ministry of Finance	• Provide financial support to malaria prevention and control, supervision and expenditure management

5.2.1.3 Management of large international activities

In the case of a large international event, it is important to pay attention to preparation against disease spreading in a large population, which include following activities:

1. Risk assessment, make monitoring plan in advance;
2. Management of training to clinicians, also personnel and volunteer training;
3. Strengthening cooperation for disease control and prevention with health care institutions;
4. Health education to improve the knowledge on malaria prevention and control;
5. Reserves of necessary antimalarial drugs, diagnostic reagents and equipment, mosquito instruments, etc. (Zheng et al., 2013);
6. Preparations for contingency plans;
7. Learning the experience of other countries through international collaboration.

5.2.1.4 Emergency preparedness

Emergency preparedness for possible malaria epidemics should be part of the general organization of emergency health services, which in turn should be an integral part of every health system. Under certain circumstances, preparedness plans for malaria epidemics and for emergency health services should be included in the national disaster preparedness plan, particularly in areas where there is a recognized risk of natural disasters (DaSilva et al., 2004).

Preparedness for malaria epidemics is taken based on an understanding of the epidemiology of malaria and of the epidemic risk factors. The more complete the understanding and the more developed information system and monitoring of risk factors are, the higher will be the preparedness, and the more accurate the forecasting and the more adequate the preventive response (Tambo et al., 2014). An immediate response should be efficient if the appropriate human resources, supplies, equipment and logistical arrangements are ready to be brought into action. Preparedness should therefore include the identification of these resources and the mechanisms required for their rapid mobilization. In general, the establishment of stocks of insecticides, spraying equipment and antimalarial drugs is highly desirable (Salam et al., 2014).

5.2.2 Diagnosis and treatment

At present, there are several methods for malaria diagnosis, but blood film examination continues to be the most affordable and commonly used in many places of China, despite its limitations for low parasite densities and failure to reveal infection in its latent stage. Whenever possible, the serological and parasitological information should be collected and evaluated together.

For imported *P. vivax* treatment, since chloroquine resistance was discovered in several regions (Yang et al., 2005; Guan et al., 2005; Lu et al., 2006), the government adopted 8 days with chloroquine plus primaquine to be used as the current treatment. For radical treatment in resting stage, oral primaquine for 8 days with total dosage of 180 mg was adopted.

For uncomplicated *P. falciparum* malaria treatment, oral artemisinin-based combination therapy (ACT) is the standard treatment as recommended by WHO. Currently, the drugs used in China include artesunate/amodiaquine, artemether/lumefantrine, and dihydroartemisinin/piperaquine, etc. (Huang et al., 2012). Another important drug, pyronaridine phosphate, which was developed by the National Institute of Parasitic Diseases, Chinese Center for Disease Control and Prevention, was used to treat *P. falciparum* that has developed resistance to chloroquine or piperaquine (Poravuth et al., 2011; Chen, 2014).

Injection-type artemisinin monotherapy can only be used in severe malaria cases (via IV or IM) as a medical emergency (WHO, 2006). While rescuing severe cases in medical units at or above the county level, the artesunate intravenous injection was suggested, and generally needs to increase the dosage in the treatment.

5.2.3 Surveillance

A malaria vigilance system needs to be carefully planned and well managed to ensure early recognition and prompt response to imported cases that can give rise to an epidemic (Bergquist, et al., 2012; Zhou et al., 2013). This is a very important component of malaria vigilance in areas free of malaria transmission in China. The results can form the basis for measuring the magnitude of the malariogenic potential of an area. Assessment of the risk of malaria epidemics should become a regular preoccupation of services in all areas with high malariogenic potential. Such assessment might be possible if based on an information system capable of identifying not only high-risk areas, but also high-risk periods and the population groups at risk.

To alleviate the consequences of imported malaria in P.R. China, a sentinel surveillance network (China Information System for Disease Control and Prevention V2.0.) has been established. A thorough description of the epidemiological and clinical aspects of each malaria case has been recorded, and this was shown to be very helpful. Information has been collected prospectively since 2004. It is possible to identify the areas from which importation of *P. falciparum* malaria is particularly high, thus indicating a high risk for travellers to that region. Another system, an annual reporting system, the data including total cases, autochthonous cases and imported cases etc. collected from all provinces, could easily detect how many cases of imported and autochthonous. Since the chloroquine-resistant cases were discovered in some areas of Yunnan Province, it is maintaining a sharp vigilance with prompt and clear diagnosis and treatment (Yang et al., 1999; Zhang et al., 2008; Bi & Tong, 2014)).

5.3 Preparation for malaria resurgence through the imported cases

5.3.1 Training

During the elimination stage, low malaria incidence made it less of a public health burden, but it still requires significant financial and human resources based on the marginal efficiency principle. Gaining the political support to sustain significant funding will be challenging (Zofou, et al., 2014). Significant communication and education are necessary to ensure that populations understand the ongoing risk and support case management activities although malaria cases are less frequent. Continuous retraining of health workers will also be necessary to ensure that health workers are alert in preventive and curative measures. It is necessary for us to keep (1) its continuing ability to rapidly diagnose malaria through microscopy and/or RDTs, and provide the recommended treatment, particularly when cases are rare; and (2) ensuring that workers have constant access to diagnostics and drugs for rapid treatment.

The training of medical and paramedical personnel is subject to the following considerations:

1. Integrate malaria services and staff into the general health system, which might involve the assignment of staff to more comprehensive duties for which special training that may be required;
2. Attach importance to a disease that no longer exists in the country among staff;
3. Public demand for other types of health care and the establishment of other priorities.

Proper training of medical personnel, including clinical physicians, rural doctors and laboratory workers, is indispensable to prevent the reintroduction of malaria (Harvey et al., 2008). When developing the curricula for such training, attention should be given based on the past. Training programmes should also be provided for health administrators, to ensure that they are alert to the need for malaria vigilance, further development of the health services, particularly rural health services, and consequent administrative and budgetary implications.

On the other hand, it would also be responsible for training students and teachers in order to keep the public aware of the dangers of the reintroduction of malaria (Okeke and Uzochukwu, 2009). Training in malaria should be practical, emphasizing not theory but individual protection and case detection, treatment, and notification.

For physicians in malaria-endemic and malaria-free but threatened countries, malaria training should begin in medical schools, which should retain malaria in the curriculum and among the subjects for examination. Special seminars are provided to the physicians who are to work in areas where malaria cases are likely to occur, or in health services at ports, airports, and border-crossing posts. As part of the routine examination of patients, physicians will ask specific questions on travel abroad. They should enquire where the patient has been during the last weeks, months and years.

5.3.2 Lessons learned

Experience has shown that it is very difficult to prevent the entry of *Plasmodium* parasite carriers into malaria-free areas. In P.R. China, a case definition of suspected malaria has been established, in detail, for patients who do not have clinical symptoms but meet the assessment criteria of imported malaria, and this can be strictly followed by the staff of the general health services for taking blood smears for microscopic examination.

For the initial screening of immigrants, it is essential to know the status of malaria endemicity in the areas from which they come. Detailed information should be collected including place of origin, areas where they have travelled, whether or not they have had malaria or any symptoms and current health status. This information is of great help for medical staff or laboratory workers to decide whether laboratory examination, chemotherapy or follow-up is necessary. The risk of overlooking malaria when its symptoms are obscured by a more obvious infection of other origin should be stressed.

In P.R. China, the activities to deal with the imported malaria are summarized as follows:

1. Information sharing between sectors and departments, e.g. health community and quarantine service, immigration services or other sectors, has been effective for information gathering and sharing (Bergquist, et al., 2012; Zhou et al., 2013);
2. Regular meetings that were held at the China–Myanmar border to share issues related to imported malaria have been useful to implement coordinated preventive measures on both sides of the border, such as the prevention of invasion of *An. arabiensis* or the strengthening of sentinel surveillance (Yu et al., 2013);

3. Surveillance based on multi-purpose home visits has been necessary in some areas. Weekly, monthly, seasonal and annual reports have been supplied in order to inform the malaria situation. These reports have proved to be very useful to management of the imported cases;
4. Confirmed malaria cases are given radical treatment through the whole country, and epidemiological investigations are made in the surrounding residents. These investigations include examination of blood films from an appropriate number of people, and entomological study of the vector in the area;
5. '1-3-7 target elimination mode' has been effectively executed for the management of the imported malaria cases in the malaria elimination phase.

In preparing for the malaria resurgence through imported malaria, it is necessary for all the medical communities to adopt the following measures:
1. Local measures against mosquitoes in general, and antianopheline measures in areas of high receptivity in particular, both antilarval and imagicidal;
2. Effective activities should be executed at airports, seaports and other popular tourist spots that are subject to the threat of reintroduction of imported malaria (Zhang et al., 2011);
3. Strengthening cooperation in the identification of malaria cases between general health services in the country and laboratory staff. The responsibility for adequate diagnostic support should be shared between the medical staff and the laboratory services;
4. Apart from the public health implications of imported malaria, it is imperative to deal with the severe illness because a delay in diagnosis may result in the death of the patient;
5. Adequate briefing for travelers and the availability of recommended antimalarials is most important. This aspect, however, has been somewhat neglected in many areas;
6. Personal protection methods such as mosquito-proofing of houses, use of repellents, vaporizing mats and aerosols, should be encouraged as a supplementary effort to reduce contact between humans and mosquitos (Moore et al., 2008).

The experience, for example, in Mauritius shows that high-risk populations should target and monitor, especially around the border (WHO, 2012b). In addition, surveillance need to proactively screen migrant groups coming back from malaria-endemic countries in the passenger-screening programme. More importantly, the reactive case detection and response system in Mauritius closely monitors positive cases, which ensures successful treatment and screens contacts and neighbours to identify additional infections in order to prevent local transmission. The strategy for preventing reintroduction also includes routine island-wide larviciding based on entomological surveillance to maintain low levels of *Anopheline* breeding and therefore diminish receptivity. As sporadic introduced cases have been reported in the country during the past decade, ongoing vigilance is critical to prevent indigenous transmission from those cases.

Another example from the Eastern Mediterranean in applying preventive measures has revealed that malaria transmission may be re-established by the entry of infected persons, the importation of infected *Anopheline* mosquitoes or by undetected parasite carriers.

6. CASE STUDIES
6.1 Treatment of an imported severe *P. falciparum* case

Male patient, 38 years old, worked in Pakistan, with a history of mosquito bites.

Treatment procedures are as follows:
1. Giving artesunate intravenous injections for 14 days;
2. Transfused with red blood cells and albumin;
3. Appropriately use furosemide (diuretics);
4. Strengthen the liver protection, anti-infection treatment;
5. Give bicarbonate for neutralization;
6. Give dexamethasone to reduce intravascular hemolysis when necessary (Figure 9.8).

Lessons learned:
1. Delayed treatment. Medical staff should promptly consider malaria disease after acquiring the travel history and seeing clinical symptoms of fever;
2. Sensitivity to malaria was not enough for laboratory testing personnel, training needed to establish attention to the prevention and treatment of imported malaria, which may induce misdiagnosis in diagnosis.

6.2 Treatment of imported mixed *P. falciparum* and *P. vivax* case

Male patient, 27 years old, worked in the forest of Myanmar. Four of his colleagues died due to severe malaria. Clinical symptoms were fever, chills and sweats, nothing changed after oral administration with pyronaridine phosphate and intramuscular injection with quinine. The patient was diagnosed as *P. vivax* malaria by microscopy.

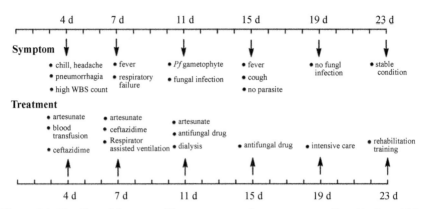

Figure 9.8 *Timeline of the event.* The symptom and treatment were listed in the middle part, the case is from Tang, 2010.

Treatment procedures are as follows:
1. Chloroquine with 1.0 g per meal for the first dose, 0.5 g per meal for the second and third days. Primaquine was taken 39.6 mg per meal for 8 days;
2. The patient caught high fever at day 8 and the spleen was enlarged. After careful examination, *P. falciparum* with chloroquine resistance was confirmed;
3. Artemether was taken by intramuscular injection of 320 mg at day 10. The patient was injected with 160 mg artemether after achieving normal body temperature. Symptoms were resolved at day 12;
4. The patient had a reoccurrence of malaria after 3 months.

Lessons learned:
1. Misdiagnose. When diagnosing imported malaria cases from endemic areas with mixed *P. vivax* and *P. falciparum*, the blood films should be carefully checked after one species is found in microscopy. Molecular technology such as PCR could be applied to reconfirm in order to prevent missing other species in the mixed infection;
2. Training is important for local medical staff. The drug usage should be standardized according to WHO recommendations, otherwise the disease may reoccur due to irregular administration (Figure 9.9).

7. CHALLENGES

The major challenges to ensure adequate response to control the reintroduction of malaria transmission are as follows:
1. Inadequate and untimely information exchange on the status of imported malaria among different areas;

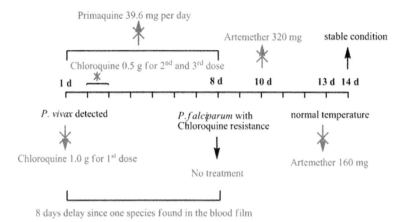

Figure 9.9 *Lesson learned from the case.* The red arrow and characters represent the wrong treatment or what we should learn from this case. The case is from Tang, 2010. (For interpretation of the references to color in this figure legend, the reader is referred to the online version of this book.)

2. Inadequate collaborative action among malaria-endemic countries, especially around China–Myanmar border, which often do not provide relevant information to persons leaving their territory;
3. Surveillance information including epidemic data, resident population and floating population should be exchanged between the countries on a regular monthly and annual basis, and there should be cross-notification of disease outbreaks;
4. Collaboration in monitoring susceptibility of malaria vectors to various insecticides as well as drug-resistant parasites in border areas (Yan et al., 2013);
5. Lack of awareness of malaria reintroduction into areas free of malaria among the general health services staff;
6. Inadequacies in malaria training. Collaboration in training activities among existing training centres and facilities in the regions need to be strengthened. Suitable epidemic warning and reporting systems as part of their epidemiological surveillance systems have to be developed to monitor and recognize early signs of impending epidemics. Malaria outbreaks/epidemics need to list as notifiable diseases in each participating country;
7. Immature country-level partnership in malaria control, which has to develop further to take more effective and sustainable malaria control efforts.

In regard to the imported *P. falciparum* malaria, it has spread to more provinces in the malaria-free areas in P.R. China. Compared to the imported malaria cases in 16 non-endemic provinces in 1998 (Tang, 1999), the imported cases were reported in 31 provinces in 2012. Increased population movement was one of major factors resulting in the spread of imported *P. falciparum* malaria in P.R. China (Lu, 2008). Meanwhile, due to the gradual decline of autochthonous cases, the increasing imported malaria, have brought the potential risk of outbreak in previous malaria endemic regions, and also the big pressure of the malaria diagnosis, treatment and mortality. For example, the imported malaria cases in Guangxi Province have increased by 95.53% in 2012, as for the imported *P. falciparum* malaria, the reported cases were 179.05% higher than the number reported in 2011.

In order to achieve the best possible preparedness for malaria reemerging, the risk of re-transmission in epidemic-prone areas and transmission intensity have been documented under the following 3 fields:
1. risk districts or areas with an originally high malaria endemic potential;
2. risk population moving from non- and low-endemic areas to highly malarious areas;
3. Unusual meteorological conditions.

The control strategies are carried out to focus on three types of people, such as: (i) nationals returning from long-term residence in malarious areas abroad, (ii) arriving foreigners and international travelers, and (iii) various groups of potential importers of malaria, particularly seasonal workers and refugees. The latter was considered as the major source of imported cases in many countries. What is also needed is to cooperate between the health services and immigration and quarantine services, customs and police, which is very useful for acquiring information on the entry of possibly infected sources. Vigilance is necessary to step up at the airports and other ports which malaria cases pass through Additional steps taken to prevent malaria resurgence include local measures against mosquitoes in general, and antianopheline measures in areas of high receptivity. Others than this, adequate briefings for travellers and the availability in the market of recommended antimalarials are also required.

8. THE WAY FORWARD

The number of imported malaria cases has declined but the proportion has increased gradually, particularly in recent years. Therefore, prompt and proper strategies will take place to avoid malaria resurgence. In addition, malaria deaths have become a high risk due to the imported *P. falciparum* malaria, especially from migrant labourers from Africa. Given this reason, all hospitals and local authorities in P.R. China will improve the capability in case detection and diagnosis of malaria. More importantly, travel history should be required for all patients with fever through cooperative information sharing with other sectors. Local authorities will further strengthen their verification processes concerned with gathering information about imported malaria cases.

ACKNOWLEDGEMENTS

The project was supported by the National S & T Major Programme (grant no. 2012ZX10004220), by the National S & T Supporting Project (grant no. 2007BAC03A02) and by China UK Global Health Support Programme (grant no. GHSP-CS-OP1).

REFERENCES

Bartoloni, A., Zammarchi, L., 2012. Clinical aspects of uncomplicated and severe malaria. Mediterr. J. Hematol. Infect. Dis. 4, e2012026.

Bergquist, R., Whittaker, M., 2012. Control of neglected tropical diseases in Asia Pacific: implications for health information priorities. Infect. Dis. Poverty 1, 3.

Bhutta, Z.A., Salam, R.A., Das, J.K., Lassi, Z.S., 2014. Tackling the existing burden of infectious diseases in the developing world: existing gaps and the way forward. Infect. Dis. Poverty 3, 28.

Bi, Y., Tong, S., 2014. Poverty and malaria in the Yunnan province, China. Infect. Dis. Poverty 3, 32.

Butcher, C.A., 2004. Malaria: a parasitic disease. AAOHN J 52, 302–309.

Chen, C., 2014. Development of antimalarial drugs and their application in China: a historical review. Infect. Dis. Poverty 3, 9.

Chen, G.W., Zhao, J.Y., Zhao, X.T., Zhao, H.Y., Sun, W.J., Bao, Z.R., Bao, X.S., Tian, G.Q., 2007. Joint malaria control effect in China-burma boader area of Cangyuan county. Parasit. Dis. Infect. Dis. 5, 80–81 (in Chinese).

Chu, Q.P., Chen, Y.E., Shao, S.C., Ren, J.Q., 2007. Characteristics of prevalence and control effect in Xuzhou city, China. Chin. J. Etiol. Bio. 6, 6–7 (in Chinese).

Cao, C.Q., Wang, W.M., 2009. Analysis of malaria situation in Nantong city. Chin. J. Schisto. Control 21, 555–556 (in Chinese).

Chen, C.R., Liu, L.C., Lai, J., Wang, L.Y., Zhou, F.K., 2011. Incidence trend of malaria in Taizhou municipality in Zhejiang province. Dis. Detection 26, 867–869 (in Chinese).

Centers for Disease Control and Prevention, 2004. Multifocal autochthonous transmission of malaria–Florida, 2003. MMWR Morb. Mortal. Wkly. Rep. 53, 412–413.

Cao, J., Zhou, S.S., Zhou, H.Y., Yu, Y.B., Tang, L.H., Gao, Q., 2013. Malaria from control to elimination in China: transition of goal, strategy and interventions. Chin. J. Schisto. Control 25, 439–443 (in Chinese).

Ding, J., 2012. Current endemic situation research in China. Chin. Pub. Health 28, 717–718 (in Chinese).

Diouf, G., Kpanyen, P.N., Tokpa, A.F., Nie, S., 2011. Changing landscape of malaria in China: progress and feasibility of malaria elimination. Asia Pac. J. Pub. Health 26, 93–100.

DaSilva, J., Garanganga, B., Teveredzi, V., Marx, S.M., Mason, S.J., Connor, S.J., 2004. Improving epidemic malaria planning, preparedness and response in Southern africa. Report on the 1st Southern african Regional epidemic Outlook Forum, Harare, Zimbabwe, 26–29 september, 2004. Malar. J. 3, 37.

Gao, Q., Le, Y.S., Gu, C.Z., 2002. Present situation of malaria in central part of China. Chin. J. Parasit. Dis. Control 15, 193–194 (in Chinese).
Gu, Z.C., Zheng, X., 2001. Malaria situation in the People's Republic of China in 2000. Chin. J. Parasitol, Parasit. Dis. 19, 257–259 (in Chinese).
Gao, Q., Beebe, N.W., Cooper, R.D., 2004. Molecular identification of the malaria vectors *Anopheles anthropophagus* and *Anopheles sinensis* (Diptera: Culicidae) in central China using polymerase chain reaction and appraisal of their position within the Hyrcanus group. J. Med. Entomol. 41, 5–11.
Gao, Q., 2011. Opportunities and challenges of malaria elimination in China. Chin. J. Schisto. Control 23, 347–349 (in Chinese).
Guan, Y.Y., Tang, L.H., Hu, L., Feng, X.P., Liu, D.Q., 2005. The point mutations in Pfcrt and Pfmdr1 genes in *Plasmodium falciparum* isolated from Hainan province. Chin. J. Parasitol, Parasit. Dis. 3, 135–139 (in Chinese).
Huang, G.Q., Yuan, F.Y., Jin, X.L., Zhao, C.L., Su, Y.P., Shen, Y.Z., 2007. To analyze epidemic situation and control of malaria in Jiangsu, Shandong, Henan, Anhui and Hubei province. Chin. J. Vect. Dis. Control 18, 398–401 (in Chinese).
Hui, F.M., Xu, B., Chen, Z.W., Cheng, X., Liang, L., Huang, H.B., Fang, L.Q., Yang, H., Zhou, H.N., Yang, H.L., et al., 2009. Spatio-temporal distribution of malaria in Yunnan Province, China. Am. J. Trop. Med. Hyg. 81, 503–509.
Huang, F., Tang, L., Yang, H., Zhou, S., Sun, X., Liu, H., 2012. Therapeutic efficacy of artesunate in the treatment of uncomplicated *Plasmodium falciparum* malaria and anti-malarial, drug-resistance marker polymorphisms in populations near the China-Myanmar border. Malar. J. 11, 278.
Harvey, S.A., Jennings, L., Chinyama, M., Masaninga, F., Mulholland, K., Bell, D.R., 2008. Improving community health worker use of malaria rapid diagnostic tests in Zambia: package instructions, job aid and job aid-plus-training. Malar. J. 7, 160.
Jin, X.L., Gao, Q., Zhou, H.Y., Wang, W.M., Li, J.L., Gu, Y.P., Cao, J., Zhu, G.D., 2006. Current epidemic status and influencing factors of malaria in Jiangsu Province. Chin. J. Schisto. Control 18, 453–454 (in Chinese).
Jiang, T.K., Cai, J., Chen, H.P., Wang, Z., Liu, B.Y., Zeng, B.B., 2009. Malaria endemic characteristics from 2004 to 2008 in Zhejiang province. Zhejiang Preventive Med. 21, 30–31 (in Chinese).
Jiao, Y.M., Fang, Q., Xie, M., Tao, Z.Y., Wang, X.M., Xia, H., Sun, X., 2013. Time and space distribution characteristics in Anhui province between 2006 and 2010. Bengbu Acta Med. College 38, 876–878 (in Chinese).
Li, H.X., Chen, G.W., Yang, Y.C., Jiang, H., 2008. Malaria situation in yunnan province during 2001-2005. Chin. J. Parasitol. Parasit. Dis. 26, 46–49 (in Chinese).
Li, H.X., Jiang, H., Yang, Y.C., 2006. Analysis of current malaria prevalent situation in Yunnan Province from 2002 to 2004. Chin. Trop. Med. 6, 1942–1944 (in Chinese).
Lu, F., Gao, Q., Xia, H., Tao, Z.Y., Cao, J., Gu, Y.P., Zhou, H.Y., Jin, X.L., 2006. In vitro sensitivity test of *Plasmodium vivax* to chloroquine in center part of China. Chin. J. Schisto. Control 4, 265–267 (in Chinese).
Lu, Y.L., 2008. Malaria in the floating population of China. Preventive Med. Forum 14, 236–238 (in Chinese).
Li, P.S., Fan, D.H., 2005. Analysis on 3646 malaria cases in Tengchong Country, Yunnan from 1999 to 2003. Parasit. Dis. Infect. Dis. 3, 29–30 (in Chinese).
Li, H.X., Zhang, Z.X., Zhou, X.W., Bi, Y., Yang, H.L., Jiang, H., Chen, Z.W., 2003. Analysis of malaria epidemic situation of Yunnan province from 1999 to 2001. Chin. J. Parasit. Dis. Control 16, 89–92 (in Chinese).
Liu, Y.B., Cao, J., Zhou, H.Y., Wang, W.M., Cao, Y.Y., Gao, Q., 2013. Analysis of overseas imported malaria situation and implication for control in Jiangsu Province, PR China. Chin. J. Schisto. Control 25, 44–47 (in Chinese).

Lin, H., Lu, L., Tian, L., Zhou, S., Wu, H., Bi, Y., Ho, S.C., Liu, Q., 2009. Spatial and temporal distribution of falciparum malaria in China. Malar. J. 8, 130.

Liu, Y., Hsiang, M.S., Zhou, H., Wang, W., Cao, Y., Gosling, R.D., Cao, J., Gao, Q., 2014. Malaria in overseas labourers returning to China: an analysis of imported malaria in Jiangsu Province, 2001–2011. Malar. J. 13, 29.

Liu, D.Q., 2014. Surveillance of antimalarial drug resistance in China in the 1980s-1990s. Infect. Dis. Poverty 3, 8.

Liu, J., Yang, B., Cheung, W.K., Yang, G., 2012. Malaria transmission modelling: a network perspective. Infect. Dis. Poverty 1, 11.

Muentener, P., Schlagenhauf, P., Steffen, R., 1999. Imported malaria (1985–95): trends and perspectives. Bull. World Health Organ. 77, 560–566.

Ministry of Health, 2006. National Malaria Control Programme from 2006 to 2015. (internal document, in Chinese).

Ministry of Health, 2010. Action Plan of China Malaria Elimination (2010–2020). http://www.gov.cn/gzdt/att/att/site1/20100526/001e3741a2cc0d67233801.doc (in Chinese).

McMichael, A.J., 1993. Global environmental change and human population health: a conceptual and scientific challenge for epidemiology. Int. J. Epidemiol. 22, 1–8.

Ministry of Health, 2011. Malaria Control and Elimination Standards. Beijing (in Chinese).

Moore, S.J., Min, X., Hill, N., Jones, C., Zaixing, Z., Cameron, M.M., 2008. Border malaria in China: knowledge and use of personal protection by minority populations and implications for malaria control: a questionnaire-based survey. BMC Public Health 8, 344.

Nadjm, B., Behrens, R.H., 2012. Malaria: an update for physicians. Infect. Dis. Clin. North Am. 26, 243–259.

Okeke, T.A., Uzochukwu, B.S., 2009. Improving childhood malaria treatment and referral practices by training patent medicine vendors in rural south-east Nigeria. Malar. J. 8, 260.

Pavli, A., Smeti, P., Spilioti, A., Vakali, A., Katerelos, P., Maltezou, H.C., 2011. Descriptive analysis of malaria prophylaxis for travellers from Greece visiting malaria-endemic countries. Travel Med. Infect. Dis. 9, 284–288.

Poravuth, Y., Socheat, D., Rueangweerayut, R., Uthaisin, C., Pyae Phyo, A., Valecha, N., Rao, B.H., Tjitra, E., Purnama, A., Borghini-Fuhrer, I., et al., 2011. Pyronaridine-artesunate versus chloroquine in patients with acute *Plasmodium vivax* malaria: a randomized, double-blind, non-inferiority trial. PLoS One 6, e14501.

Rogers, D.J., Randolph, S.E., 2000. The global spread of malaria in a future, warmer world. Science 289, 1763–1766.

Ruan, W., Yao, L.N., Xia, S.R., Chen, H.L., Yu, K.G., 2007. Epidemiological analysis on malaria in focally supervised counties in Zhejiang Province in 2006. Dis. Detection 22, 662–664 (in Chinese).

Salam, R.A., Das, J.K., Lassi, Z.S., Bhutta, Z.A., 2014. Impact of community-based interventions for the prevention and control of malaria on intervention coverage and health outcomes for the prevention and control of malaria. Infect. Dis. Poverty 3, 25.

Schlagenhauf, P., Petersen, E., 2008. Malaria chemoprophylaxis: strategies for risk groups. Clin. Microbiol. Rev. 21, 466–472.

Schlagenhauf, P., Petersen, E., 2013. Current challenges in travelers' malaria. Curr. Infect. Dis. Rep. 15, 307–315.

Sheng, H.F., Zhou, S.S., Gu, Z.C., Zheng, X., 2003. Malaria situation in the People's Republic of China in 2002. Chin. J. Parasitol. Parasit. Dis. 21, 193–196 (in Chinese).

Sheng, C.X., Zhu, D.F., Li, H.X., 2004. Analysis on malaria epidemic situation in yunnan province from 1991 to 2000. Parasit. Dis. Infect. Dis. 2, 155–158 (in Chinese).

Shi, M., Qu, C.S., 2004. Analysis of malaria endemic trend of population in Longchuan country in Yunnan province from 1993 to 2003. Med. Animal Control 20, 673–674 (in Chinese).

Su, Y.P., Shang, L.Y., Li, D.F., He, L.J., Lu, D., 1994. Analysis of malaria situation in Henan province in 1993. Henan J. Preventive Med. 4, 232–234 (in Chinese).

Su, Y.P., Liu, Y., Zhang, H.W., Chen, J.S., Xu, B.L., 2006. Analysis of malaria situation in Henan province in 2005. Dis. Control 6, 982–983 (in Chinese).

Tambo, E., Ai, L., Zhou, X., Chen, J.H., Hu, W., Bergquist, R., et al., 2014. Surveillance-response systems: the key to elimination of tropical diseases. Infect. Dis. Poverty 3, 17.

Trampuz, A., Jereb, M., Muzlovic, I., Prabhu, R.M., 2003. Clinical review: severe malaria. Crit. Care 7, 315–323.

Teng, G., Zhuo, J., Shi, J., Wang, X., Cai, P., Li, Y.F., 2013. Epidemiology of imported malaria from 2005~2012 at Liaoning ports. Chin. J. Frontier Health Quarantine 36, 165–167 (in Chinese).

Thakor, H.G., Sonal, G.S., Dhariwa, L.A., Arora, P., Gupta, R.K., 2010. Challenges and role of private sector in the control of malaria in India. J. Indian Med. Assoc. 108, 849–853.

Tang, L.H., 1999. Research achievement in malaria control in China. Chin. J. Parasitol. Parasit. Dis. 17, 257–259 (in Chinese).

Tang, L.H., 2010. Diagnosis, Treatment and Management of Imported Malaria Cases. Shanghai Scientific and Technical Publishers Shanghai. (in Chinese).

WHO, 2006. WHO Briefing on Malaria Treatment Guidelines and Artemisinin Monotherapies. Geneva.

WHO, 2010a. Dev V: Assessment of Receptivity and Vulnerability of Development Project Sites to Malaria in Bhutan. WHO Assignment Report.

WHO, 2010b. Guidelines for the Treatment of Malaria, second ed.

WHO, 2011. World Malaria Report. Geneva.

WHO, 2012a. World Malaria Report. Geneva.

WHO, 2012b. Eliminating Malaria: Case Study 4. Preventing Reintroduction in Mauritius.

Xia, Z.G., Yang, M.N., Zhou, S.S., 2012. Malaria situation in the People's Republic of China in 2011. Chin. J. Parasitol. Parasit. Dis. 30, 419–422 (in Chinese).

Xia, Z.G., Feng, J., Zhou, S.S., 2013. Malaria situation in the People's Republic of China in 2012. Chin. J. Parasitol. Parasit. Dis. 31, 413–418 (in Chinese).

Yadav, K., Dhiman, S., Rabha, B., Saikia, P., Veer, V., 2014. Socio-economic determinants for malaria transmission risk in an endemic primary health centre in Assam, India. Infect. Dis. Poverty 3, 19.

Yin, L.Z., Zhen, H.S., Wu, C.H., 1995. The general situation of malaria imported into China. Sci. Trasl. Med. 1, 178–182 (in Chinese).

Yin, J.H., Yang, M.N., Zhou, S.S., Wang, Y., Feng, J., Xia, Z.G., 2013. Changing malaria transmission and implications in China towards national malaria elimination programme between 2010 and 2012. PLoS One 8, e74228.

Yang, L., 2001. Malaria control and research in Yunnan province. Chin. J. Parasit. Dis. Control 14, 54–56 (in Chinese).

Yang, H.L., Liu, D.Q., Huang, K.G., Yang, Y.M., Yang, P.F., Liao, M.Z., Zhang, C.Y., 1999. Assay of sensitivity of plasmodium falciparum to chloroquine, amodiaquine, piperaquine, mefloquine and quinine in Yunnan province. Chin. J. Parasitol. Parasit. Dis. 17, 43–45 (in Chinese).

Yangzom, T., Gueye, C.S., Namgay, R., Galappaththy, G.N., Thimasarn, K., Gosling, R., Murugasampillay, S., Dev, V., 2012. Malaria control in Bhutan: case study of a country embarking on elimination. Malar. J. 11, 9.

Ye, L.P., Xu, G.Z., Sun, Y.W., Zhang, J.N., Lu, F., 2004. Epidemiological analysis of 131 imported malaria cases. Chin. Pub. Health 20, 880 (in Chinese).

Yao, L.N., Xu, X., Chen, H.L., Xia, S.R., Yang, T.T., Ruan, W., Yao, S.R., Yu, K.G., 2009. Malaria surveillance in zhejiang province, 2009. Chin. J. Schisto. Control 22 (556), 561 (in Chinese).

Yuan, J.F., Wu, J.C., Bao, M.H., Gao, J.M., 2004. Longitudinal surveillance on malaria in migratory population in Wuxi city. Chin. J. Schisto. Control 14, 388–389 (in Chinese).

Yang, H.L., Yang, P.F., Li, X.L., Gao, B.H., Zhang, Z.Y., Yang, Y.M., 2005. The change of *Plasmodium falciparum* resistance to chloroquine in Yunnan province, China. Chin. J. Parasit. Dis. Control 18, 368–370 (in Chinese).

Yu, G., Yan, G., Zhang, N., Zhong, D., Wang, Y., He, Z., Yan, Z., Fu, W., Yang, F., Chen, B., 2013. The Anopheles community and the role of *Anopheles minimus* on malaria transmission on the China-Myanmar border. Parasit. Vectors 6, 264.

Yan, J., Li, N., Wei, X., Li, P., Zhao, Z., Wang, L., Li, S., Li, X., Wang, Y., Yang, Z., et al., 2013. Performance of two rapid diagnostic tests for malaria diagnosis at the China-Myanmar border area. Malar. J. 12, 73.

Zheng, Q., Vanderslott, S., Jiang, B., Xu, L.L., Liu, C.S., Huo, L.L., et al., 2013. Research gaps for three main tropical diseases in the People's Republic of China. Infect. Dis. Poverty 2, 15.

Zhou, S.S., Tang, L.H., Sheng, H.F., 2005. Malaria situation in the People's Republic of China in 2003. Chin. J. Parasitol. Parasit. Dis. 23, 385–387 (in Chinese).

Zhou, S.S., Tang, L.H., Sheng, H.F., Wang, Y., 2006a. Malaria situation in the People's Republic of China in 2004. Chin. J. Parasitol. Parasit. Dis. 24, 1–3 (in Chinese).

Zhou, S.S., Wang, Y., Tang, L.H., 2006b. Malaria situation in the People's Republic of China in 2005. Chin. J. Parasitol. Parasit. Dis. 24, 401–403 (in Chinese).

Zhou, S.S., Wang, Y., Tang, L.H., 2007. Malaria situation in the People's Republic of China in 2006. Chin. J. Parasitol. Parasit. Dis. 25, 439–441 (in Chinese).

Zhou, S.S., Wang, Y., Fang, W., Tang, L.H., 2008. Malaria situation in the People's Republic of China in 2007. Chin. J. Parasitol. Parasit. Dis. 26, 401–403 (in Chinese).

Zhou, S.S., Wang, Y., Fang, W., Tang, L.H., 2009a. Malaria situation in the People's Republic of China in 2008. Chin. J. Parasitol. Parasit. Dis. 27, 455–457 (in Chinese).

Zhou, S.S., Wang, Y., Xia, Z.G., 2011a. Malaria situation in the People's Republic of China in 2009. Chin. J. Parasitol. Parasit. Dis. 29, 1–3 (in Chinese).

Zhou, S.S., Wang, Y., Li, Y., 2011b. Malaria situation in the people's republic of China in 2010. Chin. J. Parasitol. Parasit. Dis. 29, 401–403 (in Chinese).

Zhang, J.M., Wu, L.O., Yang, H.L., Duan, Q.X., Shi, M., Yang, Z.Q., 2008. Malaria prevalence and control in Yunnan Province. Chin. J. Pathogen. Bio. 3, 950–956 (in Chinese).

Zhou, H.Y., Cao, J., Wang, W.M., Li, J.L., Gu, Y.P., Zhu, G.D., Gao, Q., 2009b. Epidemic and control of malaria in Jiangsu Province. Chin. J. Schisto. Control 6, 503–506 (in Chinese).

Zhang, X.P., Gao, Q., Wu, Z.X., Zhao, Y.J., Ge, J., 1996. Investigation of falciparum malaria control in Jiangsu province. Chin. J. Parasitol. Parasit. Dis. 9, 245–249 (in Chinese).

Zhou, H.Y., Wang, W.M., Cao, J., Zhu, G.D., Gao, Q., 2011c. Epidemiologicai analysis of malaria prevalence in Jiangsu Province in 2009. Chin. J. Schisto. Control 23, 402–405 (in Chinese).

Zhou, H.Y., Wang, W.M., Liu, Y.B., Cao, J., Gao, Q., 2012. Epidemiologicai analysis of malaria prevalence in Jiangsu Province in 2010. Chin. J. Schisto. Control 22, 622–624 (in Chinese).

Zhou, X., Li, S.G., Chen, S.B., Wang, J.Z., Xu, B., Zhou, H.J., et al., 2013. Co-infections with Babesia microti and Plasmodium parasites along the China-Myanmar border. Infect. Dis. Poverty 2, 24.

Zhou, X.N., Bergquist, R., Tanner, M., 2013. Elimination of tropical disease through surveillance and response. Infect. Dis. Poverty 2, 1.

Zhang, H.W., Su, Y.P., Zhou, G.C., Liu, Y., Cui, J., Wang, Z.Q., 2007. Re-emerging malaria in Yongcheng city of Henan province. Chin. J. Vector Bio. Control 18, 42–44 (in Chinese).

Zheng, C.J., Zhang, Q.A., Li, H.Z., 2010. Analysis on the diagnosis and report of malaria cases in China during 2005 – 2008. Chin. J. Dis. Control 14, 507–509 (in Chinese).

Zoller, T., Naucke, T.J., May, J., Hoffmeister, B., Flick, H., Williams, C.J., Frank, C., Bergmann, F., Suttorp, N., Mockenhaupt, F.P., 2009. Malaria transmission in non-endemic areas: case report, review of the literature and implications for public health management. Malar. J. 8, 71.

Zhang, M., Liu, Z., He, H., Luo, L., Wang, S., Bu, H., Zhou, X., 2011. Knowledge, attitudes, and practices on malaria prevention among Chinese international travelers. J. Travel Med. 18, 173–177.

Zofou, D., Nyasa, R.B., Nsagha, D.S., Ntie-Kang, F., Meriki, H.D., Assob, J.C., et al., 2014. Control of malaria and other vector-borne protozoan diseases in the tropics: enduring challenges despite considerable progress and achievements. Infect. Dis. Poverty 3, 1.

CHAPTER TEN

Preparation for Malaria Resurgence in China: Approach in Risk Assessment and Rapid Response

Ying-Jun Qian[1], Li Zhang[1], Zhi-Gui Xia[1], Sirenda Vong[2], Wei-Zhong Yang[3], Duo-Quan Wang[1], Ning Xiao[1,*]

[1]National Institute of Parasitic Diseases, Chinese Center for Disease Control and Prevention; Key Laboratory of Parasite and Vector Biology, Ministry of Health; WHO Collaborating Center for Malaria, Schistosomiasis and Filariasis, Shanghai, People's Republic of China
[2]World Health Organization, China Representative Office, Beijing, People's Republic of China
[3]Chinese Preventive Medicine Association, Beijing, People's Republic of China; Chinese Center for Disease Control and Prevention, Beijing, People's Republic of China
*Corresponding author: E-mail: ningxiao116@126.com

Contents

1. Background 268
2. Malaria in China 268
3. Risk Determinants of Secondary Transmission by Imported Malaria 271
   3.1 Population movement 271
   3.2 Vectorial capacity of transmission 272
   3.3 Cross-sectoral cooperation 273
4. Response to Reintroduction of Malaria 274
   4.1 Prompt diagnosis and treatment 274
   4.2 Foci investigation and intervention 276
   4.3 Prophylaxis 277
   4.4 Risk assessment 277
   4.5 Public awareness 278
5. Research Needs in the National Malaria Elimination Programme 280
6. Discussion 281
7. Conclusions 282
Acknowledgements 283
References 283

Abstract

With the shrinking of indigenous malaria cases and endemic areas in the People's Republic of China (P.R. China), imported malaria predominates over all reported cases accounting for more than 90% of the total. On the way to eliminate malaria, prompt detection and rapid response to the imported cases are crucial for the prevention of secondary transmission in previous endemic areas. Through a comprehensive literature review, this chapter aims to identify risk determinants of potential local transmission caused by the imported malaria cases and discusses gaps to be addressed to reach the elimination goal by 2020. Current main gaps with respect to dealing with potential malaria resurgence in P.R. China include lack of cross-sectoral cooperation, lack of rapid response and risk assessment, poor public awareness, and inadequate research and development in the national malaria elimination programme.

1. BACKGROUND

Although many countries are on the way to malaria elimination with sharp reduction of autochthonic malaria infections, imported cases are the main source of increasing local malaria transmission as long as the environment is favorable. The People's Republic of China (P.R. China) is one of the countries targeting elimination of malaria by 2020, and the possibility of secondary transmission induced by increasing imported cases is a high risk. This review systematically evaluates the risks of sustained malaria transmission induced by the imported cases and response efforts in P.R. China since 2010, the year the national malaria elimination programme (NMEP) was launched (Ding, 2012).

2. MALARIA IN CHINA

In P.R. China, malaria has been known for more than 4000 years and was identified as one of the top five parasitic diseases that seriously affected socioeconomic development in the twentieth century. Thanks to great efforts made by the government and whole society, the malaria status in P.R. China changed quickly in terms of the number of malaria cases which has been reduced year by year recently. In 2010, the Chinese Government announced the malaria elimination action plan with the aim to eliminate malaria by 2020. Since then, surveillance and response activities have been strengthened, which has steadily reduced indigenous malaria cases. However, with developing globalization and increasing international population migration, imported cases of malaria have become the predominant threat to the NMEP in P.R. China (Zheng et al., 2013). Among the four species of imported *Plasmodium* spp., *Plasmodium falciparum* and *P. vivax* are dominant in P.R. China.

The number of nationwide reported malaria cases in 2011 was 4479, with 29.3% reported as indigenous cases and with a reduction by 43.0% compared to that in 2010 (7855). The confirmed cases were 81.7% of the total cases, while the other 18.3% were clinically diagnosed cases (Zhou et al., 2011a).

Compared to the reduced autochthonous malaria, the number of imported malaria has been sustained at a high level, accounting for a growing share of total malaria cases between 2010 and 2012 in the mainland of China. The areas that suffered from autochthonous malaria were significantly shrinking (Yin et al., 2013), while imported malaria cases were becoming dominant and widely distributed in the whole country with the four *Plasmodium* species. It is foreseeable that this situation could bring about high risk of reintroduction of malaria transmission in the areas where malaria has effectively been under control (Xia et al., 2012). A total of 651 counties in 23 provinces were affected by the imported malaria cases in 2010, 760 counties in 26 provinces in 2011, and 598 counties in 29 provinces in 2012. Most of these cases resulted from *P. falciparum* malaria were people who came back to China from African countries or Southeast Asian countries, especially Myanmar; and only a very small proportion of them were people who migrated from some provinces inside China where local transmission still occurs (Yin et al., 2013). Therefore, it is essential to understand the importance of this situation, through a literature review, in order to respond to the potential resurgence of secondary malaria transmission induced by these imported cases in this current stage of malaria elimination in P.R. China (Tambo et al., 2014a).

We searched for terms in two large databases, PubMed (http://www.pubmed.com) for English literatures and China National Knowledge Infrastructure (CNKI) (http://www.cnki.net/) for Chinese literatures. The search terms were *malaria, imported, risk, transmission, resurgence and elimination*, as well as combinations in both English and Chinese. Because of China's shifting malaria control to elimination from 2010 (Ministry of Health, 2010), we then searched papers published from 2010 to 2013. The search yielded 8466 papers, and then a four-step screening was used to establish the final database for paper review. In the first screening, papers written by authors who are non-Chinese were excluded. Second, we went through title screening, in which we established the exclusion criteria. Third, we screened papers through detailed abstract reading, and the fourth step was full-text reading and analyzing (Figure 10.1, Table 10.1).

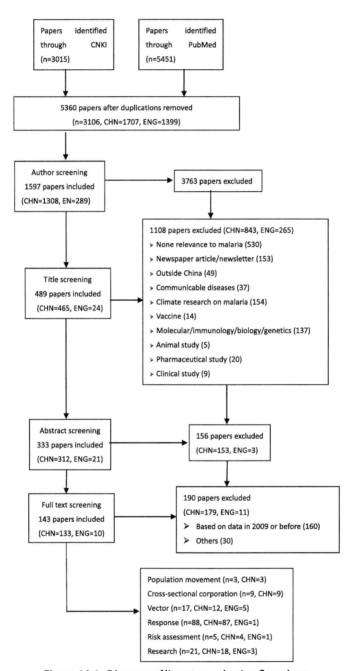

Figure 10.1 *Diagram of literature selection flow chart.*

Table 10.1 Risk determinants and their roles

Risk determinants	Roles and function	References
Mobile populations	Mobile populations' contribution to reintroduction to malaria-free areas.	Bruce-Chwatt (1968), Prothero (1977).
Vectorial capacity	Biting behavior and capacity as well as insecticide-resistance of vectors are key factors that cause potential local transmission risk.	Liu and Liu (2010), Liu et al. (2011b), Yu et al. (2013), Wang et al. (2013b).
Cross-sectoral cooperation	Inter-sectoral cooperation contributes to good information sharing and joint efforts to prevent from reintroduction.	Li et al. (2011a), Chen et al. (2012a).
Response to reintroduction	Timely and appropriate response contributes to saving cases' life and to removing any potential reservoirs.	Ministry of Health (2011), Xie and Tan (2011), Chen et al. (2013).

3. RISK DETERMINANTS OF SECONDARY TRANSMISSION BY IMPORTED MALARIA

3.1 Population movement

In the past, population movement has often contributed to the spread of disease (Prothero, 1977). Lack of consideration of this factor contributed to failure of malaria eradication campaigns in the 1950s and 1960s (Bruce-Chwatt, 1968). The movement of infected people from areas where malaria was still endemic to areas where the disease had been eliminated led to resurgence of the disease (Service, 1991). Imported malaria cases have been a major part of reported malaria in P.R. China in recent years, indicating that provinces with more mobile populations and imported cases need to pay greater attention to rapid response action (Xia et al., 2012).

Statistics showed that more and more malaria cases were found in workers who had returned from malaria endemic countries overseas. For example, in Guangdong ports, the malaria incidence among incoming passengers was 0.27 per 100,000 from 2010 to 2011, which was 2.7 times higher than reported malaria incidence in Guangdong Province in the same time period. Of all the imported cases, 57.5% were Chinese exported labors returning from Africa, which indicates that migrant labors become predominant in the imported malaria cases (Dai et al., 2012). It was also found that a great proportion of workers stay in malaria endemic areas all year round since they work for big engineering projects. For instance, in Xiamen, Fujian Province, there are around 4000 export labourers annually, and 10,000 workers stay abroad all year round (Chen et al., 2010). In Jiangdu District of Yangzhou City, Jiangsu Province, a total of 5000 people work abroad each year, and 2000 of them work in high malaria endemic areas overseas (Zhu and She, 2012). As a central governmental enterprise with 36,000 staff, over 2000 people participate in international projects abroad in the Gezhouba Group Corporation. Among 700 returnees to Xilin District, Yichang City, 14 malaria cases were reported in 2010 (Yang, 2011). Currently labor export management is not standardized in P.R. China due to the following three reasons (Zhu and She, 2012). First, many intermediaries that dispatch labourers to relevant companies have not been authorized. Second, labour export organizations or companies sometimes change the destination countries for labourers without authorization. Third, returnees always return in batches without valid information, all of which makes the management of imported malaria difficult.

3.2 Vectorial capacity of transmission

Anopheles is the vector of malaria and there are more than 60 species of anopheline mosquitoes reported in P.R. China, of which the four dominant vectors are *Anopheles sinensis*, *An. anthropophagus*, *An. minimus* and *An. dirus* (Liu and Liu, 2010). *An. sinensis* is the most widely distributed malaria vector in P.R. China except for in Qinghai and Xinjiang. *An. anthropophagus* has proven to be one vector with a high capacity to transmit *P. falciparum* malaria, which inhabits more than 30 counties of 18 provinces. *An. minimus* acts as the main vector living in the hilly areas. *An. dirus* is mainly distributing in Hainan Province, as well as in southern Yunnan Province and southwest Guangxi Zhuang Autonomous Region.

Most of the literatures on vectors in P.R. China focussed on the population and seasonal distribution, biological characteristics and flight range (Liu et al., 2011a,b; Shang, 2012; Wang et al., 2012a,b), and very little is about the resistance to insecticides (Wang et al., 2013b). For the most widely distributed *An. sinensis*, by using the mark-release-recapture (or capture-recapture) method, the flight range of *An. sinensis* is proved to be within 400 m, and 80% were captured from the center point of the range of 100 m (Liu et al., 2011a). *An. sinensis* plays a major role in the maintenance of *P. vivax* malaria transmission in P.R. China. Host-seeking behavior survey of culicine species and *An. sinensis* was conducted in *P. vivax* distributed areas and the results indicated that pigs, goats and calves were more attractive to *An. sinensis* and *Culex tritaeniorhynchus* than what dogs, humans, and chickens done (Liu et al., 2011b).

Malaria around the China–Myanmar border is an important issue in the elimination stage. As the principal malaria vector with a wide geographic distribution in this area, many studies showed anopheline mosquito is abundant with *An. minimus* being the dominant species and having a high human blood index along the China–Myanmar border (Yu et al., 2013).

3.3 Cross-sectoral cooperation

In order to promote the NMEP, in May 2010 the National Health and Family Planning Commission (NHFPC, previously called Ministry of Health, MOH) of P.R. China issued the National Malaria Elimination Action Plan (2010–2020) in collaboration with 12 other central ministries and commissions. In this document, the responsibilities of each department are clarified and close cooperation is highlighted, which include the National Development and Reform Commission, Ministry of Education, Ministry of Science and Technology, Ministry of Information Industry, Ministry of Public Security, Ministry of Finance, Ministry of Commerce, General Administration of Quality Supervision, Inspection and Quarantine, State Administration of Radio, Film and Television, National Tourism Administration, Health Department of General Logistics Department and the Logistics Department of Armed Police Force (Ministry of Health, 2010).

However, lack of close coordination between health and other sectors may delay effective response to imported malaria. In 2012, for instance, some imported malaria cases were identified by Ezhou Center for Disease Control and Prevention (CDC), Hubei Province. Active case detection, foci investigation and response were then conducted quickly. However, until the completion of the response, the local bureau of exit–entry inspection and quarantine and the relevant labor export agency did not contact CDC to provide any detailed information, which directly delayed the process of case finding (Chen et al., 2012a).

To further prevent imported *P. falciparum* cases among returnees, Yangzhou City government issued the 'Management project of imported falciparum malaria in Yangzhou' to carry out labor export surveys in 2010, and a database was then established that includes information on both labour export agencies and export labourers (Xu et al., 2011). By the end of 2012, the information on 159 agencies and 9999 export workers was put into the local disease control system; this information showed that 84.3%

(5639/6687) returnees received the health services from 2011 to 2012. Meanwhile, health education and services were provided by the team of entry–exit inspection and quarantine authorities, CDCs, health facilities and labour export companies to the export labourers. As a result, malaria awareness was greatly improved among those workers one year after the implementation of the project, and prompt diagnosis and treatment were received without deaths or complicated cases.

It is worth mentioning that besides incorporation of malaria control in the national health system, malaria elimination should consider the health system in the army. As an independent entity, the army has its own health system. Integrating information between the public health services and army system is very important in the stage of malaria elimination. In 2011, it was reported that the Air Force General Hospital treated pilots infected with *P. falciparum* malaria (Li et al., 2011a). Although timely, standardized diagnosis and treatment were applied, there was no further information on the follow-up of the case reporting. Therefore, the interaction between the hospital, public health, and entry–exit inspection and quarantine units need to be strengthened.

4. RESPONSE TO REINTRODUCTION OF MALARIA

Effective and timely responses to the reintroduction of malaria due to population movement reflect the capacity of a health system to provide timely action for public health issues (Zhou et al., 2013). Malaria is one of infectious diseases not only worldwide distribution particularly in the tropical areas, but is also characterized by transmission quickly in some environmental settings where the intervention and surveillance is week (Tambo et al., 2014a). There are at least following 5 challenges currently in the response to the reintroduction of malaria in China, indentified based on the surveillance data.

4.1 Prompt diagnosis and treatment

At the current stage of malaria elimination, prompt case identification and treatment is crucial to the follow-up activities (Ministry of Health, 2011). However, the quality of diagnosis and treatment is still need to be improved in some of malaria endemic areas. The misdiagnosis or improper treatment sometimes caused patient died of malaria, specifically occurred mainly in the imported *P. falciparum* cases before launching the NMEP. For example, some imported malaria cases were not diagnosed on a timely basis, and often were misdiagnosed as other diseases or confused with other *Plasmodium* species by health facilities at different levels (Cao, 2012; Chen et al., 2010; Chen, 2012; Chen et al., 2012b; Cui et al., 2011; Miao and Chen, 2011; Wang et al., 2013c; Yao et al., 2013; Ye and Li, 2010; Wang et al., 2013; Zhou et al., 2013a). This may delay or misdirect treatment, cause death (Chu et al., 2011; Cui et al., 2011; Gao et al., 2013; Ke et al., 2011), and result in continuous transmission (Chen et al., 2013). In 2011, 145 imported *P. falciparum* cases were reported in Henan Province, of which the longest time interval from onset to confirmed diagnosis was up to 125 days, and only 13.1% patients were diagnosed within 24 h (Chen et al., 2012c). In Hubei Province, among 28 imported *P. falciparum* cases in 2010, only 50% of cases were correctly diagnosed by health facilities at the county level or above (Huang et al., 2012). On the other hand, due to the low educational level and poor awareness of malaria, most patients usually self-medicate after the onset of symptoms (Lu et al., 2011; Liu et al., 2011; He et al., 2011), or go to private clinics as well as other primary care health facilities, such as village clinics or township health centers (Chen et al., 2011a; Li et al., 2010; Xie et al., 2011; Feng and Jiang, 2012). In most settings, these institutions do not carry out microscopy examination for malaria confirmation and treat patients as upper respiratory tract infections, gastritis or liver disease (Cui and Fang, 2011; Wang et al., 2013a; Yu et al., 2013; Ministry of Health, 2011). In one extreme case in 2010, a patient was treated as common infection at a local health facility and then clinically diagnosed as *P. falciparum* malaria in a provincial hospital without microscopy. When the provincial CDC read the blood film, it was full of *P. falciparum* in each visual field. Because of the delayed diagnosis, the patient died finally (Xie and Tan, 2011). The same story happened in Luoyang City, Henan Province, in 2010 where two *P. falciparum* malaria cases were misdiagnosed at the village clinic, township health centre, county hospital and city hospital and finally led to death (Li, 2013). With the development of symptoms, patients were then referred to another hospital at a higher level where confirmed diagnosis could be done (Chen, 2012; Cui et al., 2011; Li et al., 2010; Ministry of Health, 2011; Rao and O, 2011; Ying et al., 2011). In case of missed or delayed diagnosis, some papers attributed the situation to that most of patients did not tell their migrant history (Ke et al., 2011; Lu and long, 2011). Nevertheless, it was the responsibility of clinic doctors who should be aware and ask whether the patients had visited malaria endemic countries (Huang et al., 2012; Ke et al., 2011; Zhou et al., 2011b).

Moreover, the literature shows that there is a serious shortage of microscopists in P.R. China (Li et al., 2013; Gao et al., 2012). In a baseline survey conducted in Yunnan Province in 2010, only 58.78% (77/131) health facilities were able to carry out malaria microscopy (Li et al., 2013). In Guizhou Province, 40.3% cases were misdiagnosed (27/67) in 2011 (Zhou et al., 2013b). Among the total reported cases, 49% had conducted microscopy after presumptive treatment. Only 25.37% blood films were qualified, among which 70.59% were qualified at county CDC, 12.50% at township hospitals and zero at village clinics. As one of the pilot sites of malaria elimination, a baseline survey was performed in Chengan County, Hebei Province, in 2010, results showed there was no staff working on microscopy for malaria examination in whole county. There was a huge shortage of professionals working on malaria diagnosis and treatment (Ya-min et al., 2012).

In terms of treatment, it was often seen that uncomplicated cases or clinically diagnosed cases were treated as inpatients and given intravenous or intramuscular injections at different levels (Chen et al., 2012d; Tong et al., 2011; Wang, 2011; Liu et al., 2011; Wang et al., 2013), which was not right protocol in accordance with the National Malaria Elimination Action Plan (2010–2020) sometimes (Li et al., 2011b). In addition, according to recommendations of World Health Organization (WHO) and the national protocol (Ministry of Health, 2009; WHO, 2010), *P. falciparum* malaria cases should take primaquine for radical treatment. However, this is not usually seen in some hospital or CDCs (Quan et al., 2011; Wang et al., 2013). In terms of treatment, some clinical doctors are confused about different malaria species or have no idea of radical and preventive treatment. Some one consider that chloroquine plus primaquine is used for radical treatment of all malaria (Wu et al., 2011), and there still accidently occurred cases that clinical doctors treated as cases with monotherapy (Yu al., 2012). Therefore, more training on standard diagnosis and treatment for malaria cases in clinics and hospitals at various levels is urgently required, in order to improve the capacity on diagnosis and treatment in the stage of malaria elimination.

4.2 Foci investigation and intervention

According to the national guidelines (Ministry of Health, 2010; China CDC, 2011), malaria cases reported through the website must be verified immediately through blood tests for confirmation of *Plasmodium* species by county-level CDCs. Epidemiological investigation should be completed within three working days, and foci investigation and treatment are to be done within one week by county CDCs. Active case detection and relevant interventions should be conducted at all households of the epidemic site, and malaria prevention and consultation services should be provided. In Xiaochang and Xiantao, local CDCs of Hubei Province performed prompt case detection and foci intervention soon after the imported cases were identified, which was a quick and adequate response (Ye et al., 2013; Zhao et al., 2012). However, foci investigation and follow-up intervention in many imported malaria episodes sometimes do not strictly adhere to these guidelines (Quan et al., 2011). For example, active case detection is only focused on those who were on the same flight instead of neighborhoods after patients went home (Chen et al., 2010, 2012a). In other cases, vector interventions were not adequately conducted even during malaria transmission season (Chen et al., 2012d), or necessary health education was neglected (Yang et al., 2013). It is essential that the response to imported cases, even one imported case, must be taken as quickly as possible within one week.

4.3 Prophylaxis

WHO recommends that international travelers may also need to take preventive medication prior to, during, and upon return from their travels. Prior to their travel to malaria-endemic countries or regions, individuals should consult their national disease control centres, or other institutions offering travel advices, for information regarding preventive measures. In the Chinese national guidelines for malaria treatment, both chloroquine and piperaquine phosphate are recommended for malaria prophylaxis for no more than four consecutive months due to side effects (Ministry of Health, 2009).

In practice, the preventive medication was taken differently in different population. Prophylaxis by taking preventive drugs was significantly higher among managers, technicians, and businesspeople than that among workers (Jiang et al., 2013). Data shows that non-prophylaxis often leads to malaria infection within 15 days or 1 month for individuals without immunity.

4.4 Risk assessment

Risk assessment is one of important components in the response to the malaria reintroduction (Agomo and Oyibo, 2013; Tambo et al., 2014b). With the implementation of the NMEP, risk assessment has been applied in those areas where malaria elimination goal is nearly to be achieved. In 2011, risk assessment for malaria elimination in Guangdong Province was conducted to explore any potential risks in achieving malaria elimination (Lin et al., 2011). Two counties were selected as pilot sites in class II and III, respectively, based on the stratification of malaria endemic areas (Ministry of Health, 2010). In comparison with a class II county, the proportion of basic equipment and diagnostic technology, human resources, skilled medical personnel armed with malaria knowledge as well as basic awareness among village residents were significantly higher than those in a class III county, which indicated that the class III counties may have a relatively higher risk of reintroducing malaria because of low capacity to diagnosis and treat malaria in local health systems.

In Yunnan Province, the risk assessment conducted in 2010 indicated that class III counties were at high risk in terms of the target of malaria elimination (Li et al., 2013). The results showed that many local residents had no knowledge about malaria prevention, especially in class III counties where malaria was not endemic for continuous 3 years, and very few health education activities were conducted. Nearly 50% of health facilities were unable to carry out malaria microscopy, and the coverage of technical training was low, especially in the class III counties.

Another risk assessment conducted in 2011 was based on mosquito surveillance (Liu et al., 2012). It was concluded that there was quite low risk of local continuous transmission due to imported malaria around Beijing Capital Airport (Li et al., 2013a).

4.5 Public awareness

Public awareness of diseases, including malaria, plays an important role in disease control and prevention. A lack of reasonable knowledge of disease leads to low detection rate, and the interruption of treatment and discrimination (Becker et al., 2002; Liu et al., 2013a). Health education aims to gradually improve populations' healthy behavior (Chang and Wang, 2008). In malaria prevention and treatment, health education helps improve public awareness and knowledge of malaria, community compliance with control measures, and facilitate collaborations between health facilities and individuals to achieve the given targets (Ministry of Health, 2007). With the development of economic globalization, management of imported malaria becomes more and more important. As a developing country with an increasing tourism economy, China is now facing big challenges of management for the imported malaria cases, particularly in non- or low- endemic areas, where the disease burden may be increased due to lack of necessary awareness (Du and Wu,

2010). Dynamics of influence on malaria by social factors have been widely reported, anthropological characteristics such as age and gender, and community knowledge and awareness of malaria have proven to be strong indices of malaria infection (Beck and Davies, 1981). Medication and professional health services are not the effective way to control parasitic diseases. Instead, the best way is to improve social and economic conditions, health education, health policy and necessary medical services (Jiao and Meng, 2006). In the current settings, large numbers of export labourers returning to China after the closure of abroad projects often result in malaria outbreaks or re-transmission within two months. High awareness and knowledge of malaria among the mobile population, local residents, as well as qualified skills of malaria management in health practitioners undoubtedly improve the timely identification of malaria cases and deter the reintroduction or continued transmission of malaria (Beck and Davies, 1981).

Since the issue of the National Malaria Control Programme (2006–2015), the malaria awareness rate among residents and school-aged children in endemic areas has become one of the key indicators for assessing the programme (Ministry of Health, 2006). In 2010, the National Malaria Elimination Action Plan (2010–2020) was issued to achieve the goal of eliminating malaria by 2020 throughout China. This document lays out a huge challenge to achieve an awareness rate of 80% and 85% among residents and school-aged children, respectively, in endemic areas by 2015 (Ministry of Health, 2010).

Through door-to-door investigations by uniformed surveyors, the public awareness of malaria could be assessed at different levels (Cao et al., 2011; Wu et al., 2012; Jiang et al., 2013; Lian, 2012; Zhang et al., 2012). In general, the awareness of malaria, symptoms, prevention and treatment were higher among school-aged children than local residents, although some differences remained in certain areas. However, awareness was lower than expected in the action plan in these areas (Cao et al., 2011; Jiang and Zhang, 2011; Jiang et al., 2013; Zhang et al., 2012). In Chongqing, for example, only 42.02% and 27.69% of respondents were aware of the serious consequences of *P. falciparum* malaria, and the best way to prevent malaria (Wu et al., 2012). As a class III county, no malaria has occurred in Chengan of Hebei Province for the past 30 years. Consequently, low malaria awareness was identified in local residents with only 2.86%, 5.71% and 6.19% for the knowledge on malaria endemic areas, transmission route and prevention skills, respectively (Ya-min et al.,

2012). In Hainan, where *P. falciparum* malaria was once hyper-endemic, only 20% residents were aware of the free treatment policy for malaria, and the ownership of insecticide-impregnated nets per family was as low as 20%. Reports showed that the proportion of doctor-seeking behavior and blood examination, as well as individual protection awareness after going back were not satisfied among exported labourers (Jiang et al., 2013; Liu et al., 2013b). It shows doctor-seeking behavior was more popular in both elderly and higher educated people, while the awareness of malaria was lower among younger people (Jiang et al., 2013). In terms of awareness among school-aged children, a survey conducted in Changyang County, Hubei Province, in 2010 showed that it was significantly higher among children living in cities than those in villages (Jiang and Zhang, 2011).

In Fujian Province, researchers conducted a survey on malaria awareness targeting health practitioners, including doctors and public health staff (Lian, 2012). The total awareness rate was as low as 51.5%, with the rate in clinical doctors (49.2%) much lower than that among public health practitioners (68.3%). Not surprisingly, public health practitioners tended to better understand basic malaria knowledge, clinic knowledge and prevention. The results revealed a huge gap between the requirements of the national malaria elimination programme and the current status. It also revealed that malaria training is the key impact factor in awareness, indicating training is an effective way to improve the awareness among health practitioners (Lian, 2012).

5. RESEARCH NEEDS IN THE NATIONAL MALARIA ELIMINATION PROGRAMME

In recent years, due to the decline in malaria cases and lower infection, it is urgent to conduct confirmed diagnosis promptly for each case, but it seems that traditional microscopy cannot meet the requirement. More sensitive diagnostic tools are urgently needed to meet the demands of the NMEP (Zheng et al., 2013). Most domestic researchers are focusing on the development of new diagnostic tools, including new diagnostic reagents, PCR and LAMP technology (Zhang et al., 2013; Shen et al., 2011; Wang et al., 2012b; Yi et al., 2010). Currently, malaria blood film microscopy and rapid diagnostic test strips are widely used at the grass roots level. Microscopy is the gold standard for malaria diagnosis, which requires skilled personnel. But it may cause misdiagnosis or improper judgment between species if the parasite density is lower than 50 /μL or mixed infection occurs. The malaria rapid diagnostic test strip is simple, fast, and has high sensitivity and specificity in the diagnosis of *P. falciparum* and *P. vivax* infections, but may provide false-negative diagnosis in detection of *P. ovale* and *P. malariae* infections. Nested PCR is of high sensitivity and accuracy in detections, but requires special equipment with skilled operation and high detection costs for each sample. For instance, the performance assessment on a domestic malaria diagnosis kit showed high agreement with the results of microscopy, and it was proved to be suitable for grass-roots application in malaria-endemic areas (Shen et al., 2011). In response to the increasing imported malaria, nested PCR could be used as an effective complement to microscopy and display its advantages of high sensitivity and specificity (Zhang et al., 2013; Li et al., 2013b). Shi YX established a nested PCR method that proved to be useful in the detection and identification of malaria species compared to microscopy (Shi et al., 2011). In Zhejiang Province, five samples from the imported *P. vivax* malaria cases misdiagnosed initially were retested by nested PCR for identification (Yao et al., 2013). Results showed four infections with *P. ovale* malaria and one mixed infection with *P. ovale* and *P. vivax* malaria. However, the rapid diagnosis test results were all negative. With the development of loop-mediated isothermal amplification technology (LAMP), it was found LAMP is suitable to replace nested PCR in the field, since it is a rapid, sensitive and promising diagnostic tool for *P. falciparum* and *P. viviax* detection (Chen et al., 2011b; Wang et al., 2012b). Yi et al. (2010) established a simple, rapid and highly sensitive fluorescent quantitative LAMP, which showed high sensitivity (90.00%), specificity (93.33%) and correction rate (91.67%), which took only 30 min.

Besides the research and development of new diagnostic tools, new antimalarial drugs, and the surveillance tools on drug resistance and insecticide resistance are also needed in the stage of malaria elimination (Fu, 2013; Wang et al., 2013b; Liu, 2014). As the broad distributed vector, it has been found that *An. sinensis* was resistant to both deltamethrin and DDT, and resistance to pyrethroid has risen strikingly recently.

6. DISCUSSION

In this chapter, the following four gaps have been identified in the national malaria elimination programme in P.R. China, through the literature review. First, in the current stage of malaria elimination, lack of cross-sectoral cooperation is a major gap. It seems that limited information is delivered at the grass-roots CDC level, resulting in the aggregation of individual imported cases, as well as delayed responses. Therefore, it is urgent to tighten and strengthen the interactive activities among the multi-sectors, including departments of inspection and quarantine, health, commerce, foreign affairs, and export labour agencies, to jointly involve them in the national malaria elimination programme (Tambo et al., 2012).

Second, rapid response and risk assessment are necessary to protect the population from potential reintroduction of malaria transmission locally, after imported malaria cases are identified (Liu et. al., 2012; Tambo et al., 2014a,b). However, there is lack of routine risk assessment activities and the response is always delayed or inadequate with poor diagnosis and treatment, which is so crucial in the stage of malaria elimination. Moreover, a significant number of clinically diagnosed but unconfirmed cases still exist in the malaria elimination stage, which need to be addressed through strengthening local laboratory diagnostic capabilities (Xia et al., 2012).

Third, improvement of the public awareness level is urgent, and it is particularly necessary to enhance health education among low educated populations, and clinical doctors (Bhutta et al., 2014; Salam et al., 2014). Misdiagnosis and overtreatment are frequent in grass-roots hospitals due to lack of malaria training. In addition, private sectors should be paying greater attention to malaria diagnosis and treatment.

Fourth, research and development is the backbone of the NMEP, due to the innovative tools that are able to accelerate and sustain the achievement of malaria elimination (Zheng et al., 2013). New diagnostic tools are widely developed and evaluated at various levels in P.R. China, but the development of antimalarial drugs is limited in the context of potential artemisinin resistance spreading from the Greater Mekong Subregion (Cui et al., 2012; Chen, 2014).

7. CONCLUSIONS

The impact factors of reintroduction or continuous malaria transmission induced by imported cases include population movement, cross-sectoral coordination and corporation, vector biology, health services and response to imported malaria, as well as research achievements. Some factors, such as vector biology and population movement may not change in a short term. However, others could be greatly improved through many efforts, such as strengthening prompt, qualified diagnosis and treatment, and enhancing public awareness. It is very crucial that an effective management mechanism should be made and a nationwide surveillance system be established to monitor returning exported workers (Salam et al., 2014; Zofou et al., 2014). It is important to have adequate coverage with well-qualified laboratory and clinical services for malaria detection, treatment and case management. People, especially clinical doctors, need to be trained about malaria prevention, diagnosis and treatment. It is urgent that these gaps be addressed if the goal of malaria elimination is to be achieved by 2020.

ACKNOWLEDGEMENTS

The project was supported by the National S & T Major Programme (grant no. 2012ZX10004220), by the National S & T Supporting Project (grant no. 2007BAC03A02) and by China UK Global Health Support Programme (grant no. GHSP-CS-OP1) and by WHO (2012/269948-0).

REFERENCES

Agomo, C.O., Oyibo, W.A., 2013. Factors associated with risk of malaria infection among pregnant women in Lagos. Nigeria Infect. Dis. Poverty 2, 19.

Beck, J.W., Davies, J.E., 1981. Medical Parasitology. C.V. Mosby Company, Saint Lonis.

Becker, G.J., McClenny, T.E., Kovacs, M.E., Raabe, R.D., Katzen, B.T., 2002. The importance of increasing public and physician awareness of peripheral arterial disease. J. Vasc. Interv. Radiol. 13, 7–11.

Bhutta, Z.A., Sommerfeld, J., Lassi, Z.S., Salam, R.A., Das, J.K., 2014. Global burden, distribution, and interventions for infectious diseases of poverty. Infect. Dis. Poverty 3, 21.

Bruce-Chwatt, L.J., 1968. Movements of populations in relation to communicable disease in Africa. East Afr. Med. J. 45, 266–275.

Bureau of Disease Control, MoH, 2007. Operational Manual of Malaria Prevention. People's Medical Publishing House, Beijing (in Chinese).

Cao, C.Q., 2012. Analysis of construction of microscopy ability for malaria in Nantong City, 2011. Chin. J. Schisto. Control 700–702 (in Chinese).

Cao, X.B., Wang, X.J., Gu, G.M., Li, L., Cao, Y., Chen, H.X., 2011. Health education needs for malaria control in rural residents in Haian County. Chin. J. Schisto. Control 704–707 (in Chinese).

Chang, Q., Wang, Z.P., 2008. The key points of health education in infectious diseases control. Endem. Dis. Bull. 23, 2 (in Chinese).

Chen, C., 2014. Development of antimalarial drugs and their application in China: a historical review. Infect. Dis. Poverty 3, 9.

Chen, X.Z., 2012. Case report of an imported falciparum malaria. Lab. Med. Clin. 754–755 (in Chinese).

Chen, G.W., Lin, M.Z., Li, J.Y., Lu, X.W., 2010. Investigation and analysis of an imported falciparum malaria case. Sci. Travel Med. 45–46 (in Chinese).

Chen, X.L., Zhang, N.K., Zhao, L.M., 2011a. Malaria situation in some certain city. Guide China Med. 303 (in Chinese).

Chen, Q., Zhang, G.Q., Zhang, B., Shen, Y.J., Tian, Z.G., Cao, J.P., Xu, Y.X., Feng, X.P., He, Y.P., Tang, L.H., 2011b. Using loop-mediated isothermal amplification to detect *Plasmodium falciparum* infection in China. J. Pathog. Biol. 269–272 (in Chinese).

Chen, J.Y., Zhao, J.Y., Zhao, M., 2012a. A report on the response to an imported falciparum malaria. Zhejiang Prev. Med. 41–42 (in Chinese).

Chen, H., Han, C.X., Zhou, L., Lin, Y., Ren, C.L., 2012b. Case report on imported falciparum malaria. Chin. J. Clin. Lab. Sci. 556 (in Chinese).

Chen, W.Q., S, Y.P., D, Y., Zhang, H.W., 2012c. Epidemiological analysis of imported malaria in henan province in 2011. Chin. J. Parasitol. Parasit. Dis. 387–390 (in Chinese).

Chen, W.J., Dong, X.B., Hu, C.Y., Wang, L.X., 2012d. Investigation and response to family clustering malaria in Xiaonan District, Xiaogan City. J. Public Health Prev. Med. 64 (in Chinese).

Chen, S., Tao, L., Zhu, Y.L., Gu, W.P., 2013. Epidemiological and clinical analysis of two cases of imported pernicious malaria. Lab. Med. Clin. 1097–1098 (in Chinese).

China CDC, 2011. National technical guideline for malaria elimination. In: China (in Chinese).

Chu, W.Y., Pan, M., Wu, Q., 2011. Report on hemolytic uremic syndromecaused by imported falciparum malaria. Int. J. Lab. Med. 32, 1267–1268.

Cui, X.B., Fang, Z.B., 2011. A death due to imported falciparum malaria. Prev. Med. Trib. 245–246 (in Chinese).

Cui, X.B., K, H., Kong, Q.P., Hu, L.Q., 2011. An investigation of imported case infected with dengue and falciparum. J. Public Health Prev. Med. 78–79 (in Chinese).

Cui, L., Yan, G., Sattabongkot, J., Chen, B., Cao, Y., Fan, Q., Parker, D., Sirichaisinthop, J., Su, X.Z., Yang, H., et al., 2012. Challenges and prospects for malaria elimination in the Greater Mekong Subregion. Acta Trop. 121, 240–245.

Dai, J., Hong, Y., Zhang, X.G., Zhang, W., Deng, J., Wu, H.M., Huang, L., Pan, D.G., Huang, J.C., Shi, Y.X., 2012. The analysis of surveillance status and prevention measures on imported malaria cases at Guangdong Ports, 2010–2011. Chin. J. Dis. Control Prev. 517–520 (in Chinese).

Ding, J., 2012. Current malaria situation and research progress. Chin. J. Public Health 28 (5), 717–718 (in Chinese).

Du, J.W., Wu, K.S., 2010. Global malaria control and elimination: report of a technical review. China Trop. Med. 10, 965–969 (in Chinese).

Feng, Y.X., Jiang, Z.H., 2012. Analysis of imported malaria infections in Hechi City in 2010. China Trop. Med. 283–284 (in Chinese).

Fu, F.Y., 2013. Bioassasy Deltamethrin Resistance and Preliminary Study on Its Mechanism of *Anopheles sinensis* Populations. Master of degree. Chongqing Normal University (in Chinese).

Gao, Y.M., Li, X.J., Wu, S.M., Wei, M.L., Liu, Y.M., 2012. Baseline survey results about pilot malaria eradication in Cheng'an County of Hebei Province. Occup. Health 1740–1741 (in Chinese).

Gao, S.T., Li, X.H., Xie, X., Mei, S.J., 2013. Epidemiological investigation on malaria death of an imported case. Chin. J. PHM 376–377 (in Chinese).

He, Z.Y., Wu, B.Q., Liang, Y., Wang, X.M., Dou, X.F., Jia, L., Li, X.Y., Wang, Q.Y., 2011. Case report of two deaths due to *Plasmodim falciparum*. J. Pathog. Biol. 804 + 782 (in Chinese).

Huang, G.Q., Hu, L.Q., Li, K.J., Lin, W., Sun, L.C., Zhang, H.X., Pei, S.J., Dong, X.R., Liu, J.Y., Yuan, F.Y., 2012. Morbidity situation of imported falciparum malaria in Hubei. J. Trop. Med. 1016–1018 (in Chinese).

Jiang, S.Q., Zhang, B., 2011. Survey on awareness rate of knowledge related to malaria among the primary and secondary students in Changyang County of Hubei Province. Chin. J. Health 863–864 (in Chinese).

Jiang, J., Cheng, B., Liu, J.H., Zhang, H., 2013. Survey on knowledge, attitude and practice about malaria among migrant workers in yichang. Pract. Prev. Med. 773–776 (in Chinese).

Jiao, Y., Meng, Q.Y., 2006. Analysis on situation, policy intervention and challenge of malaria control in China. Chin. Primary Health Care 20, 20–22 (in Chinese).

Ke, H., Chai, L., Liu, Z.J., 2011. Analysis of malaria death due to an imported case. Chin. J. Misdiagn 150 (in Chinese).

Li, S.H., Cai, Y.F., Yuan, F.Y., Pei, S.J., Hu, L.Q., Shi, B.F., Wang, H., Shi, X.H., Huang, L.R., 2010. Case report of imported falciparum malaria in Zhongxiang City of Hubei Province. Chin. J. Parasitol. Parasit. Dis. 478 (in Chinese).

Li, P., Pan, L.H., Lin, F., Li, S.F., 2013. Epidemiological analysis of imported malaria in 2011. J. Med. Pest Control 925–926 (in Chinese).
Li, B.F.,Yang,Y.M., Xu, J.W., Chen, G.W., Zhou, S., Zhao, X.T.,Yang, R.,Yang, R., Shen, J.Y., Lv, Q., Huan, G.Z., et al., 2013. Baseline investigation of the National Malaria Elimination program and projects funded by the Global Malaria Fund in Yunnan Province. J. Pathog. Biol. 448–450 + 472.
Li, L., Zhou, P., Dong, S.H.,Wang, C.W., Li, X.J., Si, H.Y., Sun,Y., 2011a. Clinical care of complicated imported falciparum malaria. Nurs. J. Chin. PLA 28 (1B), 55–56 (in Chinese).
Li, S.J., Zheng, S.L., Yu, Y.M., Yang, Y.J., Ren, N., Wang, J., 2011b. Clinical analysisi of 41 imported falciparum malaria. China Med. Eng. 94–95 (in Chinese).
Li, X.Y., Zhang, S.J., Zhao, X., Li, C.Q.,Wang, X.M., He, Z.Y., Tian, L.L., He, J., Pang, X.H., He, X., et al., 2013a. Epidemiological analysis of mosquito monitoring around the capital international airport in Beijing, China. Chin. Prev. Med. 86–88 (in Chinese).
Li, K., Zhou, S.S., Huang, F., Xia, Z.G., Zheng, X., 2013b. Comparison of falciparum in low parasitemia infection by 3 PCR methods. J. Pathog. Biol. 331–335 (in Chinese).
Li,Y.X., 2013. Report on 2 death cases due to mis-diagnosed imported falciparum malaria. China Trop. Med. 1043–1044 (in Chinese).
Lian, M.M., 2012. Investigation Report of the Grassroots Medical Workers Awareness Rate and Its Influening Factors of Malaria Preventive Knowledge in Fujian Province. Master. Fujian Medical University (in Chinese).
Lin, R.X., Zhang, X.C., P,B., L,W.C.,Wei, H.X., Qiu,W.W., Xie, J.H., 2011.A feasibility study of malaria elimination in Guangdong province. J.Trop. Med. 93–95 + 108 (in Chinese).
Liu, D.Q., 2014. Surveillance of antimalarial drug resistance in China in the 1980s-1990s. Infect. Dis. Poverty 3, 8.
Liu, J.,Yang, B., Cheung, W.K.,Yang, G., 2012. Malaria transmission modelling: a network perspective. Infect. Dis. Poverty 1, 11.
Liu, Q.Y., Liu, X.B., 2010. Prevention and control of vector Anopheles: a key approach for malaria elimination in China. Chin. J.Vector Biol. Control 409–413 (in Chinese).
Liu, Q.Y., Liu, X.B., Zhou, G.C., Ren, D.S., Jiang, J.Y., Guo,Y.H., Zheng, C.J., Li, H.S., Liu, J.L., Chen,Y., et al., 2011a. Primary study on the flight range of *Anopheles sinensis* based on the mark-release-recapture method in Yongcheng city, Henan province. Chin. J. Vector Biol. Control 201–204 (in Chinese).
Liu, X.B., Liu, Q.Y., Guo,Y.H., Jiang, J.Y., Ren, D.S., Zhou, G.C., Zheng, C.J., Zhang,Y., Liu, J.L., Li, Z.F., et al., 2011b. The abundance and host-seeking behavior of culicine species (Diptera: Culicidae) and *Anopheles sinensis* in Yongcheng city, People's Republic of China. Parasit.Vectors 4, 221.
Liu, H., Li, M., Jin, M., Jing, F., Wang, H., Chen, K., 2013a. Public awareness of three major infectious diseases in rural Zhejiang province, China: a cross-sectional study. BMC Infect. Dis. 13, 192.
Liu, J., Liu, C.F., Liu, C.X., Xu,Y.Q., G, D.Y., Shi, L., Zhao, C.Z., Li, D.X., Xin, B.Q., 2013b. Study of KABP on malaria among international travelers at Shenzhen ports. Chin. Front. Health Quar. 11–14 (in Chinese).
Liu,Y., Zhang,Y., Lu, F.R., 2011. Case report of imported falciparum malaria in Haerbin. Chin. J. Parasitol. Parasit. Dis. 156 (in Chinese).
Liu,Y.,Yan, Q.Y., Zhang, H.W.,Wang, H., 2011. Analysis of imported falciparum malaria in Henan Province, 2010. Contemp. Med. 156–157 (in Chinese).
Lu, S.P., Long,Y.J., 2011. Survey report on imported pernicious malaria cases. Chin. Front. Health Quar. 167–168 + 171 (in Chinese).
Miao, P., Chen, H.X., 2011. Investigation and response to imported vivax malaria in Rudong County. Jiangsu J. Prev. Med. 22 (4), 53 (in Chinese).
Ministry of Health, 2006. National Malaria Control Programme (2006–2015). In: Beijing: Ministry of Health (in Chinese).

Ministry of Health, 2009. National protocol for anti-malaria treatment. In: Beijing (in Chinese).
Ministry of Health, 2010. National Action Plan for Malaria Elimination (2010–2020). In: Beijing (in Chinese).
Ministry of Health, 2011. National Technical Guideline for Malaria Elimination in China. In: Beijing (in Chinese).
Prothero, R.M., 1977. Disease and mobility: a neglected factor in epidemiology. Int. J. Epidemiol. 6, 259–267.
Quan, X.B., Lu, Y.J., Zhan, X.Y., D, Z., 2011. Survey of an imported malaria case. China Trop. Med. 11 (9), 1170 (in Chinese).
Rao, X.M., O, Y.Y., 2011. Relapse of imported malaria and literature review. ACTA Acad. Med. Zunyi 193–194 (in Chinese).
Salam, R.A., Das, J.K., Lassi, Z.S., Bhutta, Z.A., 2014. Impact of community-based interventions for the prevention and control of malaria on intervention coverage and health outcomes for the prevention and control of malaria. Infect. Dis. Poverty 3, 25.
Service, M.W., 1991. Agricultural development and arthropod-borne diseases: a review. Rev. Saude Publica 25, 165–178.
Shang, X.P., 2012. Research the Density, Ecological Habit and Resistance to Insecticides of the Main Malaria Media in Hubei Province. Master. Wuhan Uni. Sci. Technol. (in Chinese).
Shen, X., Liu, H., Li, C.F., 2011. Effect evaluation of ABON rapid diagnostic reagents for mixed malaria/ *Plasmodium falcipatum*. Parasit. Infect. Dis. 235–237 (in Chinese).
Shi, Y.X., Huang, J.C., Su, J.K., Hong, Y., Li, X.B., Zheng, S., Xing, L.Q., Guo, B.X., 2011. Nested PCR for malaria detection and plasmodium species identificaiton. Chin. J. Parasitol. Parasit. Dis. 263–266 (in Chinese).
Tambo, E., Adedeji, A.A., Huang, F., Chen, J.H., Zhou, S.S., Tang, L.H., 2012. Scaling up impact of malaria control programmes: a tale of events in Sub-Saharan Africa and People's Republic of China. Infect. Dis. Poverty 1, 7.
Tambo, E., Ai, L., Zhou, X., Chen, J.H., Hu, W., Bergquist, R., et al., 2014a. Surveillance-response systems: the key to elimination of tropical diseases. Infect. Dis. Poverty 3, 17.
Tambo, E., Ugwu, E.C., Ngogang, J.Y., 2014b. Need of surveillance response systems to combat Ebola outbreaks and other emerging infectious diseases in African countries. Infect. Dis. Poverty 3, 29.
Tong, X.C., Huang, H.C., Xu, T.M., 2011. Clinical analysis of 8 imported falciparum malaria. Mod. Med. J. China 72–73 (in Chinese).
Wang, H.J., Sl, Hu, Sl, Li, Gu, Z.C., Chen, J.S., Zhu, G.D., Huang, F., 2012a. Investigation on anopheline species in Chayu County, Linzhi Prefecture of Tibet Autonomous Region. Chin. J. Schisto. Control 333–335 (in Chinese).
Wang, Z.Y., Jiang, L., Cai, L., Wang, W.J., Zhang, Y.G., Hong, G.B., Zhang, X.P., Lu, W., 2012b. Analysis and establishment of loop-mediated isothermal amplification for the diagnosis of *Plasmodium vivax*. J. Trop. Med. 12, 157–161 (in Chinese).
Wang, Z.L., Wang, Y.R., Fu, T.X., Mao, D.H., 2013. Clinical analysis of 91 cases of imported falciparum malaria from Africa. Chin. J. Schisto. Control 324–325 (in Chinese).
Wang, W.M., Zhou, H.Y., liu, Y.B., li, J.L., Cao, Y.Y., Cao, J., 2013a. Comparison of seasonal fluctuation and nocturnal activity patterns of *Anopheles sinensis* in different regions of Jiangsu pronince. China Trop. Med. 13, 292–295 (in Chinese).
Wang, D.Q., Xia, Z.G., Zhou, S.S., Zhou, X.N., Wang, R.B., Zhang, Q.F., 2013b. A potential threat to malaria elimination: extensive deltamethrin and DDT resistance to *Anopheles sinensis* from the malaria-endemic areas in China. Malar. J. 12, 164.
Wang, X.G., Lei, Y.L., Lan, J.Q., Mei, J.H., Li, Z.H., 2013c. Laboratory testing for a case of imported *Plasmodium ovale* infection in Zhejiang Province. Chin. J. Parasitol. Parasit. Dis. 78–79 (in Chinese).

Wang, D.X., 2011. Epidemiological and clinical analyasis of 14 imported malaria cases. Guide China Med. 398–399 (in Chinese).
WHO, 2010. Guidelines for the Treatment of Malaria. World Health Orgnization, Geneva.
Wu, M., Ke, Q.X., Zhang, L.Z., Shuai, J., 2011. Analysis of mis-treatment of complicated imported falciparum malaria. Lab. Med. Clin. 1529–1530 (in Chinese).
Wu, C.G., Luo, X.J., Luo, F., Li, S.S., Xiao, B.Z., Jiang, S.G., 2012. Current status of malaria prevention in Chongqing. J. Trop. Med. 472–474 (in Chinese).
Xia, Z.G., Yang, M.N., Zhou, S.S., 2012. Malaria situation in the People's Republic of P.R. China in 2011. Chin. J. Parasitol. Parasit. Dis. 30, 419–422 (in Chinese).
Xie, X.R., Tan, H.B., 2011. Rescue of complicated falciparum malaria. Med. J. Natl. Defend. Forces Southwest China 21 (3), 328 (in Chinese).
Xu, Y.H., Gao, Y., Yang, J., Zuo, Y.P., 2011. Effect of management scheme on control of imported falciparum malaria in Yangzhou City, 2010. Chin. J. Schisto. Control 728–729 (in Chinese).
Yang, Q., 2011. The analysis of 14 imported malaria cases in Gezhouba Group Company in 2010. J. Public Health Prev. Med. 2 (5), 80 (in Chinese).
Yang, Y.L., Wu, S.R., Lu, Y.H., Cao, P.G., Guo, C.K., 2013. Epidemiological investigation of imported falciparum malaria in Longlin county, guangxi zhuang autonomous region. Chin. J. Schisto. Control 100–101 (in Chinese).
Yao, L.N., Zhang, L.L., Ruan, W., Chen, H.L., Lu, Q.Y., Yang, T.T., 2013. Species identification in 5 imported cases previously diagnosed as vivax malaria by parasitological and nested PCR techniques. Chin. J. Parasitol. Parasit. Dis. 221–223 + 234 (in Chinese).
Ye, Q.Y., Yu, K.X., Zhou, J.S., 2013. Epidemiological investigation and response to the first imported malaria in Xiaochang county, Hubei province. J. Public Health Prev. Med. 93–94 (in Chinese).
Ye, Y., Li, L., 2010. One death of imported falciparum malaria. Exp. Lab. Med. 437 + 346 (in Chinese).
Yi, H.H., Xu, B., Fang, C., Song, Y.W., Wu, P.L., Wang, Y.F., Xu, Z., Zhao, J.W., Xu, J.C., 2010. Study on the detection of *Plasmodium falciparum* by fluorescent quantitative loop-mediated isothermal amplification. China Trop. Med. 1178–1180 (in Chinese).
Yin, J.H., Yang, M.N., Zhou, S.S., Wang, Y., Feng, J., Xia, Z.G., 2013. Changing malaria transmission and implications in P.R. China towards national malaria elimination programme between 2010 and 2012. PloS One 8, e74228.
Yu, G., Yan, G., Zhang, N., Zhong, D., Wang, Y., He, Z., Yan, Z., Fu, W., Yang, F., Chen, B., 2013. The Anopheles community and the role of *Anopheles minimus* on malaria transmission on the China–Myanmar border. Parasit. Vectors 6, 264.
Yu, Y.J., Cao, H., Sen, S.R., Li, M.Y., Yang, X., Zhan, X.M., 2012. An imported falciparum malaria case in Guangdong. Chin. J. Parasitol. Parasit. Dis. 94 + 99 (in Chinese).
Zhang, H.Y., Gao, Y.H., Long, R.T., Lin, Y.Z., Lin, J., Li, W., He, S.H., 2012. Knowledge, attitude and practice on malaria among residents in Hainan province. Hainan Med. J. 128–130 (in Chinese).
Zhang, B., Tian, B., Liao, Y., Shen, X.J., Zeng, M., Liu, Y.P., Wen, L., 2013. Application of nested PCR in the diagnosis and typing of imported malaria. Pract. Prev. Med. 229–231 (in Chinese).
Zhao, S.J., Zhao, Q.P., Peng, P.Z., Liu, G.L., Zhao, J., 2012. Response to an imported falciparum malaria. J. Public Health Prev. Med. 93–94 (in Chinese).
Zheng, Q., Vanderslott, S., Jiang, B., Xu, L.L., Liu, C.S., Huo, L.L., et al., 2013. Research gaps for three main tropical diseases in the People's Republic of China. Infect. Dis. Poverty 2, 15.
Zhou, S., Wang, Y., Li, Y., 2011a. Malaria situation in the People's Republic of P.R. China in 2010. Chin. J. Parasitol. Parasit. Dis. 29 (6), 401–403 (in Chinese).

Zhou, Z.H., Wu, M.S., Wang, Q.J., 2011b. A case report of imported falciparum malaria. Chin. J. Clin. Lab. Sci. 204 (in Chinese).

Zhou, R.M., Zhang, H.W., D,Y., Qian, D., Liu,Y., Chen,W.Q.,Yan, Q.Y., S,Y.P., Zhao, X.D., Xu, B.L., 2013a. Laboratory detection on two cases with imported *Plasmodium ovale* infection. Chin. J. Parasitol. Parasit. Dis. 127–130 (in Chinese).

Zhou, G.R., Geng,Y., Li,A.M., 2013b.Verification survey of network reported malaria infection in Guizhou Province in 2011. China Trop. Med. 384–385 + 389 (in Chinese).

Zhu, X.G., She, G.S., 2012. Epidemic situation and prevention and control countermeasures of imported falciparum malaria in Jiangdu District,Yangzhou City. Chin. J. Schisto. Control 616–617 (in Chinese).

Zofou, D., Nyasa, R.B., Nsagha, D.S., Ntie-Kang, F., Meriki, H.D., Assob, J.C., et al., 2014. Control of malaria and other vector-borne protozoan diseases in the tropics: enduring challenges despite considerable progress and achievements. Infect. Dis. Poverty 3, 1.

CHAPTER ELEVEN

Transition from Control to Elimination: Impact of the 10-Year Global Fund Project on Malaria Control and Elimination in China

Ru-Bo Wang[1,2,3,*], Qing-Feng Zhang[1,2,3], Bin Zheng[1,2,3], Zhi-Gui Xia[1,2,3], Shui-Sen Zhou[1,2,3], Lin-Hua Tang[1,2,3], Qi Gao[4], Li-Ying Wang[5], Rong-Rong Wang[5]

[1]National Institute of Parasitic Diseases, Chinese Center for Disease Control and Prevention, Shanghai, People's Republic of China
[2]WHO Collaborating Centre for Malaria, Schistosomiasis and Filariasis, Shanghai, People's Republic of China
[3]Key Laboratory of Parasite and Vector Biology, Ministry of Health, Shanghai, People's Republic of China
[4]Jiangsu Provincial Institute of Parasitic Diseases, Wuxi, People's Republic of China
[5]National Health and Family Planning Commission of the People's Republic of China, Beijing, People's Republic of China
*Corresponding author: E-mail: rubo_wang@163.com

Contents

1. Introduction	290
2. Epidemiological Justification for Rounds of GFATM Malaria Programmes	291
3. General Information about Rounds	292
3.1 GFATM malaria programme in China: R1	293
3.2 GFATM malaria programme in China: R5	293
3.3 GFATM malaria programme in China: R6	293
3.4 GFATM malaria programme in China: NSA	295
3.5 GFATM malaria programme in China: R10	296
4. Programme Management	296
4.1 Management structure	297
4.2 Programme implementation mechanism	297
4.3 Monitoring and evaluation	298
4.4 Procurement and supply management	299
4.5 Financial management and audit	300
5. Programme Input	300
5.1 Fund	300
5.2 Health products, equipment and materials	301
5.3 Human resources	301

6. Main Programme Activities and Output	302
6.1 Case management	303
6.2 Vector control	304
6.3 Health education and community mobilization	306
6.4 Malaria control for populations at high risk	306
6.5 Malaria surveillance	309
6.6 Multi-sector cooperation	309
6.7 Operational research	310
7. Achievements and impacts	310
7.1 From malaria control to elimination in China	310
7.2 Filling the resource gap for malaria control and elimination	313
7.3 Improved multi-sectorial cooperation and communication	314
7.4 Contribution to policy and ability of the national malaria control and elimination	314
7.5 Enforced public awareness of malaria control and prevention	314
7.6 Cross-border cooperation mechanism for malaria control	315
8. Looking Forward	315
9. Conclusions	316
Acknowledgements	317
References	317

Abstract

The Global Fund to Fight AIDS, Tuberculosis and Malaria (GFATM) supported a project on the control and elimination of malaria in People's Republic of China which was one of the biggest-scale international cooperation programmes to control malaria in the country during the past 10 years. The project promoted the effective implementation of the Chinese national malaria control programme. On the basis of epidemiologic data, an overview of the project activities and key performance indicators, the overall impact of the GFATM project was evaluated. We also reviewed relevant programme features including technological and management approaches, with a focus on best practice, innovations in implementation and the introduction of international standards. Last, we summarised the multi-stakeholder cooperation mechanism and comments on its sustainability in the post-GFATM period. Recommendations for the future management of the Chinese national malaria elimination programme are put forward after considering the challenges, shortcomings and lessons learnt during the implementation of the GFATM project in China to sustain past achievements and foster the attainment of the ultimate goal of malaria elimination for the country.

1. INTRODUCTION

The Global Fund to Fight AIDS, Tuberculosis and Malaria (GFATM) was established in Geneva in 2002 with an aim to bundle and mobilise resources for the fight against the three diseases. The approach it adopted as the most effective way to fight these infections is to spur partnerships among government, civil society, the private sector and communities living with the diseases (Global Fund, 2002).

Since 2002, the People's Republic of China (China) successfully applied for GFATM support for 15 specific and three consolidated programmes. These included five grants dedicated to malaria, such as: Round 1 (R1), Round 5 (R5), Round 6 (R6), Round 10 (R10) and the National Strategy Application (NSA). The last grant reflected the shift from a rounds-based mode to an ever closer alignment with the National Malaria Elimination Programme (NMEP). Projects were implemented by the Centres for Disease Control and Prevention (CDCs) at different administrative levels, together with partners such as international non-governmental organizations (NGOs) and other government sectors (China PR, 2013).

Here, we critically review the activities, experiences and impacts of the GFATM-supported malaria control and elimination programme in China. We mainly rely on data obtained from the annual programme reports prepared from 2003 to 2012 and made available by the National Programme Office (NPO) and the information publicly available through the GFATM website.

2. EPIDEMIOLOGICAL JUSTIFICATION FOR ROUNDS OF GFATM MALARIA PROGRAMMES

In 2002, Yunnan and Hainan provinces still faced a serious malaria problem (Sheng et al., 2003). The number of malaria cases originating from the two provinces accounted for more than 70% of the total number in China. Migration was increasing significantly in these areas, which could result in a significant increase of malaria transmission, especially of multi-drug-resistant *Plasmodium falciparum* present on the border region of Yunnan province and in the mountainous areas of Hainan province. To control malaria in these two target provinces and limit the spread of multi-drug-resistant strains, China GFATM malaria programme R1 was implemented from 2003 to 2008.

In 2003, about 96% of the malaria cases registered in China were from six provinces: Yunnan and Hainan (both of which had falciparum malaria) and the central provinces of Anhui, Hubei, Henan and Jiangsu, where vivax malaria became an increasing problem (Liu, 2014; Zhou et al., 2005). To roll back the re-emergence of malaria in central provinces and reduce the burden of malaria across the country, the China GFATM malaria programme R5 was approved by the GFATM in 2005.

From 2001 to 2005, about 25.6% of all malaria cases registered in Yunnan province were originally from Myanmar. The surveillance and reporting systems for malaria were relatively weak in the four special administrative regions of Myanmar bordering China, and the high cross-border mobility of Chinese workers resulted in a high and difficult to manage malaria burden. In 2006, the GFATM approved the China GFATM malaria programme R6 to reduce malaria burden on the China–Myanmar border.

In 2008, the nationwide malaria situation on China met the criteria of pre-elimination recommended by the World Health Organization (WHO). Most of the remaining endemic areas were poverty-stricken counties lacking the required resources to achieve the malaria elimination goal (Bi & Tong, 2014). To achieve the stated goal of malaria elimination in China, the NSA was approved by the GFATM in 2009. Unlike the previous programmes that followed the logic of 'rounds', the NSA was in line with the NMEP. NSA focuses on malaria elimination of Type 1 and 2 counties whereas the government fund focuses on Type 3 and 4 counties (Type 1 – local infections detected in 3 consecutive years and the annual incidence was ≥1 per 10,000; Type 2 – local infections detected in the last 3 years and at least in 1 year the annual incidence was <1 per 10,000 and >0; Type 3 – no local infections reported in the last 3 years; Type 4 – non-malaria epidemic area).

Last, in 2012, the China GFATM malaria programme R10 was approved to consolidate malaria control on the China–Myanmar border with an aim to continually reduce the malaria burden in the project area and contribute to achieving malaria elimination in China by 2020.

3. GENERAL INFORMATION ABOUT ROUNDS

From 2003 to 2012, China completed R1, R5, R6 and NSA. R10 was closed by the end of 2013. Overall, approximately 116 million USD had been disbursed by the GFATM for these grants, accounting for 15.9% of the total value of all GFATM grants in China approved between 2003 and 2012 (see Table 11.1 and Figure 11.1).

In general, a GFATM grant can last for 5 years, divided into two periods – Phase I (2 years) and Phase II (3 years). The 25th GFATM board meeting in 2011 decided that the following eligibility criteria for renewal applications would become effective started in 2012: (1) *group of 20 (G20) upper middle income countries will less than an extreme disease burden will no longer be eligible for renewals of grants;* and (2) *the counterpart financing and fucus of proposal requirements under the policy on eligibility, counterpart financing and prioritization will apply.* China is a G20 member country, and the malaria burden in China is less than an extreme situation. According to the decision, NSA and R10 were closed in advance after the programme activities of Phase I finished, and the GFATM will not support malaria control and elimination in China after 2014.

The coverage of Global Fund-supported projects gradually expanded from 47 counties in 2003 to 762 counties in 20 provinces in 2010 (Figure 11.2). Most of the project areas were poverty-stricken regions with relatively weak economic conditions and a heavy malaria burden.

3.1 GFATM malaria programme in China: R1

GFATM malaria programme R1 covered 25 border counties of the Yunnan province, 10 counties of the Hainan province and 12 counties of the other eight project provinces or autonomous regions (including Henan, Hubei, Anhui, Jiangsu, Guangdong, Guangxi, Sichuan and Guizhou). The main activities included early diagnosis, appropriate treatment and effective protection; malaria management among mobile populations in border areas of Yunnan; malaria-related health education and promotion; malaria surveillance and programme management capability.

3.2 GFATM malaria programme in China: R5

The target area for R5 covered 6 provinces, namely Yunnan, Hainan, Anhui, Henan, Hubei and Jiangsu provinces. Overall, 19.14 million people at risk living in rural areas of 1813 townships in 121 counties benefited directly and a further 63.8 million benefited indirectly from the activities. The activities were integrated into the NSA in 2010. The project pursued three strategic objectives: (1) prevention, (2) diagnosis and treatment and (3) surveillance and epidemic response. Malaria control and management capacity in target areas were enhanced through the introduction of new techniques, including artemisinin-based combination therapy (ACT), rapid diagnosis tests (RDTs), long-lasting insecticide-treated nets (LLINs) and multi-sectorial participation as well as health education.

Table 11.1 Key information about the GFATM malaria programmes in China

Round	Period	Approved budget (USD)	Coverage	Beneficiaries	Overall goal
R1	2003–2008	6,347,448	47 counties in 10 provinces of China	9.3 million	Roll back malaria in target provinces and control the spread of multi-drug-resistant malaria
R5	2006–2010	31,161,319	1813 townships of 121 counties in six provinces of China	63.8 million	Roll-back re-emerging malaria in central provinces and reduce the burden of malaria in resource-poor areas of central and southern China
R6	2007–2012	11,865,704	12 border counties in Yunnan provinces of China and four special regions of Myanmar	3.5 million	Reduce the malaria burden of mobile Chinese workers crossing the common border in 12 project counties of five prefectures in Yunnan and reduce the malaria burden of local residents in four special administrative regions of Myanmar bordering with Yunnan
NSA	2010–2012	63,436,279	762 counties in 20 provinces of China	500 million	Reduce to zero of locally contracted malaria cases in China (except in some border areas in Yunnan province) until 2015
R10	2012–2013	5,080,078	Seven counties in Yunnan provinces of China and five special regions of Myanmar	2.2 million	Reduce the malaria burden in five special regions in Myanmar; monitoring drug resistance against artemisinin and its derivatives, control the number of imported infections and promote elimination of malaria in China

NSA, National Strategy Application.

Impact of the Global Fund Project on Malaria Control 295

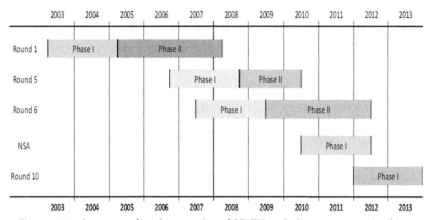

Figure 11.1 *Sequence of implementation of GFATM malaria programmes in China.*

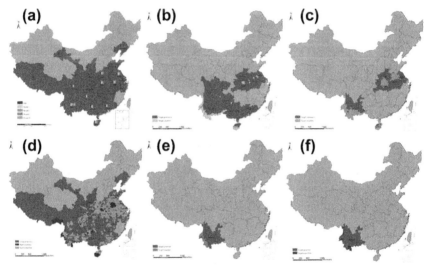

Figure 11.2 *Target area of GFATM malaria programmes in China.* (a) target provinces of R1, R5, R6, National Strategy Application and R10; (b) 47 target counties in 10 provinces of R1; (c) 121 target counties in six provinces of R5; (d) 762 target counties in 20 provinces of NSA; (e) 12 target counties in Yunnan province of R6 and (f) 7 target counties in Yunnan province of R10.

3.3 GFATM malaria programme in China: R6

GFATM malaria programme R6 covered 12 counties in Yunnan province of China and four special administrative regions in Myanmar on the China–Myanmar border. The programme activities in Yunnan province were consolidated into the NSA in 2010 whereas the activities of Myanmar ended in 2012. The main content of R6 was (1) to improve the malaria control service for frequently mobile Chinese workers (FMCWs) in Yunnan; (2) to improve the malaria service accessibility and quality for FMCW and local residents in Myanmar and (3) to develop cross-border malaria surveillance, information exchange and joint prevention mechanisms.

3.4 GFATM malaria programme in China: NSA

According to the original proposal, the NSA was intended to last 5 years. However, the programme ended ahead of schedule on June 30, 2012, after the relevant decision of the 25th GFATM board meeting. The programme covered Type 1 and Type 2 counties and its objectives were (1) to offer timely diagnosis and proper treatment to malaria patients, (2) to improve malaria prevention, (3) to enhance malaria control of populations at high risk and (4) to further improve the malaria surveillance system.

With the principle of the one strategy plan and resource consolidation, NSA consolidated two programme grants (R5 and R6) and consolidated all of the resources on malaria elimination in China. NSA was in line with the NMEP, including the goal, objectives, indicators and main activities. The target area of NSA spread to the lower malaria-burden provinces, and the previous grants only covered the high malaria-burden provinces.

3.5 GFATM malaria programme in China: R10

GFATM malaria programme R10 covered five special regions in Myanmar on the China–Myanmar border and seven border counties in Yunnan. The target population included 586,000 local residents and 100,000 Chinese migrant workers in Myanmar as well as 1.5 million frequent border crossers. The main activities were (1) to improve access to diagnostic and treatment services; (2) to improve access to LLINs; (3) to maximise the utilization of preventive, diagnostic and treatment services through the production of information education communication (IEC)/behavioural change communication (BCC), and (4) to strengthen project implementation, management and information exchange and enhance the efficiency of cross-border malaria control.

4. PROGRAMME MANAGEMENT

In 2003, programme management was a new concept for malaria control in China. Under the leadership and supervision of the Ministry of

Health (MOH), local health bureaus of provincial governments and county governments, the GFATM malaria programme established a management system on basis of the CDC infrastructure. A series of documents, including programme agreements, work plans, and the *Handbook of Management and Technology* were issued upon launch of each programme. Monitoring and evaluation (M&E) activities, procurement activities and financial and audit activities ensured that all programmes were progressing smoothly.

4.1 Management structure

As the Principal Recipient (PR) of GFATM programmes in China, the China CDC was responsible for implementing and managing the GFATM programmes under the guidance and supervision of the Country Coordinating Mechanism for the GFATM Programmes (China CCM). The PR established nine office or departments, including Programme Offoce, Financial Department, Procurement Department, M&E Department, Human Resources Department, Audit Department, NPO for AIDS, NPO for Tuberculosis, NPO for Malaria. The three NPOs were responsible for planning and implementing specific Programmes.

The NPO for Malaria of R1, R5, R6 and NSA was facilitated by the National Institute of Parasitic Diseases (NIPD), China CDC in Shanghai to execute all of its functions whereas the R10 grant was implemented in conjunction with the Yunnan Provincial Bureau of Health. The NPO for Malaria, in cooperation with the PR, was responsible for the development and implementation of plans, disbursement requests, financial management, procurement and assets management supervision, assessment, technical assistance (TA), programme balance statements and reports, consolidation and submission as well as receiving supervision, instruction and audit initiated by relevant government authorities (Tang, 2009).

Sub-recipients (SRs) were based in local CDCs, which were responsible for conducting programme-related activities (e.g. trainings, M&E, finance management etc.) Local health bureaus of provincial governments and county governments were responsible for supervising the implementation of the programmes. Township hospitals mainly performed malaria diagnosis, treatment and health education (Figure 11.3).

4.2 Programme implementation mechanism

The programme implementation was performed in accordance with the agreement signing covering agreed work plan and M&E plan between GFATM, and PR. After an agreement had been signed with the GFATM, the

PR signed the programme agreements with the health bureaus/or CDCs of the target provinces and partners. Target provinces then signed programme agreements with the health bureaus/or CDCs of target counties.

The work plan, M&E plan and procurement and supply management plan approved by the GFATM and MOH were annexed to programme agreements. Considering the local malaria epidemiology, specific work plans were developed at lower levels to perform the project activities. Standard documents such as the *Financial Management Manual*, the *Procurement and Asset Programme Manual*, the *Audit Guideline*, and the *Programme Management and Technical Guidelines*, issued by PR, were used to guide project execution and project control.

4.3 Monitoring and evaluation

M&E allows measurement of the progress and effectiveness of a programme at all levels. All GFATM programmes in China developed an M&E plan and a performance framework to capture data on the impact of the interventions, monitor their effectiveness and ensure financial and programmatic accountability.

The M&E plan defined indicators, data management, programme supervision, etc. In 2003, the M&E plan was only used for the GFATM programme

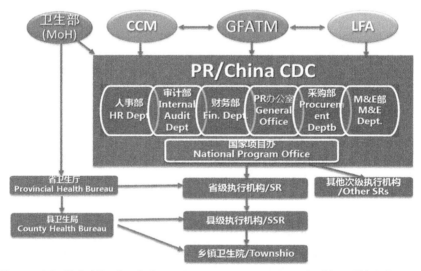

Figure 11.3 *Global Fund malaria grant management structure in china.* (CCM, Country Coordinating Mechanism for the GFATM Programmes; GFATM, Global Fund to Fight AIDS, Tuberculosis and Malaria; LFA, Local Fund Agents; PR, Principal Recipient).

in China. In 2010, the M&E plan and framework were enhanced to support both the NMEP and the NSA.

The numerator and denominator of indicators, collection methods, frequency and coverage of data were definitely prescribed in the M&E plan. For the malarial programme, the indicators were classified as impact indicators, outcome indicators and progress indicators. Totally, there were 14, 15, 19 and 22 indicators in R1, R5, R6, NSA, respectively. The proportion of impact and outcome indicators per round increased over rounds (Zhao et al., 2011).

The baseline surveys to establish indicators were completed at the beginning of each programme, and the results were semi-annually measured and reported through the programme process reporting system. All data were verified at all levels.

The PR and NPO conducted supervision of each province each year, the Provincial Programme Office conducted supervision of counties twice per year, the Prefectural Programme Office conducted supervision of townships twice per year and township hospitals conducted supervision of villages 1–2 times per year. The problems found during supervision were followed up to be rectified. Meanwhile, R5 received a data quality audit (DQA) mandated by the GFATM in 2008. The result of the DQA indicated that the data generated by the malaria programme were of high quality.

To evaluate the effects of the programme in more detail, R1, R6 and NSA mandated experts of universities and research institutes to exteranally evaluate the programme implementation. The evaluation mainly focused on the completion of work plans, the results of M&E indicators, programme implementation capabilities, investment in malaria control and prevention and the degree of satisfaction of people who were to benefit from the programme. Several suggestions and recommendations of these programme evaluations were used for the national malaria control and elimination programme. The NSA also received a diagnostic review, conducted by the GFATM in 2011, which found no major problems.

4.4 Procurement and supply management

All health products, equipment and material used or distributed in the frame of the GFATM-supported projects were procured through public bidding. The procurement procedure was open, transparent and equitable in accordance with relevant laws of China. New malaria control products such as LLINs, RDTs and ACTs were successfully introduced to China. In accordance with GFATM policy, only LLINs approved by WHO's Pesticides Evaluation Scheme (WHOPES) (WHO, 2007) were procured, the RDTs were recommended by WHO and anti-malarials were pre-qualified by WHO. In 2006, only one sort of 'WHOPES-approved' LLINs was registered by the Ministry of Agriculture in China, and no RDTs were registered by the China Food and Drug Administration (CFDA). Thus, RDTs were initially only procured through the reimbursable procurement of WHO. However, in 2012, three 'WHOPES-approved' LLINs and one RDT had been registered in China, and many anti-malarial manufacturers had applied for pre-qualification.

An Inventory Management System was set up in programme areas, and an inventory officer was responsible for inventory information management at each level. The inventory information was collected and analysed to avoid stock-outs and waste. All CDCs at each level had set up warehouses, and products were quickly distributed to township hospitals through the CDCs system. Every distribution was recorded in detail, with signature. The allocation of products was adjusted according to the progress of programme activities, and buffer-stocks of products were to be kept.

4.5 Financial management and audit

The performance-based funding mechanism was introduced since the beginning of the programme in China in 2003. The NPO semi-annually submitted a disbursement request for programme funds to the GFATM on the basis of the semi-annual performance reports of the SRs. Programme funds were then disbursed to SRs within 10 working days after receipt from the GFATM. A *Financial Management Manual* was issued, financial management training was held once every year and financial supervision was performed once every year. The semi-annual financial reports were submitted to the GFATM through the 'China Global Fund Financial Management System'.

To ensure that expenditures were real, legitimate and reasonable, external and internal audits at each level were performed once a year. External and internal audits reviewed the internal financial control system, budget fund allocation, budget implementation, completion of the main indicators, accounting of current accounts, monetary funds, inventories, fixed assets accounting and so on. Any rectifications were performed in a timely manner according to the findings contained in the audit report.

5. PROGRAMME INPUT

5.1 Fund

As of November 2012, the GFATM had disbursed a total of USD 116 million to support the Chinese malaria control and elimination programmes. These funds played an important role in helping China to achieve its malaria control goals. The peak of disbursements was reached in 2010 because the funding of NSA in 2010 was more than the other grant (Figure 11.4).

5.2 Health products, equipment and materials

From 2003 to 2012, the GFATM-supported malaria programs procured and distributed 5516 microscopes, 6655 sprayers, 967,560 RDTs, 2,399,069 person-doses of drugs, 1,823,153 LLINs and 178,074 L of pesticide. In addition, vehicles and office equipment improved the hardware of CDCs and facilitated programme activities and malaria control (Table 11.2).

Data from the GFATM database including 1514 entries from 79 countries show that the costs for LLINs and RDTs dropped significantly from 2005 to 2012 (Wafula et al., 2013). In China, the cost for purchasing a single LLIN decreased from 38 RMB in 2006 to 28 RMB in 2012 and that of RDTs maintained at approximately 0.7 USD.

5.3 Human resources

CDC staff were responsible for most of the programme activities such as diagnosis, treatment, vector control, surveillance, data collection, data analysis and management. The salaries of CDC staff were provided by governments at each level, but township hospital and village doctors were provided with incentives to improve case management and reporting. Because of

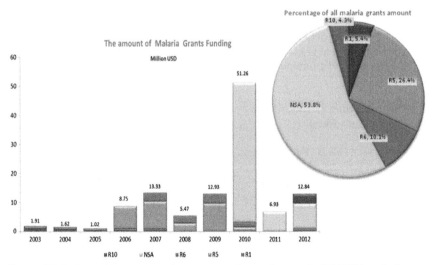

Figure 11.4 *Annual amount and percentage by each round of GFATM malaria programme from 2003 to 2012 in China.*

human resources limitations, some partners recruited new staff to perform malaria programme activities.

For programme staff, annual training in M&E, financial management, procurement and supply was held at each level. The management capacity at each level was enhanced to ensure that the programme was correctly implemented. (Re-) training courses on diagnosis, treatment, vector control, surveillance etc. were provided to a total of 444,941 trainees with support of the NSA. All township microscopists and 98% of the village doctors in programme areas were (re-)trained. A sample survey of 27,499 township and village doctors in 2012 showed that 94.9% of them were satisfied with the training of the programme.

With support of R5, approximately 15 persons gained their Master's degree in public health management and technical specialties, some of which were related to the technology or management of malaria control in China.

6. MAIN PROGRAMME ACTIVITIES AND OUTPUT

Over the 10 years the GFATM malaria programme was implemented, approximately 1.40 million malaria cases (including suspected cases) were treated, 1.10 million patients infected in the prior year were treated, 2.80 million bed-nets were treated with insecticide and 1.80 million LLINs were distributed. Various local IEC/BCC materials were developed and distributed, and health education activities were performed, which improved the residents' awareness of malaria prevention and treatment.

Table 11.2 Number of health products and equipment procured for the malaria control programme in China by the GFATM from 2003 to 2012

Health product and equipment	Equipment
Medicine: 2,399,069 (person-doses)	Cars: 538 (units)
Insecticide: 178,074 (litres)	Motorbikes: 1908 (units)
RDTs: 967,560 (units)	Computers: 1461 (units)
LLINs: 1,823,153 (nets)	Cameras: 1012 (units)
Microscopes: 5516 (units)	Projectors: 802 (units)
Sprayers: 6655 (units)	Small refrigerators: 1600 (units)

RDTs, rapid diagnosis tests; LLINs, long-lasting insecticide-treated nets.

6.1 Case management

For diagnosis, microscopy remained the gold standard method throughout the programme. The diagnostic options increased from only one method (microscopy) at the time of the R1 to three methods including microscopy, RDTs and polymerase chain reaction (PCR), in the NSA period. In 2005–2010, 81 countries used RDTs for malaria diagnosis in project supported by the GFATM, improving the case management of acute febrile illness of children (Zhao et al., 2012). RDTs were the primary diagnostic tool in the remote villages in southern China in 2008–2012. Microscopy and RDTs have lower sensitivities than PCR (Yan et al., 2013). PCR was mainly used for case verification in provincial CDC and NIPD because of the lack of PCR equipment in county CDCs and townships.

The quality assurance (QA) system was updated by establishing a network of reference laboratories, setting up microscopy oversight teams and (re-)training microscopists. Microscopes and related materials were provided to township hospitals and periodic double-checking of malaria diagnosis was provided.

A total of 13 provincial reference laboratories were established, and all 762 counties in 20 provinces and autonomous regions set up microscopy monitoring groups for quality control. RDTs were adopted to detect malaria in Yunnan, Hainan, Guizhou and Xizang provinces or autonomous regions. A total of 12,485,598 fever patients received microscopic tests in the frame of R1, R6 and NSA (Table 11.3).

Before 2010, case detection mainly relied on blood examination for fever patients in hospitals at all levels. Since the implementation of NSA, in addition to blood examination, active case detection (ACD) was performed to find infection sources.

Treatment competence was strengthened through special training and the provision of appropriate anti-malarial drugs. In China, patients with *P. vivax* received chloroquine/primaquine (CQ/PQ) treatment and patients with *P. falciparum* received artemisinin therapy before the implementation of R5. In 2006, ACT for *P. falcipaum* and copackaged CQ/PQ for *P. vivax* treatment were first used in the programme areas of R5. Then, the national malaria treatment policy was revised to use ACT as the first-line therapy for *P. falciparum* case treatment in line with international policy. Artemisinin monotherapy for uncomplicated malaria was no longer to be used, and injectable artesunate was used to treat severe or complicated malaria. In addition, the treatment with primaquine for the radical cure of *P. vivax* cases ('spring treatment') was provided in R1, R5 and NSA during pre-transmission season in March or April.

The follow-up for case treatment was provided by village doctors. The CFDA performed routine anti-malarial drug quality testing and QA at all levels, from manufacturing over distribution to storage. A total of 1,504,613 confirmed and suspected malaria cases were treated between 2003 and 2012 (Table 11.4). Since 2004, each malaria case was reported on time through the national notifiable infectious diseases reporting system network.

6.2 Vector control

LLINs are a key malaria control strategy of relevant programmes worldwide (Ghebreyesus et al., 2008). In the highly endemic areas of Yunnan, Hainan and Guizhou provinces, LLINs were distributed and nets were treated with insecticide (ITNs). Meanwhile, in four special regions of Myanmar, the

Table 11.3 Malaria diagnosis in the GFATM malaria programmes in China

Round	Microscopic test (number of fever patients received)	RDTs	PCR
R1	Case detection and verification (2,530,245)	X	X
R5	Case detection and verification (not available)	*P. falciparum* diagnosis in remote villages of Yunnan and Hainan province	X
R6	Case detection in malaria diagnosis and treatment post (548,872)	Case detection by mobile medical team	X
NSA	Case detection and verification (9,406,481)	• Primary method of diagnosis limited to remote villages mainly in Type 1 counties • Backup in programme counties at the township level • Febrile patients among travelers screening at most international ports of entry (air, sea and land)	Case verification
R10	Case detection and verification (not available)	Case detection by mobile medical team	X

RDTs, rapid diagnosis tests; PCR, polymerase chain reaction.

local residents and mobile population also received LLINs through R6. The distribution of LLINs and ITNs were free of charge, with exception for the LLINs distributed through social marketing in R5 in eight counties of Yunnan province.

LLINs were distributed based on the malaria incidence in the programme areas and the economic conditions of the households. All nets in the programme area were treated once per year immediately before malaria transmission season before 2010. Only nets of villages with an incidence <1% in the previous year were treated with insecticide after 2010, and LLINs were distributed in the villages with an incidence ≥1%; 1.8 million LLINs were distributed and 3.56 million ITNs were distributed (Figure 11.5).

Net ownership can be significantly improved in a short time, but net use is not always proportionally increased (Shargie et al., 2010). The percentage of households at risk receiving at least one LLIN/ITN in the last 12 months was 88.3% in China in 2012, and the fraction of the population at risk who reported sleeping under LLINs/ITNs at the previous night was 65.1%.

Indoor residual spraying (IRS) for vector control was only used in confirmed transmission foci. In 2012, 3941 foci received IRS, and the effectiveness was assessed to ensure that residents received better protection from the malaria infections.

Table 11.4 Malaria treatment in GFATM malaria programmes in China

Round	Number of cases (confirmed and suspected)	Treatment scheme	Spring treatment
R1	517,220	• CQ/PQ treatment for patients with *Plasmodium vivax* malaria • Artemisinin therapy for patients with *Plasmodium falciparum* malaria	√
R5	738,462	• Co-packaged CQ/PQ treatment for patients with *P. vivax* malaria • ACT for patients with *P. falciparum* malaria	√
R6	211,286	The same as R5	X
NSA	37,645	The same as R5	√
R10	No reported	The same as R5	X
Total	1,504,613		

CQ/PQ, chloroquine/primaquine; ACT, artemisinin-based combination therapy; Spring treatment, or radical treatment of *Plasmodium vivax* malaria cases with combined use of chloroquine and primaquine to prevent the relapse of malaria.

6.3 Health education and community mobilization

All provinces and partners performed investigations on health education needs and developed locally appropriate IEC/BCC methodologies and materials in conjunction with the various target groups. For example, in Yunnan province, the materials printed consisted of local minority language calendars, notebooks, picture posters, videos and so on. These were each targeted for the different population needs of students, residents and cross-border mobile workers (Table 11.5).

BCC activities were performed by NGOs, the Ministry of Education (MOE) and CDCs by way of IEC material distribution, television, radio, short messaging service (SMS), websites, health education courses, community activities, face-to-face communication and so on. Approximately 400 million people received anti-malaria health education. Since 2005, the day of April 26th has been defined as a 'Chinese Malaria Day', immediately after the April 25th 'World Malaria Day'. During 'Chinese Malaria Day', many IEC/BCC activities are held all over the country (Table 11.6).

6.4 Malaria control for populations at high risk

In contrast to highly endemic areas of Africa, where malaria mostly affects pregnant women and children, the population at highest risk in China

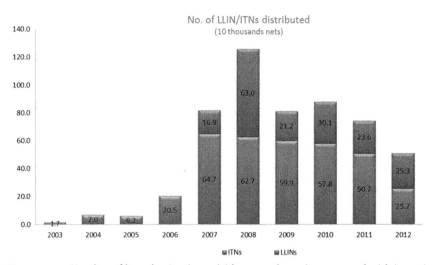

Figure 11.5 *Number of long-lasting insecticide-treated nets/nets treated with insecticide (LLINs/ITNs) distributed by GFATM programmes from 2003 to 2012 in China.*

was exposed to risk factors related to work activities and location of residence. For example, the forest-goers in Hainan province were the high risk population of malaria infections. Some ethic minority people of Yunnan, Hainan and Guizhou provinces as well as international migrants are at high

Table 11.5 IEC/BCC activities in the frame of the GFATM malaria programmes in China

Round	IEC/BCC activities
R1	• 259,640 anti-malaria posters exhibited in public areas • 12,644 anti-malaria IEC courses held for primary and secondary school students • Malaria programmes over radio or TV broadcasted 532 times for target counties
R5	• Village doctor training CDs distributed by NPO • Various locally appropriate IEC/BCC materials targeting different populations distributed such as booklets, aprons, leaflets and fans; flyers on correct use of bed-nets, materials and calendars for international travelers • Participatory community health education activities (e.g. dramas and song contests) • Malaria counseling rooms, bulletin boards and multi-media screens in border ports • TV and radio health education advertisements as well as knowledge lectures • 10,758,415 primary and middle school students received education on malaria control
R6	• 276,617 families reached by face-to-face community activities • 133,455 primary/middle school students and 5025 villagers trained • 2079 village committees for malaria control established • 267,000 calendars and 7000 posters delivered
NSA	• 306.81 million pieces of IEC materials distribution or exhibited • Health education activities for students of 4663 primary and middle schools in seven provinces • Community health education activities in 13 counties of Yunnan and 12 counties of Guizhou • 176,879 families visited face-to-face by a health worker in 3115 natural villages of four counties of Yunnan • 100,000 people received consultation, 1.3 million IEC materials distributed at entry and exit ports
R10	• IEC/BCC materials distributed • IEC/BCC activities conducted at Chinese Malaria Day

IEC, information education communication; BCC, behavioural change communication; NPO, National Programme Office.

Table 11.6 Amount of IEC materials distributed in the frame of the NSA

Category	Name	Number (x 10,000)
Publicity materials	1. Poster (poster, brochure, flyer)	1320
	2. Calendar	32
	3. Microscopy map	4
	4. Promotional display panel	0.1
Household materials	5. Teacup (glass, disposable cup)	645
	6. Shopping bag, apron	139
	7. Anti-mosquito product (i.e. electric mosquito swatter)	1
	8. Clothing (T-shirt, cuff)	19
	9. Umbrella	10
	10. Facial tissue	327
	11. Travel bag	1
	12. Plastic fan	33
	13. Towel	4
	14. Poker	48
	15. Flashlight	4
School materials	16. Exercise notebook (stationery box, visual chart)	336
	17. CD-ROM	6
	18. Health education book	2
Office materials	19. Pens (pen holder, advertising pen)	125
	20. Mouse pad	1
	21. Office notebook	11
Total		3068.1

IEC, information education communication.

risk of malaria infections as well. Meanwhile, the higher malaria incidence across the south border of China affects the malaria transmission patterns in Yunnan province (Chen et al., 2010). The population at highest risk mainly included the mobile workers on the China–Myanmar border of Yunnan province.

The key activities for populations at high risk were to expand and strengthen coverage of integrated malaria services through distributing LLINs or 'malaria packs contained with LLINs, piperaquine and IEC materials' as well as providing IEC/BCC materials in cooperation with partners. Meanwhile, travelers back from Africa and other endemic countries received IEC/BCC materials, and RDT screening was conducted at all ports. On the China–Myanmar border, eight mobile medical teams and 73 malaria consultation posts were established in R6 to provide malaria diagnosis and treatment services to the mobile workers crossing the border. Malaria control activities among mobile workers who involved in road construction or working on plantations and in forest areas were conducted in the six provinces where covered by GFATM malaria programme of R5.

6.5 Malaria surveillance

With the further improvement of national malaria surveillance system the epidemic response capability at the county level was strengthened through training and the provision of equipment, RDTs and insecticide for IRS. Surveillance technical skills trainings were held to improve the quality of case reporting (Zhou, et al., 2013).

An anti-malarial drug resistance surveillance network was set up in R5, R6 and NSA, including *in vivo* and *in vitro* monitoring of ACT efficacy and *in vivo* study of chloroquine efficacy. Anti-malarial drug quality testing in the private and public sector was conducted in Yunnan. Mosquito density surveillance and insecticide resistance surveillance activities were performed in 200 sentinel counties in the frame of the NSA. *In vitro* testing of sensitivity to anti-malarial drugs was performed on the China–Myanmar border with support of the GFATM because *P. falciparum* have high resistance to chloroquine in that area (Zhang et al., 2012), as well as artemisinin resistance for *P. falciparum* infection has been suspected in the areas on the China–Myanmar border (Huang et al., 2012).

Since 2010, every reported case was followed up, resulting in 226,254 ACD activities involving 1,933,444 people surveyed which contributed to the clearance of infection sources timely.

6.6 Multi-sector cooperation

Through the programme, multi-sector cooperation on malaria control and elimination was established and continued to be improved. Before 2003, no NGOs took part in malaria control, but with the launch GFATM malaria programmes in China, some NGOs such as Health Unlimited/ Health Poverty Action (HU/HPA), Humana People to People (HPP), and Population Service International (PSI) conducted many malaria control activities as partners. At first, they only took part in temporary activities such as monitoring and providing TA in R1. Six partners conducted activities as SRs, such as health education and promotion, LLIN distribution, anti-malarial quality control and malaria screening to the mobile population at border ports in R5. In R6, 70% of the funding was used for activities performed by two NGOs. Especially in the Myanmar border areas, malaria control was mainly performed by HU/HPA. For the NSA, these partners continued to implement malaria elimination activities. Since R1, WHO provided TA to the Chinese malaria control and elimination efforts and the China GFATM malaria programme. Meanwhile, especially in elimination areas, the private sector was engaged in malaria diagnosis, treatment and reporting, etc. (Table 11.7).

6.7 Operational research

The operational research of GFATM grants focused on the scientific problems of malaria control and elimination in China. NPO developed operational research bidding guidance. Through public bidding, operational research proposals were selected and contracted. From 2007 to 2012, more than 40 operational researches were supported by the GFATM malaria programme in China. On the basis of the reviewing comments of malaria experts from universities and research institutes in China, a total of 28 studies of the NSA were granted in the seven fields, namely genetic screening (seven studies), vector control (seven studies), malaria diagnosis (six studies), surveillance (three studies), imported malaria (three studies), treatment (one study) and economic evaluation (one study). Some research results have been used for fighting malaria and providing technical support to malaria control and elimination activities (e.g. the *P. vivax* sporozite-carrying mosquitoes detected through loop-mediate isothermal amplification).

7. ACHIEVEMENTS AND IMPACTS

In China, great achievements on malaria control has been gained, for example, the malaria incidence decreased from 4 per 10,000 to below 1 per 10,000 over the past 10 years. As the biggest international cooperation programme focusing on malaria in China, the GFATM malaria programmes was instrumental in this success, which has produced tremendous impacts in the following six fields.

7.1 From malaria control to elimination in China

The GFATM malaria programmes including R1, R5 and R6 contributed to reduction of malaria transmission in these years. With the support of R1, the malaria burden in Yunnan and Hainan, the two provinces most severely affected by the disease, significantly reduced. The number of reported malaria cases fell from 6357 to 1844 in Hainan and from 13,816 to 4027 in Yunnan in 2003–2008. As late as 2004–2005, the

Table 11.7 The multidisciplinary partners involved in the GFATM malaria programme in China

Round	HU(HPA)	HPP	PSI	RCSC	MOE	CFDA	CIQ	WHO
R1	M&E	X	X	M&E	X	x	x	TA
R5	Community health education	X	LLINs social marketing	LLINs distribution	Student health education	QA of anti-malarials	Malaria screening in ports	TA
R6	Malaria control in Myanmar	Face to face health education	X	X	X	x	x	TA
NSA	Community health education	Face to face health education	X	LLINs distribution	Student health education	QA of anti-malarials	Malaria screening in ports	TA
R10	Malaria control in Myanmar	X	X	X	X	x	Malaria screening in ports	TA

HU/HPA, Health Unlimited/Health Poverty Action; HPP, Humana People to People; PSI, Population Service International; RCSC, Red Cross Society of China; MOE, Ministry of Education; CFDA, China Food and Drug Administration; CIQ, China Entry-Exit Inspection and Quarantine; LLINs, long-lasting insecticide-treated nets; NSA, National Strategy Application; WHO, World Health Organization; QA, quality assurance; TA, technical assistance; M&E, monitoring and evaluation.

malaria incidence was increasing in central China; with the support of R5, the 26,873 cases reported in 2008 were concentrated in a few provinces, and *P. falciparum* was confined to Yunnan and Hainan provinces. While the NSA began to be implemented in 2010 when the Action Plan of the National Malaria Elimination (2010–2020) was issued, resulting in the number of malaria cases continue to decrease sharply from 7855 in 2010 to 2718 in 2012. Under the joint efforts of Global Fund and China government, malaria incidence (1/100,000) decreased from 3.91 in 2003 to 0.2 in 2012 (Figure 11.6).

Much lower transmission of malaria has been observed in 2011, the annual reported malaria incidence in 88% of the 75 Type 1 counties was less than 1 per 10,000, and 21.33% of Type 1 counties and 86.17% of the 687 Type 2 counties reported zero locally transmitted malaria cases.

Over the 10-year period, most notably in the frame of R1, R5 and NSA, *P. falciparum* malaria was eliminated in Hainan province. The incidence in target counties of Hainan was 0.4 per 10,000 in 2007, and in 2011 there were no locally transmitted *P. falciparum* malaria cases. A case in point is Wanning city, which previously had been a highly endemic area and where transmission has dropped to zero since 2010 (Lin et al., 2013). In Yunnan, the *P. falciparum* malaria incidence dropped significantly year by year and was close to 0.12 per 10,000 in 2011.

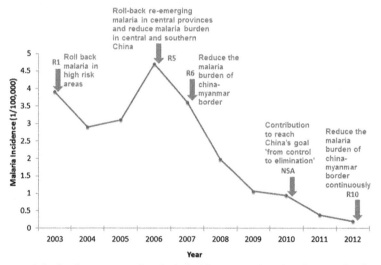

Figure 11.6 *Reduction pattern of malaria incidence correlated to the rounds of GFATM malaria programmes in 2003 - 2012.*

In central China, the re-emergence of malaria was rolled back with the support of R5. From 2000 to 2006, a steady increase in malaria cases resulting from the re-emergence of *P. vivax* malaria in Henan, Hubei, Jiangsu and especially Anhui province had been recorded. In 2009, the outbreak in Anhui province had largely been contained, and the incidence in only 10 counties more than 1 per 1000. In Henan province, the incidence of malaria has decreased from 0.375 per 1000 in 2006 to 0.047 per 1000 in 2010 (Liu et al., 2011). With the implementation of the GFATM grants in Suining county of Jiangsu province, the number of malaria cases decreased from 37 cases in 2006 to 1 case in 2012, and the incidence decreased from 0.28 per 10,000 to 0.01 per 10,000, with the decline rate of 96.43% (Tang, 2013).

In the China–Myanmar border area, R6 and R10 were implemented from 2007 to 2012. The malaria burden of FMCWs and local residents in four special regions of Myanmar was significantly reduced, for example, the prevalence of malaria decreased by 94% among local residents of Myanmar last 5 years based on the surveillance data of HPA.

7.2 Filling the resource gap for malaria control and elimination

The malaria control and elimination funds of the governments at different levels mainly covered human resources, logistics, overheads, special malaria control and elimination funding. The programme areas were mostly poor and rural, with support focused on weak links in malaria elimination such as diagnosis, treatment, monitoring and awareness for malaria prevention (Bi & Tong, 2014). The existing funding gap was approximately 20% of the total malaria control and elimination budget in China, and it was promptly filled by the GFATM. It is important to note that the GFATM successfully leveraged the government at each level to increase malaria control funds. For example, from 2003 to 2012, the central government invested approximately 193 million RMB in malaria control and elimination, increasing from 5 million RMB in 2003 to 46 million RMB in 2012.

The health products and materials distributed by the GFATM greatly eased the lack of resources for malaria control and elimination in China. Health products such as microscopes, insecticide and RDTs filled the gap in malaria diagnosis and treatment in township hospitals, and LLINs boosted prevention of transmission. The procurement and distribution of materials such as vehicles and office equipment also provided necessary help for the programme implementation and malaria elimination.

7.3 Improved multi-sectorial cooperation and communication

The multi-sectorial cooperation and communication greatly benefited malaria control in China. Many government sectors and NGOs played important roles in malaria control. In the past, no NGOs had been engaged in malaria control in China. In addition to CDCs, NGOs (HU/HPA, HPP and PSI), other government sectors (MOE, Food and Drug Administration and China Entry-Exit Inspection and Quarantine) and the Red Cross were involved in malaria control and elimination. Coordination meetings between CDCs and these partners were held regularly, and departments of tourism and commerce were also invited to attend these meetings. The number of provinces with multi-sectorial cooperation increased from 6 to 20 nationwide over the programme period. Together, these partners received 10% of the total funds from the GFATM malaria programmes in China.

7.4 Contribution to policy and ability of the national malaria control and elimination

Malaria control and elimination policy in the fields of diagnosis, treatment and vector control were revised over the course of implementation of the GFATM malaria programmes. As new malaria control techniques such as RDTs, LLINs and ACT were introduced to China, these new techniques were integrated into the national malaria control policy and spread to the entire country.

In terms of technical trainings, all microscopists, 98.02% of the village doctors and 63.41% of the medical workers in township hospitals were (re-)trained. Nearly half of the microscopy staff working with grass-root organizations was trained for several times. Thus, the programme improved the comprehensive control and prevention abilities at all levels. The test scores related to blood slide preparation and reading were significantly higher in GFATM-supported provinces than those not covered by the programme (Fu et al., 2012).

The programme staff benefited from its exposure to the management concepts of the GFATM and improved its ability to manage programmes better, such as handling of M&E, procurement, financing and audit (Li et al., 2009).

7.5 Enforced public awareness of malaria control and prevention

Health education materials targeting different groups were developed, and various health education and improvement activities were performed to improve the awareness of malaria control and prevention in the trargted areas. For example, more than 60% of the students were aware of malaria among 44,519 students surveyed in 94 counties of 20 provinces in 2010 (Yin et al., 2013). After implementation of a GFATM grant in the four counties of Chongqing City, the malaria awareness rate of primary school students increased from 58.94% to 89.96%, that of middle school students from 52.83% to 86.06% and that of local residents from 56.74% to 83.89% (Wu et al., 2013). With constantly improving awareness of malaria control and prevention, the residents could go to the hospital immediately upon getting infected and actively took part in malaria control and prevention activities.

7.6 Cross-border cooperation mechanism for malaria control

A cross-border cooperation mechanism with NGOs was established through R6 and R10. On the basis of the platform of R6, information about malaria epidemic was exchanged regularly between border counties of Yunnan in China and the special regions of Myanmar. TA in the fields of malaria diagnosis and treatment was provided to the special regions of Myanmar in the form of human resources, training and M&E. A malaria control network was established and maintained in the four special regions in Myanmar with the help of R6 and R10. Sixty-eight malaria control consultation and service posts in Yunnan province and 73 health posts in Myanmar were established on the border. These health posts were equipped with microscopes and anti-malaria drugs. Seventy percent of the funds were used by the NGOs for malaria diagnosis and treatment and health education in Myanmar. Meanwhile, there was strengthened mobilization and coordination between local health bureaus of special regions in Myanmar and those of the border counties in Yunnan province, which improved the management and technical capacity of the officers from both countries.

8. LOOKING FORWARD

From 2014 onward, the GFATM will no longer support malaria control or elimination activities in China. However, it remains challenges, especially in rural areas of the country, such as the lack of funding for vector control, diagnostic testing and treatment. It has been argued that in addition to GFATM investments, international and domestic funds should be provided if malaria elimination targets are to be reached (Katz et al., 2011). The scale-up in funding should be sustainable, and governments should invest resources to support robust, systematic and regular malaria control activities (Nahlen and Low-Beer, 2007). The central and local governments of China took timely measures to increase funding to fill the financing gap created by the termination of the GFATM in China.

At present, the incidence of malaria is at a very low level in most areas except on the China–Myanmar border of Yunnan province. In this area, malaria control and elimination remains a challenge, not least because of the difficulty of controlling a disease in the mobile population. Therefore, cross-border cooperation mechanisms among governments of China and Myanmar, together with with the NGOs, need to be enhanced, and more funds should be invested for malaria control in this area (Tambo et al., 2014).

China has acquired solid malaria control and elimination experience in the fields of diagnosis, treatment, vector control, surveillance and capacity building (Liu, 2014). In addition to its domestic work, China in able to provide train facilities, including control experience and expertise, for Africa and the Greater Mekong sub-region on malaria control and elimination. In addition, Chinese malaria experts are able to take the responsibility in implementing malaria control and elimination in Africa under multi-lateral or bi-lateral cooperation mechanisms. Meanwhile, institutions in China need to further strengthen their cooperation with international agencies and foundations.

9. CONCLUSIONS

The GFATM malaria programme in China was the biggest international cooperation project in malaria control in the country. Over a decade, the GFATM programme effectively promoted the implementation of the NMCP (2006–2015) and the NMEP (2010–2020) (Zheng, et al., 2013). By the end of 2012, the number of reported malaria cases had dropped below 3,000 annually, a record that is the lowest number in the Chinese history.

On the basis of epidemiological data, project activities and performance indicators, the impacts of the GFATM project have been summarised in six fileds. The experiences in different fields, including introduction of technology and management, best practice at community level, innovation in implementation, introduction of international standards and evaluation of the performance of multi-stakeholder cooperation mechanisms, were reviewed.

All rounds of the GFATM malaria programme, including R1, R5, R6 and NSA, had already ended successfully before 2013. After R10 closed in 2014, the GFATM completely stopped providing financial support for China malaria elimination. The efforts in the post-GFATM era were proposed, such as achievements made over the past decade have to be maintained, and the NMEP (2010–2020) should be implemented as planned. It was recommended that the country is looking forward to continuing to cooperate with the GFATM on malaria control and elimination in both African countries and border countries, such as Myanmar.

ACKNOWLEDGEMENTS

This work was supported by China Global Fund Malaria Programme R1 (Grant No. CHN-102-G02-M-00), R5 (Grant No.CHN-506-G07-M), R6 (Grant No. CHN-607-G09-M), NSA (Grant No. CHN-S10-G13-M), R10 (Grant No.CHN-011-G15-M) and by the China U.K. Global Health Support Programme (Grant no. GHSP-CS-OP1).

REFERENCES

Bi,Y., Tong, S., 2014. Poverty and malaria in the Yunnan province, China Infect. Dis. Poverty 3, 32.

Chen, G.W., Wang, J., Huang, X.Z., Li, Y.P., Hou, Z.S., Li, H.X., Xu, S.Y., Wei, C., Zhang, Z.X., 2010. Serological detection of malaria for people entering China from 19 ports of entry covering 8 border prefectures of Yunnan. Chin. J. Parasitol. Parasit. Dis. 28, 54–57 (in Chinese).

China PR, 2013. http://www.chinaglobalfund.org/en/index.html (assessed on 25.02.14.).

Fu, Q., Li, S.Z., Wang, Q., Zhang, L., Liu, W., Zheng, X., Zhang, S.S., Xia, Z.G., Zhou, S.S., Chen, Z., Wang, L.Y., Zhou, X.N., 2012. Report of analysis of national technique competition for diagnosis of parasitic diseases in 2011 II Analysis of capabilities of plasmodium detection. Chin. J. Schisto. Control 24, 274–278 (in Chinese).

Ghebreyesus, T.A., Lynch, M.C., Coll-Seck, A.W., 2008. The Global Malaria Action Plan for a Malaria-Free World. Roll Back Malaria Partnership, Switzerland.

Global Fund, 2002. History of the Global Fund. The Global Fund to fight AIDS, Geneva. Tuberculosis and Malaria. http://www.theglobalfund.org/en/about/history/.

Huang, F., Tang, L.H., Yang, H.L., Zhou, S.S., Sun, X.D., Liu, H., 2012. Therapeutic efficacy of artesunate in the treatment of uncomplicated *Plasmodium falciparum* malaria and anti-malarial, drug-resistance marker polymorphisms in populations near the China-Myanmar border. Malar. J. 11, 278.

Katz, I., Komatsu, R., Low-Beer, D., Atun, R., 2011. Scaling up towards international targets for AIDS, tuberculosis, and malaria: contribution of global fund-supported programs in 2011–2015. PLoS One 6, e17166.

Li, L., Mao, S.L., Xiao, N., Xu, G.J., Lei, Y., Zhang, R.J., Yan, D.P., 2009. Discussion on pattern of global fund malaria control project in Sichuan province. J. Prev. Med. Inf. 25, 993–996 (in Chinese).

Lin, M.H., Wen, L., Weng, S.W., Li, C.Y., Tao, Z., Zhu, D.C., Zeng, W., Huang, S.L., Zhang, L.X., Chen, X., 2013. Effect in implementation of Global Fund Malaria Project in previously high malaria-endemic area of Wanning City. China Trop. Med. 13, 4.

Liu, D.Q., 2014. Surveillance of antimalarial drug resistance in China in the 1980s-1990s. Infect. Dis. Poverty 3, 8.

Liu, Y., Guo, X.F., Zhang, H.W., Su, Y.P., Chen, J.S., 2011. Evaluation of the fifth round of projects supported by the global fund to fight HIV/AIDS, tuberculosis, and malaria in Henan province. J. Pathog. Biol. 6, 754–756.

Nahlen, B.L., Low-Beer, D., 2007. Building to collective impact: the Global Fund support for measuring reduction in the burden of malaria. Am. J. Trop. Med. Hyg. 321–327 6 Supply.

Shargie, E.B., Ngondi, J., Graves, P.M., Getachew, A., Hwang, J., Gebre, T., Mosher, A.W., Ceccato, Endeshaw, T., Jima, D., Tadesse, Z., Tenaw, E., Reithinger, R., Emerson, P.M., Richards, F.O., Ghebreyesus, T.A., 2010. Rapid increase in ownership and use of long-lasting insecticidal nets and decrease in prevalence of malaria in three regional States of Ethiopia (2006–2007). J. Trop. Med. http://dx.doi.org/10.1155/2010/750978.

Sheng, H.F., et al., 2003. Malaria situation in the People's Republic of China in 2002. Chin. J. Parasitol. Parasit. Dis. 21, 193–196 (in Chinese).

Tang, L.H., 2009. China global Fund Malaria Program: Malaria Control in High Transmission Area. Jiuzhou press, Beijing (in Chinese).

Tang, Y.E., 2013. Effect of execution of global fund malaria project in Suining county. Chin. J. Schisto. Control 25, 633–635 (in Chinese).

Wu, C.G., Luo, F., Jiang, S.G., Luo, X.J., Xie, J., Li, S.S., 2013. Effect evaluation of the awareness of knowledge about malaria: the global fund malaria project in Chongqing. Chin. J. Evid-based Med. 13, 1409–1412 (in Chinese).

Wafula, F., Agweyu, A., Macintyre, K., 2013. Regional and temporal trends in malaria commodity costs: an analysis of Global Fund data for 79 countries. Malar. J. 12, 466.

WHO, 2007. WHO Pesticide Evaluation Scheme (WHOPES). http://www.who.int/whopes/en/.

Yan, J., Li, N., Wei, X., Li, P., Zhao, Z., Wang, L., Li, S., Li, X., Wang, Y., Li, S., Yang, Z., Zheng, B., Zhou, G., Yan, G., Cui, L., Cao, Y., Fan, Q., 2013. Performance of two rapid diagnostic tests for malaria diagnosis at the China-Myanmar border area. Malar. J. 12, 73.

Yin, J.H., Wang, R.B., Xia, Z.G., Zhou, S.S., Zhou, X.N., Zhang, Q.F., Feng, X.Y., 2013. Students' awareness of malaria at the beginning of national malaria elimination programme in China. Malar. J. 12, 237.

Zhang, C.L., Zhou, H.N., Wang, J., Liu, H., 2012. In vitro sensitivity of *Plasmodium falciparum* isolates from China-Myanmar border region to chloroquine, piperaquine and pyronaridine. Chin. J. Parasitol. Parasit. Dis. 30, 41–44 (in Chinese).

Zhao, J.K., Lama, M., Korenromp, E., Aylward, P., Shargie, E., Filler, S., Komatsu, R., Atun, R., 2012. Adoption of rapid diagnostic tests for the diagnosis of malaria, a preliminary analysis of the Global Fund program data, 2005 to 2010. PLoS One 7, e43549.

Zhao, J.K., Lama, M., Sarkar, S., Atun, R., 2011. Indicators measuring the performance of malaria programs supported by the global fund in Asia, progress and the way forward. PLoS One 6, e28932.

Zheng, Q., Vanderslott, S., Jiang, B., Xu, L.L., Liu, C.S., Huo, L.L., Duan, L.P., Wu, N.B., Li, S.Z., Xia, Z.G., et al., 2013. Research gaps for three main tropical diseases in the People's Republic of China. Infect. Dis. Poverty. 2, 15.

Zhou, S.S., Tang, L.H., Sheng, H.F., 2005. Malaria situation in the People's Republic of China in 2003. Chin. J. Parasitol. Parasit. Dis. 23, 385–387.

Zhou, X.N., Bergquist, R., Tanner, M., 2013. Elimination of tropical disease through surveillance and response. Infect. Dis. Poverty. 2, 1.

CHAPTER TWELVE

China–Africa Cooperation Initiatives in Malaria Control and Elimination

Zhi-Gui Xia[1], Ru-Bo Wang[1], Duo-Quan Wang[1], Jun Feng[1], Qi Zheng[1], Chang-Sheng Deng[2], Salim Abdulla[3], Ya-Yi Guan[1], Wei Ding[1], Jia-Wen Yao[1], Ying-Jun Qian[1], Andrea Bosman[4], Robert David Newman[4], Tambo Ernest[5], Michael O'leary[6], Ning Xiao[1,*]

[1]National Institute of Parasitic Diseases, Chinese Center for Disease Control and Prevention; Key Laboratory of Parasite and Vector Biology, MOH; WHO Collaborating Centre for Malaria, Schistosomiasis and Filariasis; Shanghai, People's Republic of China
[2]Guangzhou University of Traditional Chinese Medicine, Guangdong, People's Republic of China
[3]Ifakara Health Institute, Dar es Salaam, Tanzania
[4]Global Malaria Programme, World Health Organization, Geneva, Switzerland
[5]Centre for Sustainable Malaria Control, Faculty of Natural and Environmental Science; Center for Sustainable Malaria Control, Biochemistry Department, Faculty of Natural and Agricultural Sciences, University of Pretoria, Pretoria, South Africa
[6]World Health Organization, China Representative Office, Beijing, People's Republic of China
*Corresponding author: E-mail: ningxiao116@126.com

Contents

1. Background	320
2. Existing China–Africa Collaboration on Malaria Control	322
3. The Challenges and Needs for Malaria Control in Africa	323
3.1 Epidemiology of malaria	323
3.2 Malaria control and elimination	325
3.3 Malaria programmes	325
3.4 Challenges and gaps	326
4. Potential Opportunity and Contribution to Enhance the Partnership	328
4.1 Quality control in diagnosis of malaria cases (test)	329
4.2 Drug delivery system (treat)	330
4.3 Malaria information reporting system (track)	330
5. Collaborative Research Scopes	332
5.1 Effectiveness evaluation on comprehensive control strategy	332
5.2 Early warning system for disease surveillance	332
5.3 Resistance surveillance	332
5.4 Transition of traditional medicines	332
6. The Way Forward	333
Annex 1: Antimalaria Centers in Africa Established by Chinese Government in 2007–2009	334
Acknowledgements	335
References	335

Advances in Parasitology, Volume 86
ISSN 0065-308X
http://dx.doi.org/10.1016/B978-0-12-800869-0.00012-3

Copyright © 2014 Elsevier Ltd.
All rights reserved.

Abstract

Malaria has affected human health globally with a significant burden of disease, and also has impeded social and economic development in the areas where it is present. In Africa, many countries have faced serious challenges in controlling malaria, in part due to major limitations in public health systems and primary health care infrastructure. Although China is a developing country, a set of control strategies and measures in different local settings have been implemented successfully by the National Malaria Control Programme over the last 60 years, with a low cost of investment. It is expected that Chinese experience may benefit malaria control in Africa. This review will address the importance and possibility of China–Africa collaboration in control of malaria in targeted African countries, as well as how to proceed toward the goal of elimination where this is technically feasible.

1. BACKGROUND

Malaria, one of the most important human parasitic diseases globally, has a significant impact on not only the health of affected populations, but also the social and economic development of the areas where it is present (Guinovart et al., 2006). In Africa, many countries have faced severe challenges in controlling malaria, in part due to major limitations in public health systems and poor infrastructure. According to the *World Malaria Report 2013* published by the World Health Organization (WHO), in 2012, there were an estimated 207 million cases of malaria (uncertainty interval 135–287 million), which caused approximately 627,000 deaths (uncertainty interval 473,000–789,000). An estimated 3.4 billion people continue to be at risk of malaria, mostly in Africa and southeast Asia. Approximately 80% of malaria cases and 90% of malaria deaths occurred in Africa (WHO, 2013).

Malaria once affected the social, economic and political development in the People's Republic of China (P.R. China). According to the statistics, more than 30 million malaria cases were recorded annually during the 1940s and the mortality rate was about 1%. After the establishment of P.R. China in 1949, the government invested significant resources in the control of malaria which was listed as one of the five top parasitic diseases in the country. The reduction of malaria leading to its elimination in P.R. China, the evidence-based prevention strategies, and related malaria prevention and control policies have been synthesised for the National Malaria Control Programmes (NMCPs). As a result, a network of scientific research and surveillance has been established, thousands of professional technical teams have been trained, and the coordination of malaria control within specific zones has been created for the past six decades. Through these interventions, a remarkable achievement of the NMCPs has taken place since the early 1950s. From the 1970s to the 1990s, the annual number of malaria cases significantly declined from 2.4 million to tens of thousands, the endemic regions were sharply shrunk, and falciparum malaria has been eliminated in all regions except for Yunnan and Hainan provinces. By 2009, the national number of malaria cases had been reduced to 14,000. Among 24 malaria-endemic provinces, the malaria incidence in 95% of the counties has dropped below 1 per 10,000 population with only 87 counties over 1 per 10,000. The data indicate that P.R. China has transitioned from malaria control phase to elimination phase (Tang, 1999; Tang, 2009).

To effectively improve public health, promote economic development, achieve the health-related Millennium Development Goals (MDG), and contribute to the ultimate global goal of malaria eradication (Butler, 2012; Bhutta et al., 2014), the Chinese government embarked upon the National Malaria Elimination Action Plan in 2010, with a goal of eliminating malaria by 2015 in a majority of regions with the exception of the border region in Yunnan province, and to completely eliminate malaria from P.R. China by 2020 (National Health and Family Planning Commission of P.R. China, 2009). In the implementation of this action plan from 2010 to 2012, the total number of malaria cases was reduced from about 8,000 to about 2,700. The number of local cases has been reduced significantly to less than 200 annually in the country, and locally transmitted falciparum malaria has not been reported in Hainan Province since 2010 (Yin et al., 2013).

A set of control strategies and measures for NMCPs in different local settings has been implemented successfully over the last 60 years in P.R. China. This experience has included collaboration between China and Africa in the field of malaria control, in the larger context of China's contribution to development in Africa as a major trading and investment partner. In July 2012, the WHO Director-General Dr. Margaret Chan emphasised the importance of health cooperation between Africa and China during her visit to P.R. China, and the Chinese leader of the central government gave a positive response. In early 2013, the government of P.R. China has promised to continue increasing its investment in Africa, which includes medical assistance as well as promotion of sustainable development (Wang, 2013). Recently, the National Health and Family Planning Commission (previously Ministry of Health) and the Ministry of Commerce of P.R. China have proposed to provide technical support for malaria control and elimination in Africa, which would further promote South–South cooperation through mutual exchanges in the field of malaria control and elimination.

The cooperation between China and Africa will not only strengthen China's capacity in engagement in global health, but will also provide a platform to share health products (e.g. quality control in production and delivery of antimalarial drugs), techniques (e.g. diagnostics) and intervention strategies. Based on multilateral communication, P.R. China intends to develop malaria prevention and control projects in several African countries together with international experts working in Africa. As one component of China–Africa health collaboration, this review will address the importance and possibility to strengthen the control of malaria in targeted Africa countries, as well as to proceed toward the goal of elimination where technically feasible.

2. EXISTING CHINA–AFRICA COLLABORATION ON MALARIA CONTROL

Fifty years ago, multilateral and bilateral medical collaboration between China and Africa was limited. However, over the years, an amicable relationship has developed between China and countries in Africa in the areas of medicine and health. During the 50 years since the government sent the first medical-aid team to Africa in 1963, P.R. China has sent about 20,000 medical doctors to 51 African countries, which has sustained a longstanding friendship with Africa (Zuo, 2013). China–Africa medical and health cooperation including in the area of malaria has grown through sending medical teams, training programmes, donations of medicines and medical equipment, joint research and academic exchanges.

In the competitive setting of global pharmaceutical production and distribution, some of China's enterprises have been successful in the process of pharmaceutical sale registration in dozens of countries in Africa, which provides an opportunity for Chinese-produced medicines to be used in the African market. The donation of antimalarial medicines, diagnostic kits and equipment, and the provision of experts and on-site training have demonstrated the increasing commitment on the part of the Chinese government to the malaria control and elimination programme in Africa. An outstanding example was Chinese government support in building antimalaria centres in 30 African countries and donating antimalarial medicines through bilateral cooperation since 2006 (Annex 1). Antimalaria centres have played not only a role in diagnosis and treatment, but also have contributed to communication and capacity building, including improvement of capacity in local medical research and treatment.

Experiences with the malaria control centres are variable, e.g. some have been successful, while others are faced with the challenges of financing and the lack of local contribution. Another successful example was in Comoros, where a pilot malaria elimination programme had achieved a significant reduction in malaria transmission with the support of China (Chinese Medical Cooperation in Africa, 2011).

The P.R. China has also provided academic exchanges and training programmes for African officials and technical personnel each year during the last 10 years. The National Institute of Parasitic Diseases (NIPD) of Chinese Center for Disease Control and Prevention (CDC) launched five training workshops over the last three years such as 'Infectious diseases prevention and control' and 'Prevention and control of malaria and schistosomiasis in developing counties'. More than 150 technical staff members and officials from more than 20 African countries have been trained in P.R. China in strategies and measures for malaria prevention and control.

Recently, more institutions in P.R. China have strengthened cooperation with various international agencies and foundations (e.g. the Bill and Melinda Gates Foundation; WHO; the Global Fund to Fight AIDS, Tuberculosis and Malaria (GFATM); Canada's International Development Research Centre; the Department for International Development of the United Kingdom) under multilateral or bilateral mechanisms, which have provided more opportunities to explore new ways of control and prevention of malaria. Building on these experiences and the renewed interest provided by the Forum for China–Africa Cooperation, there are further opportunities to strengthen the China–Africa cooperation in malaria control and elimination through multilateral partnerships, supported by national and international agencies.

3. THE CHALLENGES AND NEEDS FOR MALARIA CONTROL IN AFRICA

3.1 Epidemiology of malaria

According to the *World Malaria Report 2013* (WHO, 2013), approximately 80% of malaria deaths in 2012 were estimated to occur in just 17 countries, about 80% of cases in 18 countries, and most of them in Africa. The Democratic Republic of the Congo and Nigeria together accounted for 40% of the global total of estimated malaria deaths and 32% of cases. International targets for reducing cases and deaths will not be attained unless considerable progress can be made in these countries.

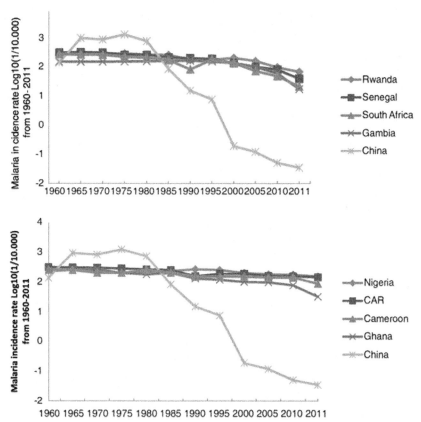

Figure 12.1 Trend in malaria incidence rate Log 10 (1/10,000) in selected African countries and P.R. China from 1960 to 2011. (1) Substantial scaling-up impact on malaria incidence in P.R. China and some African countries. (2) Scaling-up impact on malaria incidence in P.R. China and low/moderate outcomes in some African countries (Tambo et al., 2012).

When analysing the trends in malaria events from 1960 to 2011, highlighting the achievements, progress and challenges in research for moving malaria from epidemic status towards elimination in both sub-Saharan Africa and P.R. China, it was found that the gap in malaria incidence between selected African countries and China will widen in the absence of enhanced intervention in Africa (Figure 12.1). Political commitment and financial investment of stakeholders in sustaining the scaling-up impact of malaria control interventions, and networking between African and Chinese scientists and their Western partners, are urgently needed to uphold the recent gains, and to translate lessons learnt from the Chinese malaria control achievements and successes into practical interventions in malaria endemic countries in Africa and elsewhere (Tambo et al., 2012).

3.2 Malaria control and elimination

According to the *World Malaria Report 2013* (WHO, 2013), of 43 countries in the African Region with ongoing malaria transmission, eight countries (Botswana, Cabo Verde, Eritrea, Namibia, Rwanda, Sao Tome and Principe, South Africa and Swaziland) and the island of Zanzibar (United Republic of Tanzania) are on track to achieve reductions in reported malaria case incidence or malaria admission rates of 75% or more. A further two countries (Ethiopia and Zambia) are projected to achieve reductions in malaria admission rates of 50–75% by 2015, and one country (Madagascar) by <50%. An increase in locally acquired cases, from 35 in 2000 to 59 in 2012, was reported from Algeria. Cabo Verde is in the pre-elimination phase, and continues to progress towards eliminating malaria. Algeria is in the elimination phase. Both Algeria and Cabo Verde implement active case detection, case investigation and a quality assurance (QA) system for diagnostic testing guided by a national reference laboratory; they also provide treatment with primaquine for the radical cure of *P. vivax* and clearance of gametocytes in *P. falciparum* infections. Eight countries in southern Africa are signatories to the elimination eight (E8) regional initiative launched in March 2009, a goal of which is to achieve the eventual elimination of malaria in the region, with elimination in four countries (Botswana, Namibia, South Africa and Swaziland) by 2015. With continued investments in malaria control, it is expected that these countries will continue to progress towards elimination, although they do not yet meet the case management and surveillance criteria to be classified as being in the pre-elimination phase.

3.3 Malaria programmes

Many Africa countries have developed strategic plans for malaria control and elimination, and WHO has developed technical guidance and support tools for planning and programme reviews. NMCPs in Africa have received major support from the international community over the past decade, primarily from the GFATM, the U.S. President's Malaria Initiative, US Agency for International Development, the United Kingdom of Great Britain and Northern Ireland's Department for International Development and the World Bank, allowing for a rapid scaling-up of lifesaving malaria prevention and control interventions. This core support has been supplemented with additional assistance from a number of foundations, as well as bilateral and multilateral partners (WHO, 2013).

To support countries in their efforts to achieve the health-related MDG indecies and other global targets as the 2015 MDG deadline nears, WHO in 2012, along with the Roll Back Malaria (RBM) and other partners, launched a malaria situation room (MSR) to provide focused strategic support to 10 high-burden countries in sub-Saharan Africa, namely Burkina Faso, Cameroon, Côte d'Ivoire, the Democratic Republic of the Congo, Ghana, Mozambique, Niger, Nigeria, Uganda and the United Republic of Tanzania. The MSR experts collated and synthesised malaria-related information on financial flows, commodities, intervention coverage and disease trends; track challenges and progress; and identify bottlenecks that hinder country scale-up of malaria control interventions (Zofou, et al., 2014). Relevant partners are then approached to help resolve the problems identified, and progress in bottleneck resolution is monitored (WHO, 2013).

3.4 Challenges and gaps

There are several reasons why Africa bears an overwhelming proportion of the malaria burden. Most malaria infections in Africa south of the Sahara are caused by *P. falciparum*, the most severe and life-threatening form of the disease. This region is also home to the most efficient, and therefore deadly, species of mosquitoes that transmit the disease. Moreover, in many African countries, especially in rural areas, poor infrastructure, substandard sanitation, and lack of funding for vector control, diagnostic testing and treatment remain challenges (Agomo et al., 2013; Butler, 2012). Many countries lacked the infrastructure and resources necessary to mount sustainable campaigns against malaria; as a result, few benefited from historical efforts to eradicate malaria. In Africa, malaria is understood to be both a disease of poverty and a cause of poverty (Ajayi, et al., 2013; Bi & Tong, 2014). Annual economic growth in countries with high malaria transmission has historically been lower than in countries without malaria. Malaria also has a direct impact on Africa's human resources. Not only does malaria result in lost life and lost productivity due to illness and premature death, but malaria also hampers children's schooling and social development through both absenteeism and permanent neurological and other damage associated with severe episodes of the disease (Malaria in Africa; Chima et al., 2003).

The current funding gap is affecting campaigns for long-lasting insecticidal nets (LLIN), and there is a risk of malaria resurgence in several Africa countries; moreover, there are major needs to improve surveillance as well as access to diagnostic tests and treatment which are influenced by certain factors (Aidoo, 2013; Hill et al., 2013; Noor et al., 2007; Pell et al., 2011). According to the *World Malaria Report 2013* (WHO, 2013), an assessment of trends was not possible in 32 countries in Africa owing to insufficiently consistent data, and the scale-up of diagnostic testing through rapid diagnostic tests (RDTs) and microscopy remains incomplete in the public sector and to an even greater extent in the private sector. In 17 countries of Western Africa, the proportion of the population with access to an insecticide-treated mosquito net (ITN) within their household was estimated to exceed 50% in 10 countries, the use of indoor residual spraying (IRS) has increased but coverage remains relatively low, and only seven countries reported delivering sufficient antimalarial medicines to treat all patients attending public health facilities. In 10 countries of Central Africa, the recent increase in malaria cases and admissions in Sao Tome and Principe may be related to brief disruptions to spraying activities and supply of artemisinin-based combination therapy (ACTs); the proportion of the population with access to an ITN within their household was estimated to exceed 50% in five countries; and Angola and Burundi reported delivery of sufficient ACTs to treat >50% patients attending the public health facilities. In 11 countries of eastern and southern Africa, the proportion of the population with access to an ITN in their household was estimated to exceed 50% in nine countries; only Mozambique distributed insufficient ACTs to treat all patients attending public health facilities. In five low-transmission southern Africa countries, the number of people with access to an ITN in their household was estimated to exceed 50% in three countries. IRS is the primary vector control measure in South Africa, and South Africa and Swaziland reported inadequate access to antimalarial medicines.

At present, drug resistance has become a priority issue in malaria control in Africa (Sa et al., 2011; Sridaran et al., 2010). Chloroquine-resistance occurs in all of Africa (Frosch et al., 2011), and there is widespread resistance to sulfadoxine-pyrimethamine (Ako et al., 2012). Artemisinin-based drugs in Africa have played a role in malaria treatment and almost all malaria-endemic countries in Africa have changed their policies to adopt ACTs for the treatment of *P. falciparum* malaria (WHO, 2013). Unfortunately, many populations of Africa apply local traditional medicines to meet their primary health care needs with no systematic generation of safety and efficacy data through clinical and pharmacological studies, nor quality control data to support formal registrations and use for treatment and control of malaria (Addae-Mensah et al., 2011). Moreover, countries in sub-Saharan Africa are of significant concern due to the high levels of malaria transmission and widespread reports of insecticide resistance (WHO, 2012a).

From the perceptions and viewpoints on proceedings of the Africa Leaders Malaria Alliance, Africa governments should address the following critical challenges:

1. Securing adequate domestic and external funding for sustained commitment to malaria elimination;
2. Strengthening national malaria control programmes in the context of strengthening the broader health system;
3. Ensuring free access to malaria diagnosis and treatment for vulnerable groups such as children and pregnant women, in addition to universal access to long-lasting insecticide-treated nets;
4. Strengthening human resource capacity at central, district, and community levels; and
5. Establishing strong logistics, information and surveillance systems as well as early detection mechanisms for response to malaria epidemics and other public health threats (Sambo et al., 2011).

On 6–7 May 2013, the Fourth International Roundtable on China–Africa Health Cooperation was held in Gaborone, Botswana to explore new opportunities to strengthen an innovative partnership between China and Africa to cooperate on health issues. The roundtable brought African and Chinese leaders together to share experiences, discuss progress and lessons learned, and to make recommendations on how China–Africa health cooperation could be strengthened. During this roundtable, the major challenges of malaria control and elimination in Africa were identified as follows:

1. Poor infrastructure and substandard health services and coverage;
2. Poor distribution and access to diagnostic services and effective treatment;
3. Limited funding for vector control, diagnostic testing, treatment and surveillance;
4. Cross-border malaria control;
5. Threat of drug and insecticide resistance; and
6. The vicious cycle of poverty and malaria.

4. POTENTIAL OPPORTUNITY AND CONTRIBUTION TO ENHANCE THE PARTNERSHIP

Based on the aforementioned challenges and gaps, the opportunities for scaling-up Chinese involvement in malaria control in Africa may include but not limited to the following:

1. Funding for equipment and essential health commodities;
2. Extending the scope of China medical teams;
3. Training, provided language barriers can be overcome;
4. Collaborative research, also including traditional medicines;

5. Technology transfer for production of pharmaceutical products;
6. Health systems strengthening (laboratory services and IT networks); and
7. Building information system and surveillance systems.

For an accelerated and sustainable impact on malaria control and elimination in Africa, China–Africa cooperation efforts in malaria should be aligned with sound, evidence-based, and costed local strategic plans for malaria control and elimination, with a focus on identifying needs, filling gaps and leveraging all national, bilateral and multilateral as well as private sector opportunities. The P.R. China has accumulated experience, which could be shared with countries in Africa, in developing health policy leading to successful interventions but Chinese contribution should not seek to address every aspect of malaria control. It should start with the Test, Treat, and Track (T3) initiative that was advocated by WHO and all partners, and is also of great demand in Africa. To align with the T3 approach, focussing on scaling up diagnostic testing, treatment and surveillance, T3 and the corresponding surveillance manuals were launched in Namibia by the WHO Director-General in April 2012. Each suspected case of malaria will be confirmed with a rapid diagnostic test, treated with an antimalarial medicine, and tracked using a surveillance system (WHO, 2012b). Joint country efforts could be made on scaling-up access to malaria diagnostic testing, access to treatment at the community level in Africa with quality artemisinin-based antimalarial compound medicines, including lifesaving injectable artesunate, and strengthening the malaria surveillance system. China's strengths at employing these three components are described in the following sections.

4.1 Quality control in diagnosis of malaria cases (test)

During the malaria elimination stage, one of the key components is to further improve the malaria diagnosis system through quality control and assurance. According to the national strategy, to ensure the quality of malaria diagnosis in China, all positive slides or RDTs must be rechecked at the county level. All positive and 10% of randomly-selected negative results have to be verified by microscopy at an upper level, such as the prefecture, provincial or national level, within a month. Unknown species or suspect specimens with clinical symptoms will be diagnosed and confirmed by polymerase chain reaction (PCR) as well.

In line with the WHO T3 initiative launched in 2012, the national malaria diagnosis assurance system is to be updated continuously to guarantee qualified diagnostic testing so that each malaria case can be identified and traced. The diagnosis system is composed of laboratories at national, provincial and county levels. At present, the national malaria diagnosis laboratory is based in the NIPD under China CDC, and 19 provincial malaria diagnosis laboratories have been set up, all of which were recognised as reference laboratories that are responsible for malaria verification in their own province.

Therefore, P.R. China can provide expert support on the establishment of malaria diagnostic laboratories with quality assurance in local African countries. The specialists can help train the laboratory staff on malaria diagnostic techniques.

4.2 Drug delivery system (treat)

The health system in China has the advantage of a strong network of public health facilities, which is the basis for the delivery of malaria services, such as diagnosis and treatment. The rural health service delivery system is based on facilities at county, township and village levels. To ensure free, prompt, safe and effective treatment, malaria drugs are stocked at the different levels according to the different malaria burden and monthly update of drug stocks.

In China, malaria drugs are administrated free of charge to malaria patients after blood tests. G6PD deficiency is tested before giving primaquine to some ethnic minority groups in some areas. To ensure drug compliance and the safety of patients, directly observed therapy is implemented, and patients receive daily follow-up by village doctors or other health staff. To ensure the quality of antimalarial drugs, the State Food and Drug Administration strengthens the implementation of good manufacturing practices, performs routine drug quality testing (including through supporting laboratory quality improvements in rural and remote areas), and improves QA at all levels, from manufacturing to distribution to storage.

Therefore, Africans in endemic malaria areas can access and obtain practical and effective services, including antimalarial medicines, through an improved drug delivery system such as in P.R. China. Moreover, China has accumulated a lot of treatment experience using artemisinin-based therapies and can improve African capabilities toward the cure of severe *P. falciparum* malaria.

4.3 Malaria information reporting system (track)

Timely reporting, effective analyses and rapid distribution of surveillance data can assist in detecting the aberration of disease occurrence and further facilitate a timely response. In the aftermath of the 2003 outbreak of severe acute respiratory syndrome, the Chinese government strengthened its public health disease surveillance system, taking advantage of modern information technology to build an integrated, effective and reliable disease reporting system, the national Notifiable Infectious Diseases Reporting Information System (NIDRIS). This system gathers real-time epidemiologic information on approximately 40 infectious diseases, including malaria, across every province from every hospital and clinic. Based on NIDRIS, a new nationwide web-based automated system for outbreak detection and rapid response, namely the China Infectious Disease Automated-alert and Response System (CIDARS), was developed in 2008. In CIDARS, three aberration detection methods are used to detect the unusual occurrence of 28 notifiable infectious diseases at the county level and transmit information either in real time or on a daily basis. The Internet, computers and mobile phones are used to accomplish rapid signal generation and dissemination, timely reporting and reviewing of the signal response results, which greatly improved the timeliness of malaria case or epidemic detection and response (Yang et al., 2011). In 2010, when P. R. China started NMEP another web-based reporting system specific to malaria was set up to provide more detailed information on malaria patients and antimalarial interventions.

The information systems have the advantages of real-time data reporting and aggregation, national system structure and protocols to eliminate noncompatibility, enabling local CDCs and other health care organisations to aggregate data in a number of ways at any time and in any region. The systems allow for routine reporting of surveillance data and monitoring of malaria programme, which can help monitor focal malaria prevalence and predict trends due to changing transmission factors as well as to provide strategic guidance on malaria control or elimination activities, including malaria testing and treatment.

Thereafter, in Africa with the selection of pilot countries, China is interested in collaborating on capacity building and health system strengthening, with specific emphasis on monitoring/evaluation and improving health information systems. The core malaria indicators recommended by WHO and demanded in African countries could be designed in such a malaria information system. The combination of advanced technology (e.g. spatial technology) with the reporting system based on Chinese experiences could benefit local malaria detection, treatment and response. Training courses on information techniques and data management could be provided to ensure the system functions appropriately.

5. COLLABORATIVE RESEARCH SCOPES

5.1 Effectiveness evaluation on comprehensive control strategy

There will be three areas of possible research for the collaboration in the field of effectiveness evaluation of comprehensive control strategy as follows:

1. Demographic, epidemiological and environmental impact factors for malaria transmission by a comprehensive comparison in the fields of population prevalence, diagnostic capacity and effects of vector control between demonstration areas and the control areas;
2. Socioeconomic impact factors and health system status involved in a national malaria control/elimination programme, including the local social situation, economy and health services system; and
3. Strategy and measures assessment, including evaluating the control effect of a related policy and strategy applied in the demonstration areas and further exploring the successful models.

5.2 Early warning system for disease surveillance

According to the experiences in establishment of the information management system for malaria control/elimination programmes in P.R. China, a warning system of disease surveillance in Africa countries could be set up to help timely evaluate risks for malaria transmission and control effects (Tambo et al., 2014a; Tambo et al., 2014b).

5.3 Resistance surveillance

Resistance studies on both insecticide and antimalarial drugs are proposed, focussing on the possible mechanisms for resistance occurrence and transmission, which could provide evidence-based strategies for resistance containment in Africa and even worldwide (Liu, 2014).

5.4 Transition of traditional medicines

The African continent bears the greatest burden of types II and III diseases, including the so-called neglected diseases, malaria, tuberculosis, HIV/AIDS, diarrhoea and respiratory diseases. As well, China still suffers from a number of neglected tropical diseases, and emphasis is already being placed in the use of different dosage and administration in combination therapy for control and management of such diseases, including the contribution of traditional medicines (Helmstadter and Staiger, 2013; Odonne et al., 2011;

Pohlit et al., 2013; Zofou et al., 2013), and many experiences can be drawn upon for African countries in treating malaria. In addition, use of and access to available drugs are hindered by resistance, safety and compliance issues as well as cost. As many pharmaceutical companies in China have certain capabilities for developing new antimalarial drugs (Chen, 2014), there is an urgent need for the transformation of current approaches that address these candidate traditional medicines in Africa, in a manner that strengthens South–South collaborations and translates stagnating African discoveries into usable and accessible health tools. This China–Africa partnership presents a real opportunity to advance this area and build relevant capacity (Dawson et al., 2014; Mboya-Okeyo et al., 2009; Nwaka et al., 2010; Thorsteinsdottir et al., 2010; Uwimana et al., 2012).

6. THE WAY FORWARD

Over the past 60 years, P. R. China has made remarkable achievements in malaria control in terms of very limited local transmission and is proceeding with malaria elimination. Although malaria incidence and mortality was reduced by 31% and 49% respectively between 2000 and 2012, sub-Saharan Africa remains currently the highest malaria burden in the world (WHO, 2013). Many international efforts are making contributions, but there is further opportunity to strengthen China-Africa cooperation in Africa on malaria control and elimination, based on Chinese long-term experience in malaria control and China-Africa health cooperation.

Such China-Africa Cooperation efforts should be aligned with sound, evidence-based, and costed strategic plans for local malaria control and elimination, with a focus on identifying needs and filling gaps, and leveraging all national, bilateral and multilateral as well as private sector opportunities. Although a series of challenges and needs were recognized, scaling up and sustaining access to qualified malaria diagnosis, treatment and case-reporting in all public sectors, for-profit and informal health facilities across sub-Saharan Africa is central to current global strategies for malaria control and elimination. Therefore, it is the opportunity for further China-Africa collaboration to apply Chinese strengths into African improvements of quality assurance in malaria diagnosis system, drug delivery system, information reporting system, as well as the related capacity building. To ensure the successful employment of these efforts, the corresponding operation researches with high priority are also indispensable, including control effect evaluation, early warning, resistance surveillance of antimalarial drugs and insecticides, as well as transformation of traditional medicines. It is encouraged to collaborate on research and development of new products based on technical expertise and support systems from China, and building on Technical Assistance platforms and technical resource networks/Centers in Africa.

According to Chinese experience, the success of large scale campaign must generate from pilot trial. The China-Africa Cooperation efforts in malaria should be directed on an initial group of 6-8 countries, half of which with high burden and maximal potential for saving lives and half with potential to progress towards malaria pre-elimination & elimination. The criteria for selecting countries may include: 1) high level political commitment for malaria control; 2) government resources (financial or human) contributing to the initiative; 3) existing political trade and economic cooperation with Chinese government; and 4) minor language barriers to assure efficient cooperation. Chinese medical team, anti-malaria center and competitive pharmaceutical enterprises in the selected countries as well as malaria experts sent from China would have great contributions.

Simultaneously, these efforts should be in line with malaria resolutions and initiatives of United Nations (UN), WHO, Africa Union (AU) and Regional Economic Communities and in coordination with national and international agencies under bilateral and multilateral partnerships to foster collaboration effect and conformity superiority in the region.

ANNEX 1: ANTIMALARIA CENTERS IN AFRICA ESTABLISHED BY CHINESE GOVERNMENT IN 2007–2009

Year	Country	Location (city or hospital name)
2007	Senegal	Dakar
2007	Liberia	Monrovia
2007	Chad	NDjamena (hospital)
2008	Burundi	Bujumbura
2008	Uganda	Kampala (hospital)
2008	Congo	Brazzaville
2008	Gabon	Libreville (hospital)
2008	Benin	Lokosa
2008	Guinea	Bissau
2008	Ethiopia	Aoluo Mo, Naz.
2009	Togo	Lome

Year	Country	Location (city or hospital name)
2009	Mali	Bamako(hospital)
2009	Cameroon	Yaounde(hospital)
2009	Cote d'Ivoire	Abidjan
2009	Rwanda	Ki Gu Ki Jo
2009	Equatorial Guinea	Bata
2009	Zambia	Lusaka
2009	Comoros	Moheli
2009	Madagascar	Infectious Disease (hospital)
2009	Sudan	Ad Damazin
2009	Central African	Bangui
2009	Garner	Ackner
2009	Angola	Luanda
2009	Guinea	Kona Correa Steen (hospital)
2009	Mozambique	Maputo
2009	Sierra Leone	Friston(Hospital)
2009	Tanzania	Leah Amana hospital

ACKNOWLEDGEMENTS

This working paper was produced by the malaria team that was convened on 5–6 April, 2013 in Beijing, P.R. China and updated after the Fourth Roundtable on China–Africa Health Cooperation on 6–7 May, 2013 in Gaborone of Botswana by the malaria team from the National Institute of Parasitic Diseases, China CDC, based in Shanghai, and WHO. We extend our sincere thanks to all the participants who tirelessly contributed to this document. The project was supported by China UK Global Health Support Programme (grant no. GHSP-CS-OP1, OP2, OP3).

REFERENCES

Addae-Mensah, I., Fakorede, F., Holtel, A., Nwaka, S., 2011. Traditional medicines as a mechanism for driving research innovation in Africa. Malar. J. 10 (Suppl. 1), S9.

Agomo, C.O., Oyibo, W.A., 2013. Factors associated with risk of malaria infection among pregnant women in Lagos, Nigeria. Infect. Dis. Poverty 2, 19.

Aidoo, M., 2013. Factoring quality laboratory diagnosis into the malaria control agenda for sub-Saharan Africa. Am. J. Trop. Med. Hyg. 89, 403–406.

Ajayi, I.O., Jegede, A.S., Falade, C.O., Sommerfeld, J., 2013. Assessing resources for implementing a community directed intervention (CDI) strategy in delivering multiple health interventions in urban poor communities in Southwestern Nigeria: a qualitative study. Infect. Dis. Poverty 2, 25.

Ako, B.A., Offianan, A.T., Johansson, M., Penali, L.K., Nguetta, S.P., Sibley, C.H., 2012. Molecular analysis of markers associated with chloroquine and sulfadoxine/pyrimethamine resistance in *Plasmodium falciparum* malaria parasites from southeastern Cote-d'Ivoire by the time of Artemisinin-based Combination Therapy adoption in 2005. Infect. Drug. Resist 5, 113–120.

Bhutta, Z.A., Sommerfeld, J., Lassi, Z.S., Salam, R.A., Das, J.K., 2014. Global burden, distribution, and interventions for infectious diseases of poverty. Infect. Dis. Poverty 3, 21.

Bi, Y., Tong, S., 2014. Poverty and malaria in the Yunnan province, China. Infect. Dis. Poverty 3, 32.

Butler, C.D., 2012. Infectious disease emergence and global change: thinking systemically in a shrinking world. Infect. Dis. Poverty 1, 5.

Chen, C., 2014. Development of antimalarial drugs and their application in China: a historical review. Infect. Dis. Poverty 3, 9.

Chima, R.I., Goodman, C.A., Mills, A., 2003. The economic impact of malaria in Africa: a critical review of the evidence. Health Policy 63, 17–36.

Chinese Medical Cooperation in Africa, 2011. Nordiska Afrikainstitut, Uppsala. http://nai.diva-portal.org/smash/get/diva2:399727/FULLTEXT02.

Dawson, A., Brodie, P., Copeland, F., Rumsey, M., Homer, C., 2014. Collaborative approaches towards building midwifery capacity in low income countries: a review of experiences. Midwifery 30 (4), 391–402.

Frosch, A.E., Venkatesan, P.M., Laufer, M.K., 2011. Patterns of chloroquine use and resistance in sub-Saharan Africa: a systematic review of household survey and molecular data. Malar. J. 10, 116.

Guinovart, C., Navia, M.M., Tanner, M., Alonso, P.L., 2006. Malaria: burden of disease. Curr. Mol. Med. 6, 137–140.

Helmstadter, A., Staiger, C., 2013. Traditional use of medicinal agents: a valid source of evidence. Drug Discov. Today 19, 4–7.

Hill, J., Hoyt, J., van Eijk, A.M., D'Mello-Guyett, L., Ter Kuile, F.O., Steketee, R., Smith, H., Webster, J., 2013. Factors affecting the delivery, access, and use of interventions to prevent malaria in pregnancy in sub-Saharan Africa: a systematic review and meta-analysis. PLoS Med. 10, e1001488.

Liu, D.Q., 2014. Surveillance of antimalarial drug resistance in China in the 1980s-1990s. Infect. Dis. Poverty 3, 8.

Malaria in Africa. http://www.rollbackmalaria.org/cmc_upload/0/000/015/370/RBMInfosheet_3.htm.

Mboya-Okeyo, T., Ridley, R.G., Nwaka, S., 2009. The african network for drugs and diagnostics innovation. Lancet 373, 1507–1508.

Action Plan of China Malaria Elimination (2010–2020). http://www.nhfpc.gov.cn/cmsresources/mohjbyfkzj/cmsrsdocument/doc8571.doc.

Noor, A.M., Amin, A.A., Akhwale, W.S., Snow, R.W., 2007. Increasing coverage and decreasing inequity in insecticide-treated bed net use among rural Kenyan children. PLoS Med. 4, e255.

Nwaka, S., Ilunga, T.B., Da Silva, J.S., Rial Verde, E., Hackley, D., De Vre, R., Mboya-Okeyo, T., Ridley, R.G., 2010. Developing ANDI: a novel approach to health product R&D in Africa. PLoS Med. 7, e1000293.

Odonne, G., Berger, F., Stien, D., Grenand, P., Bourdy, G., 2011. Treatment of leishmaniasis in the Oyapock basin (French Guiana): a K.A.P. survey and analysis of the evolution of phytotherapy knowledge amongst Wayapi Indians. J. Ethnopharmacol. 137, 1228–1239.

Pell, C., Straus, L., Andrew, E.V., Menaca, A., Pool, R., 2011. Social and cultural factors affecting uptake of interventions for malaria in pregnancy in Africa: a systematic review of the qualitative research. PLoS One 6, e22452.

Pohlit, A.M., Lima, R.B., Frausin, G., Silva, L.F., Lopes, S.C., Moraes, C.B., Cravo, P., Lacerda, M.V., Siqueira, A.M., Freitas-Junior, L.H., Costa, F.T., 2013. Amazonian plant natural products: perspectives for discovery of new antimalarial drug leads. Molecules 18, 9219–9240.

Sa, J.M., Chong, J.L., Wellems, T.E., 2011. Malaria drug resistance: new observations and developments. Essays Biochem. 51, 137–160.

Sambo, L.G., Ki-Zerbo, G., Kirigia, J.M., 2011. Malaria control in the African Region: perceptions and viewpoints on proceedings of the Africa Leaders Malaria Alliance (ALMA). BMC Proc. 5 (Suppl. 5), S3.

Sridaran, S., McClintock, S.K., Syphard, L.M., Herman, K.M., Barnwell, J.W., Udhayakumar, V., 2010. Anti-folate drug resistance in Africa: meta-analysis of reported dihydrofolate reductase (dhfr) and dihydropteroate synthase (dhps) mutant genotype frequencies in African *Plasmodium falciparum* parasite populations. Malar. J. 9, 247.

Tambo, E., Adedeji, A.A., Huang, F., Chen, J.H., Zhou, S.S., Tang, L.H., 2012. Scaling up impact of malaria control programmes: a tale of events in Sub-Saharan Africa and People's Republic of China. Infect. Dis. Poverty 1, 7.

Tambo, E., Ai, L., Zhou, X., Chen, J.H., Hu, W., Bergquist, R., Guo, J.G., Utzinger, J., Tanner, M., Zhou, X.N., 2014a. Surveillance-response systems: the key to elimination of tropical diseases. Infect. Dis. Poverty 3, 17.

Tambo, E., Ugwu, E.C., Ngogang, J.Y., 2014b. Need of surveillance response systems to combat Ebola outbreaks and other emerging infectious diseases in African countries. Infect. Dis. Poverty 3, 29.

Tang, L.H., 1999. Chinese achievements in malaria control and research. Chin. J. Parasitol. Parastic. Dis. 17, 3 (in Chinese).

Tang, L.H., 2009. Malaria in China: from control to elimination. Int. Med. Parasitic. Dis. 36, 8 (in Chinese).

Thorsteinsdottir, H., Melon, C.C., Ray, M., Chakkalackal, S., Li, M., Cooper, J.E., Chadder, J., Saenz, T.W., Paula, M.C., Ke, W., et al., 2010. South-South entrepreneurial collaboration in health biotech. Nat. Biotechnol. 28, 407–416.

Uwimana, J., Zarowsky, C., Hausler, H., Jackson, D., 2012. Engagement of non-government organisations and community care workers in collaborative TB/HIV activities including prevention of mother to child transmission in South Africa: opportunities and challenges. BMC Health Serv. Res. 12, 233.

WHO, 2012a. Global Plan for Insecticide Resistance Management in Malaria Vectors (GPIRM). World Health Organization, Geneva.

WHO, 2012b. Test, Treat, Track: Scaling up the Fight against Malaria. World Health Organization, Geneva.

Wang, X.Y., 2013. Chinese achivements and challenges in aid to africa. Int. Res. Ref. 7, 6 (in Chinese).

WHO, 2013. World Malaria Report 2013. World Health Organization, Geneva.

Yang, W., Li, Z., Lan, Y., Wang, J., Ma, J., Jin, L., Sun, Q., Lv, W., Lai, S., Liao, Y., Hu, W., 2011. A nationwide web-based automated system for outbreak early detection and rapid response in China. Western Pac. Surveill. Response J 2, 10–15.

Yin, J.H., Yang, M.N., Zhou, S.S., Wang, Y., Feng, J., Xia, Z.G., 2013. Changing malaria transmission and implications in China towards National Malaria Elimination Programme between 2010 and 2012. PLoS One 8, e74228.

Zofou, D., Ntie-Kang, F., Sippl, W., Efange, S.M., 2013. Bioactive natural products derived from the Central African flora against neglected tropical diseases and HIV. Nat. Prod. Rep. 30, 1098–1120.

Zofou, D., Nyasa, R.B., Nsagha, D.S., Ntie-Kang, F., Meriki, H.D., Assob, J.C., Kuete, V., 2014. Control of malaria and other vector-borne protozoan diseases in the tropics: enduring challenges despite considerable progress and achievements. Infect. Dis. Poverty 3, 1.

Zuo, Y., 2013. Contributions and challenges of chinese Foreign aid medical team. Int. Economic Collaboration 11, 3 (in Chinese).

INDEX

Note: Page numbers followed by "b", "f" and "t" indicate boxes, figures and tables, respectively.

A

Active malaria screening methods, 124–125
Anhui province, 246–247
Anopheles
  An. anthropophagus, 141, 189–192
  An. dirus, 4–5, 9–10
  An. kunmingensis, 52, 70–71
  An. minimus, 193
  An. minus, 4–5
  An. sacharovi, 4–5
  An. sinensis, 4–5, 9, 141, 222–223
Anti-CSP antibodies, 125
Artemisinin-based combination therapy (ACT), 253

B

Baseline survey
  investigation, 149
  malaria incidence, 148–149
  NMEP, 148
  quality control, 149
  sampling method, 148
  vector survey, 149

C

China-Africa cooperation, malaria control
  ACTs, 326–327
  antimalaria centres, 322–323, 334–335
  challenges, 327–328
  comprehensive control effect evaluation, 332
  diagnostic capability, 329–330
  drug delivery system, 330
  drug resistance, 327
  elimination, 325
  epidemiology, 323–324, 324f
  funding gap, 326–327
  ITN, 326–327
  malaria information reporting system, 330–331
  malaria programs, 325–326
  NIPD, 323
  NMCPs, 320–321
  pilot programme, 2014-2020, 333–334
  resistance surveillance, 332
  T3 approach, 329
  traditional medicines, 332–333
  warning system of disease surveillance, 332
China Information System for Disease Control and Prevention (CISDCP), 330–331
China National Knowledge Infrastructure (CNKI), 113–114, 114f
Chinese Centers for Disease Control and Prevention (CDC), 86
Chloroquine (CQ), 123, 221–222
Colloidal gold immunochromatographic assay (GICA), 120

D

Data quality audit (DQA), 299
Drug resistance surveillance, 93

E

Elimination Eight (E8), 325
Entomological surveillance, 90–93, 92f
Enzyme-linked immunosorbent assay (ELISA), 125

F

Feasibility analysis, ME
  with geographic variations, 33–35, 35f
  MECI, 31–33, 32f, 34f
  MTRI, 32f, 33, 34f
  NMEP, 31
  transmission risks, 31
Fujian, malaria post-elimination
  epidemic situation, 185–186
  febrile patients, blood examinations, 195–196, 196t
  genetic monitoring, 197

339

Fujian, malaria post-elimination
(Continued)
  geographic patterns, 194, 194t, 195f
  local cases and epidemic status, 198
  location, 184–185, 184f
  malaria cases distribution, 199, 199f–200f
  malaria monitoring sites, 200–201, 200f
  malaria vector surveillance and control, 196–197, 197t
  population surveillance, 196
  stages of
    individual malaria cases, 1977-1985, 189, 191f
    integrated malaria control programme, 1956-1976, 186–189, 188f, 190f
    malaria incidence, 186
    pilot projects, 1950-1955, 186, 187f
    post-elimination, 2000-present, 192–193, 193f
    pre-elimination, 1986-2000, 189–192

G

Global fund (GF) malaria programme, in P.R.China
  audits, 300
  case management, 304t
    chloroquine/primaquine, 303–304
    malaria treatment, 303–304, 305t
    microscopy and RDTs, 303
    PCR, 303
    quality assurance system, 303
  communication, 314
  epidemiological justification, 291–292
  financial management, 300
  funding, 300, 301f
  health education and community mobilization, 306, 307t–308t
  health products and equipment, 301, 302t
  human resources, 301–302
  malaria control and elimination, 310
    cross-border cooperation mechanism, 315
    enforced public awareness, 314–315
    incidence, in 2003-2011, 310–312, 312f
    national contribution, 314
    P. falciparum malaria cases, 312
    resource gap, 313
  malaria surveillance, 309
  management structure, 297, 298f
  materials, 301
  monitoring and evaluation, 298–299
  multi-sector cooperation, 309–310, 311t, 314
  NSA, 296
  operational research, 310
  populations risk, 307–309
  procurement and supply management, 299–300
  programme agreements, 297
  R1, 293
  R5, 293
  R6, 293–295
  R10, 296
  sequence of implementation, 292, 295f
  target area of, 293, 295f
  vector control, 304–306, 306f
  work plan, 297–298
Glucose-6-phosphate dehydrogenase deficiency (G6PDd), 124

H

Hainan Island
  antimalarial drug administration, 64
  blood testing, 63, 64f
  chemoprophylaxis, 65f, 66–67
  geography, 51
  health system, 51
  malaria control, 61–62
  malaria elimination
    malaria reintroduction, 69
    mobile population management, 68–69
  malaria transmission
    acquisition, 59–61, 60f
    behavioral factors, 53
    demographical characteristics, 57–59, 59f
    malaria prevalence, 53–54, 54f
    natural factors, 52
    plasmodium species, 54–55, 55f
    social-economic factors, 52
    spatiotemporal distribution, 55–57, 56f–58f
  population, 51
  vector control, 64–66
  WHO/TDR fund, 67

Index

Henan province, 245–246, 246f
Histidine rich protein II (HRP-2), 120
Huang-Huai Plain
  animals, role, 207–208
  Anopheles sinensis, 207–208
  climate, 207
  distribution of malaria cases, 209, 212f
  effects, 223
  imported malaria, 224
  insecticide resistance, 225
  location, 207
  long incubation time, 224–225
  malaria incidence, 209, 212f
  malaria situation, 207, 207f
  Plasmodium vivax, 206–207
  vivax malaria re-emergence and outbreak
    age, 210–211, 211t
    aggregation, 213
    An. sinensis, 213–214
    case management, 221–222
    climate warming, 215
    diagnosis, 217–218
    districts, 207f, 210, 211t
    global fund and government, 223
    land use, 215
    malaria situation, 209, 209t
    MDA, 218–221
    months, 211–213, 213t
    rainfall, 215–216
    sex and occupation, 210–211, 211t
    vector capacity, 216–217
    vector control, 222–223
    water bodies, 214
Huang-Huai River region. *See* Huang-Huai Plain
Human blood index (HBIs), 217

I

Indoor residual spraying (IRS), 64–65, 305–306
Injection-type artemisinin monotherapy, 253
Insecticide treated nets (ITNs), 64–65, 326–327

J

Jiangsu province, 244–245

L

Local potential risk for malaria transmission (LPR), 26t
Loop-mediated isothermal amplification (LAMP), 114–115

M

Malaria early warning system (MEWS), 116
Malaria elimination (ME)
  blood examination
    methods, 177–178
    three fever patient, 175–176
  in China
    active malaria screening methods, 124–125
    border malaria, 122–123
    drug resistance monitoring, 123
    GICA, 120
    G6PD deficiency, 124
    HRP-2, 120
    LAMP, 120–121
    Plasmodium parasites, 121–122
    real-time PCR, 120
    vector distribution, 125–126
  classification map, 41
  CNKI, 113–114, 114f, 126
  control activities, 142–143, 144t
  diagnostics and detection, 114–115, 115f
  disease control network, 173–174
  drugs and drug resistance, 113f–115f, 115–116
  entomology and insecticides, 113f–115f, 116–117
  epidemic situation, 140–142
  epidemiology and disease control, 113f–115f, 116
  evaluation and certification, 171–173, 172t–173t
  feasibility analysis
    with geographic variations, 33–35, 35f
    MECI, 31–33, 32f, 34f
    MTRI, 32f, 33, 34f
    NMEP, 31
    transmission risks, 31
  feasibility assessment
    data resources, 24
    indices and weighted values, 24–25, 26t

Malaria elimination (ME) *(Continued)*
    malaria transmission, in population, 25–28, 27f
    MECI, 25, 26t
    MTRI, 25, 26t
    NMEP, 23–24
  funding mechanism, 175
  GF malaria programme, in P.R.China
    audits, 300
    case management, 303–304, 304t
    communication, 314
    epidemiological justification, 291–292
    financial management, 300
    funding, 300, 301f
    health education and community mobilization, 306, 307t–308t
    health products and equipment, 301, 302t
    human resources, 301–302
    malaria control, 310–315
    malaria surveillance, 309
    management structure, 297, 298f
    materials, 301
    monitoring and evaluation, 298–299
    multi-sector cooperation, 309–310, 311t, 314
    NSA, 296
    operational research, 310
    populations risk, 307–309
    procurement and supply management, 299–300
    programme agreements, 297
    R1, 293
    R5, 293
    R6, 293–295
    R10, 296
    sequence of implementation, 292, 295f
    target area of, 293, 295f
    vector control, 304–306, 306f
    work plan, 297–298
  in Hainan province
    malaria reintroduction, 69
    mobile population management, 68–69
  health services and raising control capacity
    disease control and prevention, 163
    equipment, 163
    supplies and fund guarantee, 163
    work mechanism and network platform, 161–163
  historical transmission pattern, 28–29, 29f
  immunology and vaccines, 113f–115f, 117–118
  incidence and interventions, 22–23, 29–30
  malaria case finding, 174–175
  malaria control knowledge, 178–179
  marginal cost-benefit principle, 40
  measures
    blood examination, 170, 171t
    case and focus survey, 164–168, 169t
    health education, 170
    malaria cases, 164, 165t–167t
    mobile population management, 168–170
    NMEP, 170–171
    professional training, 168, 169t
  monitoring and evaluation, 41–42
  NMEP
    assessment and evaluation, 149–151
    baseline survey, 148–149
    case detection, 158–159, 159t–160t
    malaria epidemic situation, 158, 158t
    malaria vector, 161, 162t
    measures, 146–147
    mosquito-proof facilities, 159–161, 160t, 162t
    organization and guarantee, 151–157
    pilot counties/districts, 143–146, 145f–146f
    targets, 147
    time schedule, 147, 148f
  percentage distribution, 112, 113f
  phase-based malaria elimination
    classification, 36–37
    incidence rate, 36
    strategy formulations, 37–39, 38t, 39f
  population distribution, 39–40
  PubMed, 113–114
  reference distribution, 112, 113f
  research gap analysis, 115f, 118–120, 119f
  research priorities, 40–41
  search strategy and selection criteria, 111, 112f

sector cooperation, 174
Shanghai, 138
statistical analysis, 111
transmission stage, elimination control, 30–31
vector surveillance, 179
World Health Organisation, 22
Yunnan province
   antimalarial drug resistance, 73–74
   capability building, 72
   epidemiology and vectors, 70–71
   high malaria transmission, on the China-Myanmar border, 70
   malaria reintroduction, 72–73
   mobile population management, 71–72
Zhejiang, 138–139
Malaria elimination capacity index (MECI)
   definition, 26t
   feasibility analysis, 32f, 33, 34f
   malaria transmission risk index, 31–32, 32f
   resources capacity, 32–33
   technical capacity, 32
Malaria interventions
   antimalarial drug administration, 63–64
   case detection, 63, 64f
   chemoprophylaxis doses, 65f, 66–67
   partnership, 65f, 67–68
   surveillance, 68
   vector control, 64–66
Malaria resurgence, in China
   central China
      An. anthropophagus, 245
      Anhui province, 246–247
      An. sinensi, 245
      Henan province, 245–246, 246f
   diagnosis and treatment, 252–253
   eastern coastal regions, 244
      Jiangsu province, 244–245
      Zhejiang province, 244
   emergency preparedness, 252
   entry and exit port, 250
   epidemiological situation
      age characteristics, 239–240, 240f
      gender, 239–240, 240f
      geographical distribution, 238–239, 239f
      imported cases, 237–238, 238f
      monthly and seasonal distribution, 240–241, 241f
      morbidity and mortality, 236–237
      reported malaria cases, 234–235
      species classification, 235–236, 236f, 237t
   government duty, 249–250
   imported malaria determination, 247–248, 255–256
   large international activities, 251–252
   malaria transmission, 258
   Mauritius, 256–257
   medical and paramedical personnel training, 254
   multi-sector cooperation, 309–310, 251t
   Myanmar border, 242–243
   non-endemic areas, 248–249
   North Korea border, 243
   P. falciparum case, treatment, 257–258, 258f
   preparation for
      cross-sectoral cooperation, 273–274
      diagnosis and treatment, 274–276
      investigation, 276–277
      literature selection flow chart, 269, 270f
      NMEP, 280–281
      Plasmodium species, 268–269
      population movement, 271–272, 271t
      prophylaxis, 277
      public awareness, 278–280
      PubMed, 269
      risk assessment, 277–278
      vectorial capacity, 272–273
   P. vivax case, treatment, 257–258, 259f
   surveillance, 253–254
Malaria surveillance/response
   CDC, 86
   data collection, 87–88
   distribution, in China, 86–87, 87f
   nationwide blood examination, 87–88, 91t
   P.R. China, 85–86
   routine surveillance
      blood examination, 89–90
      case reporting and management, 88–89, 89f
   sentinel surveillance

Malaria surveillance/response *(Continued)*
  drug resistance, 93
  entomological surveillance, 90–93, 92f
  population at risk, 93–94
 world countries, 84–85
Malaria transmission
 in 2004-2012
  acquisition, 59–61, 60f
  demographical characteristics, 57–59, 59f
  malaria prevalence, 53–54, 54f
  plasmodium species, 54–55, 55f
  spatiotemporal distribution, 55–57, 56f–58f
 artemisinin, 14
 asymptomatic and low-parasitemia infections, 14
 imported malaria, 13–14, 13f
 indeterminacy, 1949-1959
  Anopheles species, 4–5
  Plasmodium species, 4
 low transmission re-emergence, 2000-2009
  An. anthropophagus, 10
  An. minus, 10
  epidemiological features, 10–11, 12t
  falciparum malaria cases, 10, 11f
  incidence, 10
  trends of, 10, 11f
 malaria cases, in 2010-2012, 12–13, 13f
 malaria vector mosquitoes, 2–3
 NMEP, 2
 outbreak and pandemic, 1960-1979
  cases and incidence, 5, 6f
  deaths, 5, 7f
  geographical stratification, 5, 8f
  vector control interventions, 5
 phases of, 3
 recommendations, 14–15
 sporadic case distribution, 1980-1999, 8–10
Mass drug administration (MDA)
 chemoprophylaxis on transmission season, 221
 chloroquine, 218–219
 NTDs, 218
 spring treatment, 219–221

ME. *See* Malaria elimination (ME)
Mefloquine, 93
Merozoite surface protein 1 (PvMSP-1), 197

N

National Institute of Parasitic Diseases (NIPD), 323
National Malaria Control Program (NMCP), 49
National Malaria Elimination Programme (NMEP)
 assessment and evaluation
  briefing and discussion, 150
  data assessment, 150
  on-site inspection, 150–151
  scores, 151, 155t
  self-evaluation and application, 149–150
 baseline survey, 148–149
 case detection, 158–159, 159t–160t
 financial guarantee, 157, 157t
 malaria epidemic situation, 158, 158t
 malaria surveillance and response
  CDC, 86
  data collection, 87–88
  distribution, in China, 86–87, 87f
  nationwide blood examination, 87–88, 91t
  P.R. China, 85–86
  response mechanism, 102–103
  routine surveillance, 88–90, 89f
  sensitivity, 101, 102f
  sentinel surveillance, 87–88, 90–94, 92f
  techniques for, 101–102
  world countries, 84–85
 malaria vector, 161, 162t
 measures, 146–147
 mosquito-proof facilities, 159–161, 160t, 162t
 pilot counties/districts, 143–146, 145f–146f
 policy guarantee, 151–156
 risk assessment
  An. sinensis, 98
  P.R. China, 97
  predicted malaria risk areas, 98, 99f

Index

section setup, 156
sector cooperation, 174
in Shanglin, 99–101, 100f
1-3-7 strategy, 94
  early detection and timely reporting, 94–95
  expanded treatment, 95
  focus investigation, 95
  health education, 95
  P.R. China, 95
  vector control, 96
  workflow, 96, 96f
supervision process, 170–171
targets, 147
technical guarantee, 156–157
time schedule, 147, 148f
WHO requirements, 84
National Strategy Application (NSA), 292, 296
Neglected tropical diseases (NTDs), 218
NMEP. *See* National Malaria Elimination Programme (NMEP)

P

People's Republic of China (P.R. China)
  global fund (*see* Global fund (GF) malaria programme, in P.R.China)
  malaria control, 82
  malaria elimination (*see also* Malaria elimination (ME)), 95
    CNKI, 113–114, 114f
    diagnostics and detection, 114–115, 115f
    drugs and drug resistance, 113f–115f, 115–116
    entomology and insecticides, 113f–115f, 116–117
    epidemiology and disease control, 113f–115f, 116
    immunology and vaccines, 113f–115f, 117–118
    percentage distribution, 112, 113f
    PubMed, 113–114
    reference distribution, 112, 113f
    research gap analysis, 115f, 118–120, 119f
    search strategy and selection criteria, 111, 112f
    statistical analysis, 111
  malaria resurgence, 268–269
  surveillance and response system, 83, 85–86
Piperaquine, 93
Plasmodium
  P. falciparum, 4, 93, 269
  P. malariae, 4, 185–186
  P. ovale, 4
  P. vivax, 4, 224–225
Polymerase chain reaction (PCR), 303
Primaquine, 219–221
PubMed, 113–114, 113f
Pyronaridine phosphate, 253

R

Rapid diagnostic test (RDT), 89, 114–115
Real-time PCR, 120

S

Sentinel surveillance
  drug resistance, 93
  entomological surveillance, 90–93, 92f
  population at risk, 93–94
Shanghai
  control activities, 142–143, 144t
  epidemic situation, 140–142
    falciparum malaria, 140
    ovale malaria, 141
    quartan malaria, 140
    transmission trends, 141–142, 142f
    transmission vector, 141
    vivax malaria, 140
  geography, 138, 139f
  gross domestic product, 138
  location, 138
  NMEP (*see also* National Malaria Elimination Programme (NMEP)), 143
  Songjiang district, 143–145, 145f
  Zhabei district, 143, 145f
1-3-7 Strategy
  definition, 94
  early detection and timely reporting, 94–95
  expanded treatment, 95
  focus investigation, 95

1-3-7 Strategy *(Continued)*
  health education, 95
  P.R. China, 95
  vector control, 96
  workflow, 96, 96f

T
Time-series cross-sectional (TSCS), 29

V
Vectorial transmission index (VTI), 26t
Vivax malaria re-emergence/outbreak, in Huang-Huai Plain
  age, 210–211, 211t
  characteristics
    aggregation, 213
    An. sinensis, 213–214
    water bodies, 214
  districts, 207f, 210, 211t
  factors
    climate warming, 215
    diagnosis, 217–218
    land use, 215
    rainfall, 215–216
    vector capacity, 216–217
  malaria situation, 209, 209t
  MDA
    chemoprophylaxis on transmission season, 221
    chloroquine, 218–219
    NTDs, 218
    spring treatment, 219–221
  months, 211–213, 213t
  sex and occupation, 210–211, 211t

Y
Yunnan province
  antimalarial drug administration, 64
  blood testing, 63, 64f
  chemoprophylaxis, 65f, 67
  geography, 51
  health system, 51
  malaria control, 62–63
  malaria elimination

antimalarial drug resistance, 73–74
capability building, 72
epidemiology and vectors, 70–71
high malaria transmission, on the China-Myanmar border, 70
malaria reintroduction, 72–73
mobile population management, 71–72
malaria transmission
  acquisition, 59–61, 60f
  behavioral factors, 53
  demographical characteristics, 57–59, 59f
  malaria prevalence, 53–54, 54f
  natural factors, 52
  Plasmodium species, 54–55, 55f
  social-economic factors, 52–53
  spatiotemporal distribution, 55–57, 56f–58f
P. falciparum, 242
population, 51
P. vivax, 242
vector control, 66
WHO, 67–68

Z
Zhejiang
  control activities, 142–143, 144t
  epidemic situation, 140–142
    falciparum malaria, 140
    ovale malaria, 141
    quartan malaria, 140
    transmission trends, 141–142, 142f
    transmission vector, 141
    vivax malaria, 140
  geography, 138–139, 139f
  gross domestic product, 138–139
  location, 138–139
  malaria resurgence, 244
  NMEP (*see also* National Malaria Elimination Programme (NMEP)), 143
    Anji County, 145–146, 146f
    Haiyan County, 145, 146f

CONTENTS OF VOLUMES IN THIS SERIES

Volume 41

Drug Resistance in Malaria Parasites of
  Animals and Man
  W. Peters

Molecular Pathobiology and Antigenic
  Variation of *Pneumocystis carinii*
  Y. Nakamura and M. Wada

Ascariasis in China
  P. Weidono, Z. Xianmin and D.W.T. Crompton

The Generation and Expression of
  Immunity to *Trichinella spiralis* in
  Laboratory Rodents
  R. G. Bell

Population Biology of Parasitic Nematodes:
  Application of Genetic Markers
  T.J.C. Anderson, M.S. Blouin and R.M. Brech

Schistosomiasis in Cattle
  J. De Bont and J. Vercruysse

Volume 42

The Southern Cone Initiative Against Chagas
  Disease
  C. J. Schofield and J.C.P. Dias

Phytomonas and Other Trypanosomatid
  Parasites of Plants and Fruit
  E.P. Camargo

Paragonimiasis and the Genus *Paragonimus*
  D. Blair, Z.-B. Xu, and T. Agatsuma

Immunology and Biochemistry of
  Hymenolepis diminuta
  J. Anreassen, E.M. Bennet-Jenkins, and
  C. Bryant

Control Strategies for Human Intestinal
  Nematode Infections
  M. Albonico, D.W.T. Crompton, and L. Savioli

DNA Vaocines: Technology and
  Applications as Anti-parasite and
  Anti-microbial Agents
  J.B. Alarcon, G.W. Wainem and D.P McManus

Volume 43

Genetic Exchange in the Trypanosomatidae
  W. Gibson and J. Stevens

The Host-Parasite Relationship in
  Neosporosis
  A. Hemphill

Proteases of Protozoan Parasites
  P.J. Rosenthal

Proteinases and Associated Genes of Parasitic
  Helminths
  J. Tort, P.J. Brindley, D. Knox, K.H. Wolfe, and
  J.P. Dalton

Parasitic Fungi and their Interaction with the
  Insect Immune System
  A. Vilcinskas and P. Götz

Volume 44

Cell Biology of *Leishmania*
  B. Handman

Immunity and Vaccine Development in the
  Bovine Theilerioses
  N. Boulter and R. Hall

The Distribution of *Schistosoma bovis* Sonaino,
  1876 in Relation to Intermediate Host
  Mollusc-Parasite Relationships
  H. Moné, G. Mouahid, and S. Morand

The Larvae of Monogenea (Platyhelminthes)
  H.D. Whittington, L.A. Chisholm, and
  K. Rohde

Sealice on Salmonids: Their Biology and
  Control
  A.W. Pike and S.L. Wadsworth

Volume 45

The Biology of some Intraerythrocytic
  Parasites of Fishes, Amphibia and
  Reptiles
  A.J. Davies and M.R.L. Johnston

The Range and Biological Activity of FMR
  Famide-related Peptides and Classical
  Neurotransmitters in Nematodes
  D. Brownlee, L. Holden-Dye, and R. Walker

The Immunobiology of Gastrointestinal
  Nematode Infections in Ruminants
  A. Balic, V.M. Bowles, and E.N.T. Meeusen

Volume 46

Host-Parasite Interactions in Acanthocephala:
  A Morphological Approach
  H. Taraschewski

Eicosanoids in Parasites and Parasitic
  Infections
  A. Daugschies and A. Joachim

Volume 47

An Overview of Remote Sensing and
  Geodesy for Epidemiology and Public
  Health Application
  S.I. Hay

Linking Remote Sensing, Land Cover and
  Disease
  *P.J. Curran, P.M. Atkinson, G.M. Foody, and
  E.J. Milton*

Spatial Statistics and Geographic Information
  Systems in Epidemiology and Public
  Health
  T.P. Robinson

Satellites, Space, Time and the African
  Trypanosomiases
  D.J. Rogers

Earth Observation, Geographic Information
  Systems and *Plasmodium falciparum*
  Malaria in Sub-Saharan Africa
  *S.I. Hay, J. Omumbo, M. Craig, and
  R.W. Snow*

Ticks and Tick-borne Disease Systems in
  Space and from Space
  S.E. Randolph

The Potential of Geographical Information
  Systems (GIS) and Remote Sensing
  in the Epidemiology and Control of
  Human Helminth Infections
  S. Brooker and E. Michael

Advances in Satellite Remote Sensing
  of Environmental Variables for
  Epidemiological Applications
  S.J. Goetz, S.D. Prince, and J. Small

Forecasting Diseases Risk for Increased
  Epidemic Preparedness in Public Health
  *M.F. Myers, D.J. Rogers, J. Cox, A. Flauhalt,
  and S.I. Hay*

Education, Outreach and the Future of
  Remote Sensing in Human Health
  *B.L. Woods, L.R. Beck, B.M. Lobitz, and
  M.R. Bobo*

Volume 48

The Molecular Evolution of Trypanosomatidae
  *J.R. Stevens, H.A. Noyes, C.J. Schofield, and
  W. Gibson*

Transovarial Transmission in the Microsporidia
  A.M. Dunn, R.S. Terry, and J.E. Smith

Adhesive Secretions in the Platyhelminthes
  I.D. Whittington and B.W. Cribb

The Use of Ultrasound in Schistosomiasis
  C.F.R. Hatz

Ascaris and Ascariasis
  D.W.T. Crompton

Volume 49

Antigenic Variation in Trypanosomes:
  Enhanced Phenotypic Variation in a
  Eukaryotic Parasite
  H.D. Barry and R. McCulloch

The Epidemiology and Control of Human
  African Trypanosomiasis
  J. Pépin and H.A. Méda

Apoptosis and Parasitism: from the Parasite to
  the Host Immune Response
  G.A. DosReis and M.A. Barcinski

Biology of Echinostomes Except *Echinostoma*
  B. Fried

Volume 50

The Malaria-Infected Red Blood Cell:
  Structural and Functional Changes
  B.M. Cooke, N. Mohandas, and R.L. Coppel

Schistosomiasis in the Mekong Region: Epidemiology and Phytogeography
S. W. Attwood

Molecular Aspects of Sexual Development and Reproduction in Nematodes and Schistosomes
P.R. Boag, S.E. Newton, and R.B. Gasser

Antiparasitic Properties of Medicinal Plants and Other Naturally Occurring Products
S. Tagboto and S. Townson

Volume 51

Aspects of Human Parasites in which Surgical Intervention May Be Important
D.A. Meyer and B. Fried

Electron-transfer Complexes in *Ascaris* Mitochondria
K. Kita and S. Takamiya

Cestode Parasites: Application of *In Vivo* and *In Vitro* Models for Studies of the Host-Parasite Relationship
M. Siles-Lucas and A. Hemphill

Volume 52

The Ecology of Fish Parasites with Particular Reference to Helminth Parasites and their Salmonid Fish Hosts in Welsh Rivers: A Review of Some of the Central Questions
J.D. Thomas

Biology of the Schistosome Genus *Trichobilharzia*
P. Horak, L. Kolarova, and C.M. Adema

The Consequences of Reducing Transmission of *Plasmodium falciparum* in Africa
R.W. Snow and K. Marsh

Cytokine-Mediated Host Responses during Schistosome Infections: Walking the Fine Line Between Immunological Control and Immunopathology
K.F. Hoffmann, T.A. Wynn, and D.W. Dunne

Volume 53

Interactions between Tsetse and Trypanosomes with Implications for the Control of Trypanosomiasis
S. Aksoy, W.C. Gibson, and M.J. Lehane

Enzymes Involved in the Biogenesis of the Nematode Cuticle
A.P. Page and A.D. Winter

Diagnosis of Human Filariases (Except Onchocerciasis)
M. Walther and R. Muller

Volume 54

Introduction — Phylogenies, Phylogenetics, Parasites and the Evolution of Parasitism
D. T.J. Littlewood

Cryptic Organelles in Parasitic Protists and Fungi
B.A.P. Williams and P.J. Keeling

Phylogenetic Insights into the Evolution of Parasitism in Hymenoptera
J.B. Whitfield

Nematoda: Genes, Genomes and the Evolution of Parasitism
M.L. Blaxter

Life Cycle Evolution in the Digenea: A New Perspective from Phylogeny
T.H. Cribb, R.A. Bray, P.D. Olson, and D.T.J. Littlewood

Progress in Malaria Research: The Case for Phylogenetics
S.M. Rich and F.J. Ayala

Phylogenies, the Comparative Method and Parasite Evolutionary Ecology
S. Morand and R. Poulin

Recent Results in Cophylogeny Mapping
M.A. Charleston

Inference of Viral Evolutionary Rates from Molecular Sequences
A. Drummond, O.G. Pybus, and A. Rambaut

Detecting Adaptive Molecular Evolution: Additional Tools for the Parasitologist
J.O. McInerney, D.T.J. Littlewood, and C.J. Creevey

Volume 55

Contents of Volumes 28–52
Cumulative Subject Indexes for Volumes 28–52
Contributors to Volumes 28–52

Volume 56

Glycoinositolphospholipid from *Trypanosoma cruzi*: Structure, Biosynthesis and Immunobiology
J.O. Previato, R. Wait, C. Jones, G.A. DosReis, A.R. Todeschini, N. Heise and L.M. Previata

Biodiversity and Evolution of the Myxozoa
E.U. Canning and B. Okamura

The Mitochondrial Genomics of Parasitic Nematodes of Socio-Economic Importance: Recent Progress, and Implications for Population Genetics and Systematics
M. Hu, N.B. Chilton, and R.B. Gasser

The Cytoskeleton and Motility in Apicomplexan Invasion
R.E. Fowler, G. Margos, and G.H. Mitchell

Volume 57

Canine Leishmaniasis
J. Alvar, C. Cañavate, R. Molina, J. Moreno, and J. Nieto

Sexual Biology of Schistosomes
H. Moné and J. Boissier

Review of the Trematode Genus *Ribeiroia* (Psilostomidae): Ecology, Life History, and Pathogenesis with Special Emphasis on the Amphibian Malformation Problem
P.T.J. Johnson, D.R. Sutherland, J.M. Kinsella and K.B. Lunde

The *Trichuris muris* System: A Paradigm of Resistance and Susceptibility to Intestinal Nematode Infection
L.J. Cliffe and R.K. Grencis

Scabies: New Future for a Neglected Disease
S.F. Walton, D.C. Holt, B.J. Currie, and D.J. Kemp

Volume 58

Leishmania spp.: On the Interactions they Establish with Antigen-Presenting Cells of their Mammalian Hosts
J.-C. Antoine, E. Prina, N. Courret, and T. Lang

Variation in *Giardia*: Implications for Taxonomy and Epidemiology
R.C.A. Thompson and P.T. Monis

Recent Advances in the Biology of *Echinostoma* species in the "revolutum" Group
B. Fried and T.K. Graczyk

Human Hookworm Infection in the 21st Century
S. Brooker, J. Bethony, and P.J. Hotez

The Curious Life-Style of the Parasitic Stages of Gnathiid Isopods
N.J. Smit and A.J. Davies

Volume 59

Genes and Susceptibility to Leishmaniasis
Emanuela Handman, Colleen Elso, and Simon Foote

Cryptosporidium and Cryptosporidiosis
R.C.A. Thompson, M.E. Olson, G. Zhu, S. Enomoto, Mitchell S. Abrahamsen and N.S. Hijjawi

Ichthyophthirius multifiliis Fouquet and Ichthyophthiriosis in Freshwater Teleosts
R.A. Matthews

Biology of the Phylum Nematomorpha
B. Hanelt, F. Thomas, and A. Schmidt-Rhaesa

Volume 60

Sulfur-Containing Amino Acid Metabolism in Parasitic Protozoa
Tomoyoshi Nozaki, Vahab Ali, and Masaharu Tokoro

The Use and Implications of Ribosomal DNA Sequencing for the Discrimination of Digenean Species
Matthew J. Nolan and Thomas H. Cribb

Advances and Trends in the Molecular Systematics of the Parasitic Platyhelminthes
Peter D. Olson and Vasyl V. Tkach

Wolbachia Bacterial Endosymbionts of Filarial Nematodes
Mark J. Taylor, Claudio Bandi, and Achim Hoerauf

The Biology of Avian *Eimeria* with an Emphasis on their Control by Vaccination
Martin W. Shirley, Adrian L. Smith, and Fiona M. Tomley

Volume 61

Control of Human Parasitic Diseases: Context and Overview
David H. Molyneux

Malaria Chemotherapy
Peter Winstanley and Stephen Ward

Insecticide-Treated Nets
Jenny Hill, Jo Lines, and Mark Rowland

Control of Chagas Disease
Yoichi Yamagata and Jun Nakagawa

Human African Trypanosomiasis: Epidemiology and Control
E.M. Févre, K. Picozzi, J. Jannin, S.C. Welburn and I. Maudlin

Chemotherapy in the Treatment and Control of Leishmaniasis
Jorge Alvar, Simon Croft, and Piero Olliaro

Dracunculiasis (Guinea Worm Disease) Eradication
Ernesto Ruiz-Tiben and Donald R. Hopkins

Intervention for the Control of Soil-Transmitted Helminthiasis in the Community
Marco Albonico, Antonio Montresor, D.W.T. Crompton, and Lorenzo Savioli

Control of Onchocerciasis
Boakye A. Boatin and Frank O. Richards, Jr.

Lymphatic Filariasis: Treatment, Control and Elimination
Eric A. Ottesen

Control of Cystic Echinococcosis/Hydatidosis: 1863-2002
P.S. Craig and E. Larrieu

Control of *Taenia solium* Cysticercosis/Taeniosis
Arve Lee Willingham III and Dirk Engels

Implementation of Human Schistosomiasis Control: Challenges and Prospects
Alan Fenwick, David Rollinson, and Vaughan Southgate

Volume 62

Models for Vectors and Vector-Borne Diseases
D.J. Rogers

Global Environmental Data for Mapping Infectious Disease Distribution
S.I. Hay, A.J. Tatem, A.J. Graham, S.J. Goetz, and D.J. Rogers

Issues of Scale and Uncertainty in the Global Remote Sensing of Disease
P.M. Atkinson and A.J. Graham

Determining Global Population Distribution: Methods, Applications and Data
D.L. Balk, U. Deichmann, G. Yetman, F. Pozzi, S.I. Hay, and A. Nelson

Defining the Global Spatial Limits of Malaria Transmission in 2005
C.A. Guerra, R.W. Snow and S.I. Hay

The Global Distribution of Yellow Fever and Dengue
D.J. Rogers, A.J. Wilson, S.I. Hay, and A.J. Graham

Global Epidemiology, Ecology and Control of Soil-Transmitted Helminth Infections
S. Brooker, A.C.A. Clements and D.A.P. Bundy

Tick-borne Disease Systems: Mapping Geographic and Phylogenetic Space
S.E. Randolph and D.J. Rogers

Global Transport Networks and Infectious Disease Spread
A.J. Tatem, D.J. Rogers and S.I. Hay

Climate Change and Vector-Borne Diseases
D.J. Rogers and S.E. Randolph

Volume 63

Phylogenetic Analyses of Parasites in the New Millennium
David A. Morrison

Targeting of Toxic Compounds to the Trypanosome's Interior
Michael P. Barrett and Ian H. Gilbert

Making Sense of the Schistosome Surface
Patrick J. Skelly and R. Alan Wilson

Immunology and Pathology of Intestinal Trematodes in Their Definitive Hosts
Rafael Toledo, José-Guillermo Esteban, and Bernard Fried

Systematics and Epidemiology of *Trichinella*
Edoardo Pozio and K. Darwin Murrell

Volume 64

Leishmania and the Leishmaniases: A Parasite Genetic Update and Advances in Taxonomy, Epidemiology and Pathogenicity in Humans
Anne-Laure Bañuls, Mallorie Hide and Franck Prugnolle

Human Waterborne Trematode and Protozoan Infections
Thaddeus K. Graczyk and Bernard Fried

The Biology of Gyrodctylid Monogeneans: The "Russian-Doll Killers"
T.A. Bakke, J. Cable, and P.D. Harris

Human Genetic Diversity and the Epidemiology of Parasitic and Other Transmissible Diseases
Michel Tibayrenc

Volume 65

ABO Blood Group Phenotypes and *Plasmodium falciparum* Malaria: Unlocking a Pivotal Mechanism
María-Paz Loscertales, Stephen Owens, James O'Donnell, James Bunn, Xavier Bosch-Capblanch, and Bernard J. Brabin

Structure and Content of the Entamoeba histolytica Genome
C.G. Clark, U.C.M. Alsmark, M. Tazreiter, Y. Saito-Nakano, V. Ali, S. Marion, C. Weber, C. Mukherjee, I. Bruchhaus, E. Tannich, M. Leippe, T. Sicheritz-Ponten, P. G. Foster, J. Samuelson, C.J. Noël, R.P. Hirt, T.M. Embley, C.A. Gilchrist, B.J. Mann, U. Singh, J.P. Ackers, S. Bhattacharya, A. Bhattacharya, A. Lohia, N. Guillén, M. Duchene, T. Nozaki, and N. Hall

Epidemiological Modelling for Monitoring and Evaluation of Lymphatic Filariasis Control
Edwin Michael, Mwele N. Malecela-Lazaro, and James W. Kazura

The Role of Helminth Infections in Carcinogenesis
David A. Mayer and Bernard Fried

A Review of the Biology of the Parasitic Copepod Lernaeocera branchialis (L., 1767) (Copepoda: Pennellidae
Adam J. Brooker, Andrew P. Shinn, and James E. Bron

Volume 66

Strain Theory of Malaria: The First 50 Years
F. Ellis McKenzie, David L. Smith, Wendy P. O'Meara, and Eleanor M. Riley*

Advances and Trends in the Molecular Systematics of Anisakid Nematodes, with Implications for their Evolutionary Ecology and Host−Parasite Co-evolutionary Processes
Simonetta Mattiucci and Giuseppe Nascetti

Atopic Disorders and Parasitic Infections
Aditya Reddy and Bernard Fried

Heartworm Disease in Animals and Humans
John W. McCall, Claudio Genchi, Laura H. Kramer, Jorge Guerrero, and Luigi Venco

Volume 67

Introduction
Irwin W. Sherman

An Introduction to Malaria Parasites
Irwin W. Sherman

The Early Years
Irwin W. Sherman

Show Me the Money
Irwin W. Sherman

In Vivo and *In Vitro* Models
Irwin W. Sherman

Malaria Pigment
Irwin W. Sherman

Chloroquine and Hemozoin
Irwin W. Sherman

Isoenzymes
Irwin W. Sherman

The Road to the *Plasmodium falciparum* Genome
Irwin W. Sherman

Carbohydrate Metabolism
Irwin W. Sherman

Pyrimidines and the Mitochondrion
Irwin W. Sherman

The Road to Atovaquone
Irwin W. Sherman

The Ring Road to the Apicoplast
Irwin W. Sherman

Ribosomes and Ribosomal Ribonucleic Acid
Synthesis
Irwin W. Sherman

De Novo Synthesis of Pyrimidines and Folates
Irwin W. Sherman

Salvage of Purines
Irwin W. Sherman

Polyamines
Irwin W. Sherman

New Permeability Pathways and Transport
Irwin W. Sherman

Hemoglobinases
Irwin W. Sherman

Erythrocyte Surface Membrane Proteins
Irwin W. Sherman

Trafficking
Irwin W. Sherman

Erythrocyte Membrane Lipids
Irwin W. Sherman

Invasion of Erythrocytes
Irwin W. Sherman

Vitamins and Anti-Oxidant Defenses
Irwin W. Sherman

Shocks and Clocks
Irwin W. Sherman

Transcriptomes, Proteomes and
Data Mining
Irwin W. Sherman

Mosquito Interactions
Irwin W. Sherman

Volume 68

HLA-Mediated Control of HIV and HIV
Adaptation to HLA
*Rebecca P. Payne, Philippa C. Matthews,
Julia G. Prado, and Philip J.R. Goulder*

An Evolutionary Perspective on Parasitism as
a Cause of Cancer
Paul W. Ewald

Invasion of the Body Snatchers:
The Diversity and Evolution of
Manipulative Strategies in Host−Parasite
Interactions
*Thierry Lefèvre, Shelley A. Adamo, David G.
Biron, Dorothée Missé, David Hughes, and
Frédéric Thomas*

Evolutionary Drivers of Parasite-Induced
Changes in Insect Life-History Traits:
From Theory to Underlying
Mechanisms
Hilary Hurd

Ecological Immunology of a Tapeworms'
Interaction with its Two Consecutive
Hosts
*Katrin Hammerschmidt
and Joachim Kurtz*

Tracking Transmission of the Zoonosis
Toxoplasma gondii
Judith E. Smith

Parasites and Biological Invasions
Alison M. Dunn

Zoonoses in Wildlife: Integrating Ecology
into Management
Fiona Mathews

Understanding the Interaction Between an
Obligate Hyperparasitic Bacterium,
Pasteuria penetrans and its Obligate
Plant-Parasitic Nematode Host,
Meloidogyne spp.
Keith G. Davies

Host−Parasite Relations and Implications for
Control
Alan Fenwick

Onchocerca−Simulium Interactions and the
Population and Evolutionary Biology of
Onchocerca volvulus
*María-Gloria Basáñez, Thomas S. Churcher,
and María-Eugenia Grillet*

Microsporidians as Evolution-Proof Agents of
Malaria Control?
*Jacob C. Koella, Lena Lorenz, and
Irka Bargielowski*

Volume 69

The Biology of the Caecal Trematode
Zygocotyle lunata
*Bernard Fried, Jane E. Huffman, Shamus
Keeler, and Robert C. Peoples*

Fasciola, Lymnaeids and Human Fascioliasis,
with a Global Overview on Disease
Transmission, Epidemiology,
Evolutionary Genetics, Molecular
Epidemiology and Control
*Santiago Mas-Coma, María Adela Valero,
and María Dolores Bargues*

Recent Advances in the Biology of
   Echinostomes
   *Rafael Toledo, José-Guillermo Esteban, and
   Bernard Fried*

Peptidases of Trematodes
   *Martin Kašný, Libor Mikeš, Vladimír Hampl,
   Jan Dvořák, Conor R. Caffrey,
   John P. Dalton, and Petr Horák*

Potential Contribution of Sero-Epidemiological Analysis for Monitoring Malaria Control and Elimination: Historical and Current Perspectives
   Chris Drakeley and Jackie Cook

Volume 70

Ecology and Life History Evolution of
   Frugivorous *Drosophila* Parasitoids
   *Frédéric Fleury, Patricia Gibert, Nicolas Ris, and
   Roland Allemand*

Decision-Making Dynamics in Parasitoids of
   Drosophil
   Andra Thiel and Thomas S. Hoffmeister

Dynamic Use of Fruit Odours to Locate Host Larvae: Individual Learning, Physiological State and Genetic Variability as Adaptive Mechanisms
   *Laure Kaiser, Aude Couty, and Raquel
   Perez-Maluf*

The Role of Melanization and Cytotoxic By-Products in the Cellular Immune Responses of *Drosophila* Against Parasitic Wasps
   A. Nappi, M. Poirié, and Y. Carton

Virulence Factors and Strategies of *Leptopilina* spp.: Selective Responses in *Drosophila* Hosts
   *Mark J. Lee, Marta E. Kalamarz,
   Indira Paddibhatla, Chiyedza Small,
   Roma Rajwani, and Shubha Govind*

Variation of *Leptopilina boulardi* Success in *Drosophila* Hosts: What is Inside the Black Box?
   *A. Dubuffet, D. Colinet, C. Anselme, S. Dupas,
   Y. Carton, and M. Poirié*

Immune Resistance of *Drosophila* Hosts Against *Asobara* Parasitoids: Cellular Aspects
   *Patrice Eslin, Geneviève Prévost,
   Sebastien Havard, and Géraldine Doury*

Components of *Asobara* Venoms and their Effects on Hosts
   *Sébastien J.M. Moreau, Sophie Vinchon,
   Anas Cherqui, and Geneviève Prévost*

Strategies of Avoidance of Host Immune Defenses in *Asobara* Species
   *Geneviéve Prevost, Géraldine Doury,
   Alix D.N. Mabiala-Moundoungou,
   Anas Cherqui, and Patrice Eslin*

Evolution of Host Resistance and Parasitoid Counter-Resistance
   Alex R. Kraaijeveld and H. Charles J. Godfray

Local, Geographic and Phylogenetic Scales of Coevolution in *Drosophila*–Parasitoid Interactions
   S. Dupas, A. Dubuffet, Y. Carton, and M. Poirié

Drosophila–Parasitoid Communities as Model Systems for Host–*Wolbachia* Interactions
   *Fabrice Vavre, Laurence Mouton, and
   Bart A. Pannebakker*

A Virus-Shaping Reproductive Strategy in a *Drosophila* Parasitoid
   *Julien Varaldi, Sabine Patot, Maxime Nardin,
   and Sylvain Gandon*

Volume 71

Cryptosporidiosis in Southeast Asia: What's out There?
   *Yvonne A.L. Lim, Aaron R. Jex, Huw V.
   Smith, and Robin B. Gasser*

Human Schistosomiasis in the Economic Community of West African States: Epidemiology and Control
   *Hélené Moné, Moudachirou Ibikounlé, Achille
   Massougbodji, and Gabriel Mouahid*

The Rise and Fall of Human Oesophagostomiasis
   *A.M. Polderman, M. Eberhard, S. Baeta, Robin
   B. Gasser, L. van Lieshout, P. Magnussen,
   A. Olsen, N. Spannbrucker, J. Ziem, and J. Horton*

Volume 72

Important Helminth Infections in Southeast Asia: Diversity, Potential for Control and Prospects for Elimination
Jürg Utzinger, Robert Bergquist, Remigio Olveda, and Xiao-Nong Zhou

Escalating the Global Fight Against Neglected Tropical Diseases Through Interventions in the Asia Pacific Region
Peter J. Hotez and John P. Ehrenberg

Coordinating Research on Neglected Parasitic Diseases in Southeast Asia Through Networking
Remi Olveda, Lydia Leonardo, Feng Zheng, Banchob Sripa, Robert Bergquist, and Xiao-Nong Zhou

Neglected Diseases and Ethnic Minorities in the Western Pacific Region: Exploring the Links
Alexander Schratz, Martha Fernanda Pineda, Liberty G. Reforma, Nicole M. Fox, Tuan Le Anh, L. Tommaso Cavalli-Sforza, Mackenzie K. Henderson, Raymond Mendoza, Jürg Utzinger, John P. Ehrenberg, and Ah Sian Tee

Controlling Schistosomiasis in Southeast Asia: A Tale of Two Countries
Robert Bergquist and Marcel Tanner

Schistosomiasis Japonica: Control and Research Needs
Xiao-Nong Zhou, Robert Bergquist, Lydia Leonardo, Guo-Jing Yang, Kun Yang, M. Sudomo, and Remigio Olveda

Schistosoma mekongi in Cambodia and Lao People's Democratic Republic
Sinuon Muth, Somphou Sayasone, Sophie Odermatt-Biays, Samlane Phompida, Socheat Duong, and Peter Odermatt

Elimination of Lymphatic Filariasis in Southeast Asia
Mohammad Sudomo, Sombat Chayabejara, Duong Socheat, Leda Hernandez, Wei-Ping Wu, and Robert Bergquist

Combating Taenia solium Cysticercosis in Southeast Asia: An Opportunity for Improving Human Health and Livestock Production Links
A. Lee Willingham III, Hai-Wei Wu, James Conlan, and Fadjar Satrija

Echinococcosis with Particular Reference to Southeast Asia
Donald P. McManus

Food-Borne Trematodiases in Southeast Asia: Epidemiology, Pathology, Clinical Manifestation and Control
Banchob Sripa, Sasithorn Kaewkes, Pewpan M. Intapan, Wanchai Maleewong, and Paul J. Brindley

Helminth Infections of the Central Nervous System Occurring in Southeast Asia and the Far East
Shan Lv, Yi Zhang, Peter Steinmann, Xiao-Nong Zhou, and Jürg Utzinger

Less Common Parasitic Infections in Southeast Asia that can Produce Outbreaks
Peter Odermatt, Shan Lv, and Somphou Sayasone

Volume 73

Concepts in Research Capabilities Strengthening: Positive Experiences of Network Approaches by TDR in the People's Republic of China and Eastern Asia
Xiao-Nong Zhou, Steven Wayling, and Robert Bergquist

Multiparasitism: A Neglected Reality on Global, Regional and Local Scale
Peter Steinmann, Jürg Utzinger, Zun-Wei Du, and Xiao-Nong Zhou

Health Metrics for Helminthic Infections
Charles H. King

Implementing a Geospatial Health Data Infrastructure for Control of Asian Schistosomiasis in the People's Republic of China and the Philippines
John B. Malone, Guo-Jing Yang, Lydia Leonardo, and Xiao-Nong Zhou

The Regional Network for Asian Schistosomiasis and Other Helminth Zoonoses (RNAS⁺): Target Diseases in Face of Climate Change
Guo-Jing Yang, Jürg Utzinger, Shan Lv, Ying-Jun Qian, Shi-Zhu Li, Qiang Wang, Robert Bergquist, Penelope Vounatsou, Wei Li, Kun Yang, and Xiao-Nong Zhou

Social Science Implications for Control of Helminth Infections in Southeast Asia
Lisa M. Vandemark, Tie-Wu Jia, and Xiao-Nong Zhou

Towards Improved Diagnosis of Zoonotic Trematode Infections in Southeast Asia
Maria Vang Johansen, Paiboon Sithithaworn, Robert Bergquist, and Jurg Utzinger

The Drugs We Have and the Drugs We Need Against Major Helminth Infections
Jennifer Keiser and Jürg Utzinger

Research and Development of Antischistosomal Drugs in the People's Republic of China: A 60-Year Review
Shu-Hua Xiao, Jennifer Keiser, Ming-Gang Chen, Marcel Tanner, and Jürg Utzinger

Control of Important Helminthic Infections: Vaccine Development as Part of the Solution
Robert Bergquist and Sara Lustigman

Our Wormy World: Genomics, Proteomics and Transcriptomics in East and Southeast Asia
Jun Chuan, Zheng Feng, Paul J. Brindley, Donald P. McManus, Zeguang Han, Peng Jianxin, and Wei Hu

Advances in Metabolic Profiling of Experimental Nematode and Trematode Infections
Yulan Wang, Jia V. Li, Jasmina Saric, Jennifer Keiser, Junfang Wu, Jürg Utzinger, and Elaine Holmes

Studies on the Parasitology, Phylogeography and the Evolution of Host−Parasite Interactions for the Snail Intermediate Hosts of Medically Important Trematode Genera in Southeast Asia
Stephen W. Attwood

Volume 74

The Many Roads to Parasitism: A Tale of Convergence
Robert Poulin

Malaria Distribution, Prevalence, Drug Resistance and Control in Indonesia
Iqbal R.F. Elyazar, Simon I. Hay, and J. Kevin Baird

Cytogenetics and Chromosomes of Tapeworms (Platyhelminthes, Cestoda)
Marta Špakulová, Martina Orosová, and John S. Mackiewicz

Soil-Transmitted Helminths of Humans in Southeast Asia—Towards Integrated Control
Aaron R. Jex, Yvonne A.L. Lim, Jeffrey Bethony, Peter J. Hotez, Neil D. Young, and Robin B. Gasser

The Applications of Model-Based Geostatistics in Helminth Epidemiology and Control
Ricardo J. Soares Magalhães, Archie C.A. Clements, Anand P. Patil, Peter W. Gething, and Simon Brooker

Volume 75

Epidemiology of American Trypanosomiasis (Chagas Disease)
Louis V. Kirchhoff

Acute and Congenital Chagas Disease
Caryn Bern, Diana L. Martin, and Robert H. Gilman

Cell-Based Therapy in Chagas Disease
Antonio C. Campos de Carvalho, Adriana B. Carvalho, and Regina C.S. Goldenberg

Targeting *Trypanosoma cruzi* Sterol 14α-Demethylase (CYP51)
Galina I. Lepesheva, Fernando Villalta, and Michael R. Waterman

Experimental Chemotherapy and Approaches to Drug Discovery for *Trypanosoma cruzi* Infection
Frederick S. Buckner

Vaccine Development Against *Trypanosoma cruzi* and Chagas Disease
Juan C. Vázquez-Chagoyán, Shivali Gupta, and Nisha Jain Garg

Genetic Epidemiology of Chagas Disease
Sarah Williams-Blangero, John L. VandeBerg, John Blangero, and Rodrigo Corrêa-Oliveira

Kissing Bugs. The Vectors of Chagas
Lori Stevens, Patricia L. Dorn, Justin O. Schmidt, John H. Klotz, David Lucero, and Stephen A. Klotz

Advances in Imaging of Animal Models of Chagas Disease
Linda A. Jelicks and Herbert B. Tanowitz

The Genome and Its Implications
Santuza M. Teixeira, Najib M. El-Sayed, and Patrícia R. Araújo

Genetic Techniques in Trypanosoma cruzi
Martin C. Taylor, Huan Huang, and John M. Kelly

Nuclear Structure of *Trypanosoma cruzi*
Sergio Schenkman, Bruno dos Santos Pascoalino, and Sheila C. Nardelli

Aspects of Trypanosoma cruzi Stage Differentiation
Samuel Goldenberg and Andrea Rodrigues Ávila

The Role of Acidocalcisomes in the Stress Response of *Trypanosoma cruzi*
Roberto Docampo, Veronica Jimenez, Sharon King-Keller, Zhu-hong Li, and Silvia N.J. Moreno

Signal Transduction in *Trypanosoma cruzi*
Huan Huang

Volume 76

Bioactive Lipids in *Trypanosoma cruzi* Infection
Fabiana S. Machado, Shankar Mukherjee, Louis M. Weiss, Herbert B. Tanowitz, and Anthony W. Ashton

Mechanisms of Host Cell Invasion by *Trypanosoma cruzi*
Kacey L. Caradonna and Barbara A. Burleigh

Gap Junctions and Chagas Disease
Daniel Adesse, Regina Coeli Goldenberg, Fabio S. Fortes, Jasmin, Dumitru A. Iacobas, Sanda Iacobas, Antonio Carlos Campos de Carvalho, Maria de Narareth Meirelles, Huan Huang, Milena B. Soares, Herbert B. Tanowitz, Luciana Ribeiro Garzoni, and David C. Spray

The Vasculature in Chagas Disease
Cibele M. Prado, Linda A. Jelicks, Louis M. Weiss, Stephen M. Factor, Herbert B. Tanowitz, and Marcos A. Rossi

Infection-Associated Vasculopathy in Experimental Chagas Disease: Pathogenic Roles of Endothelin and Kinin Pathways
Julio Scharfstein and Daniele Andrade

Autoimmunity
Edecio Cunha-Neto, Priscila Camillo Teixeira, Luciana Gabriel Nogueira, and Jorge Kalil

ROS Signalling of Inflammatory Cytokines During *Trypanosoma cruzi* Infection
Shivali Gupta, Monisha Dhiman, Jian-jun Wen, and Nisha Jain Garg

Inflammation and Chagas Disease: Some Mechanisms and Relevance
André Talvani and Mauro M. Teixeira

Neurodegeneration and Neuroregeneration in Chagas Disease
Marina V. Chuenkova and Mercio PereiraPerrin

Adipose Tissue, Diabetes and Chagas Disease
Herbert B. Tanowitz, Linda A. Jelicks, Fabiana S. Machado, Lisia Esper, Xiaohua Qi, Mahalia S. Desruisseaux, Streamson C. Chua, Philipp E. Scherer, and Fnu Nagajyothi

Volume 77

Coinfection of *Schistosoma* (Trematoda) with Bacteria, Protozoa and Helminths
Amy Abruzzi and Bernard Fried

Trichomonas vaginalis Pathobiology: New Insights from the Genome Sequence
Robert P. Hirt, Natalia de Miguel, Sirintra Nakjang, Daniele Dessi, Yuk-Chien Liu, Nicia Diaz, Paola Rappelli, Alvaro Acosta-Serrano, Pier-Luigi Fiori, and Jeremy C. Mottram

Cryptic Parasite Revealed: Improved
   Prospects for Treatment and Control
   of Human Cryptosporidiosis Through
   Advanced Technologies
   Aaron R. Jex, Huw V. Smith,
   Matthew J. Nolan, Bronwyn E. Campbell,
   Neil D. Young, Cinzia Cantacessi, and
   Robin B. Gasser

Assessment and Monitoring of
   Onchocerciasis in Latin America
   Mario A. Rodríguez-Pérez, Thomas R. Unnasch,
   and Olga Real-Najarro

Volume 78

Gene Silencing in Parasites: Current Status
   and Future Prospects
   Raúl Manzano-Román, Ana Oleaga,
   Ricardo Pérez-Sánchez, and Mar Siles-Lucas

Giardia—From Genome to Proteome
   R. C. Andrew Thompson and Paul Monis

Malaria Ecotypes and Stratification
   Allan Schapira and Konstantina Boutsika

The Changing Limits and Incidence of
   Malaria in Africa: 1939–2009
   Robert W. Snow, Punam Amratia,
   Caroline W. Kabaria, Abdisalan M. Noor, and
   Kevin Marsh

Volume 79

Northern Host – Parasite Assemblages:
   History and Biogeography on the
   Borderlands of Episodic Climate and
   Environmental Transition
   Eric P. Hoberg, Kurt E. Galbreath, Joseph A. Cook,
   Susan J. Kutz, and Lydden Polley

Parasites in Ungulates of Arctic North
   America and Greenland: A View of
   Contemporary Diversity, Ecology and
   Impact in a World Under Change
   Susan J. Kutz, Julie Ducrocq, Guilherme G.
   Verocai, Bryanne M. Hoar, Doug D. Colwell,
   Kimberlee B. Beckmen, Lydden Polley,
   Brett T. Elkin, and Eric P. Hoberg

Neorickettsial Endosymbionts of the Digenea:
   Diversity, Transmission and Distribution
   Jefferson A. Vaughan, Vasyl V. Tkach, and
   Stephen E. Greiman

Priorities for the Elimination of Sleeping
   Sickness
   Susan C. Welburn and Ian Maudlin

Scabies: Important Clinical Consequences
   Explained by New Molecular Studies
   Katja Fischer, Deborah Holt, Bart Currie, and
   David Kemp

Review: Surveillance of Chagas Disease
   Ken Hashimoto and Kota Yoshioka

Volume 80

The Global Public Health Significance of
   Plasmodium vivax
   Katherine E. Battle, Peter W. Gething,
   Iqbal R. F. Elyazar, Catherine L. Moyes,
   Marianne E. Sinka, Rosalind E. Howes,
   Carlos A. Guerra, Ric N. Price,
   J. Kevin Baird, and Simon I. Hay

Relapse
   Nicholas J. White and Mallika Imwong

Plasmodium vivax: Clinical Spectrum, Risk
   Factors and Pathogenesis
   Nicholas M. Anstey, Nicholas M. Douglas,
   Jeanne R. Poespoprodjo, and Ric N. Price

Diagnosis and Treatment of *Plasmodium vivax*
   Malaria
   J. Kevin Baird, Jason D. Maguire, and
   Ric N. Price

Chemotherapeutic Strategies for Reducing
   Transmission of *Plasmodium vivax* Malaria
   Nicholas M. Douglas, George K. John,
   Lorenz von Seidlein, Nicholas M. Anstey,
   and Ric N. Price

Control and Elimination of *Plasmodium vivax*
   G. Dennis Shanks

Volume 81

Plasmodium vivax: Modern Strategies to
   Study a Persistent Parasite's Life Cycle
   Mary R. Galinski, Esmeralda V.S. Meyer, and
   John W. Barnwell

Red Blood Cell Polymorphism and
   Susceptibility to Plasmodium vivax
   Peter A. Zimmerman, Marcelo U. Ferreira,
   Rosalind E. Howes, and
   Odile Mercereau-Puijalon

Natural Acquisition of Immunity to *Plasmodium vivax*: Epidemiological Observations and Potential Targets
Ivo Mueller, Mary R. Galinski, Takafumi Tsuboi, Myriam Arevalo-Herrera, William E. Collins, and Christopher L. King

G6PD Deficiency: Global Distribution, Genetic Variants and Primaquine Therapy
Rosalind E. Howes, Katherine E. Battle, Ari W. Satyagraha, J. Kevin Baird, and Simon I. Hay

Genomics, Population Genetics and Evolutionary History of *Plasmodium vivax*
Jane M. Carlton, Aparup Das, and Ananias A. Escalante

Malariotherapy – Insanity at the Service of Malariology
Georges Snounou and Jean-Louis Pérignon

Volume 82

Recent Developments in Blastocystis Research
C. Graham Clark, Mark van der Giezen, Mohammed A. Alfellani, and C. Rune Stensvold

Tradition and Transition: Parasitic Zoonoses of People and Animals in Alaska, Northern Canada, and Greenland
Emily J. Jenkins, Louisa J. Castrodale, Simone J.C. de Rosemond, Brent R. Dixon, Stacey A. Elmore, Karen M. Gesy, Eric P. Hoberg, Lydden Polley, Janna M. Schurer, Manon Simard, and R. C. Andrew Thompson

The Malaria Transition on the Arabian Peninsula: Progress toward a Malaria-Free Region between 1960-2010
Robert W. Snow, Punam Amratia, Ghasem Zamani, Clara W. Mundia, Abdisalan M. Noor, Ziad A. Memish, Mohammad H. Al Zahrani, Adel AlJasari, Mahmoud Fikri, and Hoda Atta

Microsporidia and 'The Art of Living Together'
Jiří Vávra and Julius Lukeš

Patterns and Processes in Parasite Co-Infection
Mark E. Viney and Andrea L. Graham

Volume 83

Iron–Sulphur Clusters, Their Biosynthesis, and Biological Functions in Protozoan Parasites
Vahab Ali and Tomoyoshi Nozaki

A Selective Review of Advances in Coccidiosis Research
H. David Chapman, John R. Barta, Damer Blake, Arthur Gruber, Mark Jenkins, Nicholas C. Smith, Xun Suo, and Fiona M. Tomley

The Distribution and Bionomics of *Anopheles* Malaria Vector Mosquitoes in Indonesia
Iqbal R.F. Elyazar, Marianne E. Sinka, Peter W. Gething, Siti N. Tarmidzi, Asik Surya, Rita Kusriastuti, Winarno, J. Kevin Baird, Simon I. Hay, and Michael J. Bangs

Next-Generation Molecular-Diagnostic Tools for Gastrointestinal Nematodes of Livestock, with an Emphasis on Small Ruminants: A Turning Point?
Florian Roeber, Aaron R. Jex, and Robin B. Gasser

Volume 84

Joint Infectious Causation of Human Cancers
Paul W. Ewald and Holly A. Swain Ewald

Neurological and Ocular Fascioliasis in Humans
Santiago Mas-Coma, Verónica H. Agramunt, and María Adela Valero

Measuring Changes in *Plasmodium falciparum* Transmission: Precision, Accuracy and Costs of Metrics
Lucy S. Tusting, Teun Bousema, David L. Smith, and Chris Drakeley

A Review of Molecular Approaches for Investigating Patterns of Coevolution in Marine Host–Parasite Relationships
Götz Froeschke and Sophie von der Heyden

New Insights into Clonality and Panmixia in *Plasmodium* and *Toxoplasma*
Michel Tibayrenc and Francisco J. Ayala

Volume 85

Diversity and Ancestry of Flatworms Infecting Blood of Nontetrapod Craniates "Fishes"
Raphael Orélis-Ribeiro, Cova R. Arias, Kenneth M. Halanych, Thomas H. Cribb, and Stephen A. Bullard

Techniques for the Diagnosis of Fasciola Infections in Animals: Room for Improvement
Cristian A. Alvarez Rojas, Aaron R. Jex, Robin B. Gasser, and Jean-Pierre Y. Scheerlinck

Reevaluating the Evidence for Toxoplasma gondii-Induced Behavioural Changes in Rodents
Amanda R. Worth, R. C. Andrew Thompson, and Alan J. Lymbery